Mechanical Damage in Fresh Horticultural Produce

Pankaj B. Pathare • Umezuruike Linus Opara
Editors

Mechanical Damage in Fresh Horticultural Produce

Measurement, Analysis and Control

Springer

Editors
Pankaj B. Pathare
College of Agricultural and Marine Sciences
Sultan Qaboos University
Al-Khod, Oman

Umezuruike Linus Opara
SARChI Postharvest Technology Research Laboratory, Africa Institute for Postharvest Technology
Stellenbosch University
Stellenbosch, South Africa

ISBN 978-981-99-7095-7 ISBN 978-981-99-7096-4 (eBook)
https://doi.org/10.1007/978-981-99-7096-4

This Springer imprint is published by the registered company Springer Nature Singapore Pte Ltd.
The registered company address is: 152 Beach Road, #21-01/04 Gateway East, Singapore 189721, Singapore

Dedicated to
My parents, wife Sneha, and two lovely daughters Parvi and Shaarvi for their kind support with their encouragement and patience throughout the project.
Pankaj B Pathare
and
My wife Gina and
daughters Ijeoma Chidi Onumarakwu Uwaezuoke and Okaraonyemma Lanuola Ulumma
for their unflinching support and encouragement toward my academic project.
Umezuruike Linus Opara

Foreword

There was a time, in our memory, where **biological and engineering** research advanced each on different, if not contradictory, paths. Though mechanization was used for centuries, and the engineering design had developed quite knowledge to enable, for example, efficient grain harvesting, only really rough produce were appropriate for mechanical harvesting and handling. Although harvesting losses were considerable, agricultural engineering was efficient in solving important problems for farmers, namely low productivity and lack of labor.

Then there came a time where **engineers involved biology-based scientists** with the "new" idea that machines could and should be adapted to the properties of the produce, mainly in their resistance to mechanical manipulation, but also in their quality. They designed selected plant varieties, and machines, in joint and groundbreaking programs toward well-worked goals, with a lot of mutual confidence and devotion, and also with significant efforts to understand their particular approaches. Fruits and vegetables are still a very important target because losses due to especially mechanical damage are high, and they add dramatically to the rate of global food waste.

Physical properties address to the interaction of a product with a source of energy, being it mechanical, optical, spectroscopic, thermal, electrical, etc. and enabling their description and quantification by physical rather than chemical means; they are also addressed as engineering properties and include size, shape, also particulate and surface features. This area of knowledge has evolved enormously and is the fundamental basis for identifying and reducing damage in fresh produce. That includes many other applications, like the detection of internal quality, chemical compounds, food process effects, also crop monitoring, and all this is now possible in practice, with high-tech equipment and advanced decision tools, which have to be appropriate in the agri-food particular sectors.

Today**, research groups** are formed by broad and well-designed teams, created on the basis of specific goals of specific projects. They are selected from diverse specialization fields, in the areas of engineering and biology, also food specialists, also by soil-water, physics, mathematics, and computer and social sciences experts.

The chapters of this book deal with the problem of **loss of fresh produce** due to mechanical aggressiveness. For any product, it is obvious that we have to study their physiology, maturation processes and ripening, pre- and postharvest environmental conditions, the conditions of handling machinery and product susceptibility, which all jointly affect their response to aggressions. The usually complex problems are best described by multiple-variable models aiming to meaningful solutions to characterize the particular product behavior. Updated modelling, eventually using artificial intelligence, is the best tool to contribute to reduction of damage and losses of fresh horticultural products.

In the goal to diminishing the effects of aggressions, which involve losses in product and quality, the key to success will be based then on the capacity for transferring innovation into the product chains, as a result of research, best-practice standards, and effective technologies. The present book is well placed toward this relevant and global target.

Universidad Politécnica de Madrid—
UPM, Madrid, Spain
May 2023

Margarita Ruiz-Altisent

Preface

Bruising is the most common type of mechanical damage, which is inevitable along the postharvest chain, including harvesting, sorting, packaging, storage, transportation, retailing, and distribution stages. It is a major cause of postharvest loss along the value chain of fresh produce. Studies show that 30–40% of fruits and vegetables produced are affected by bruising and other types of mechanical damage. Fresh horticultural produce are particularly susceptible to bruising when they impact or compress each other or a hard surface during picking and postharvest operations in the packhouse, during transport and at retail stores. It occurs when the forces of fruits and vegetables under dynamic and static loading exceed the failure stress of the tissue. The consequences of bruise damage of fresh produce are not limited to visual aspects but may also accelerate other biological processes such as microbial spoilage and aggravate the risk of microbial contamination, hence providing potential causes for produce quality losses and lower shelf-life. Postharvest rots and decay are more prevalent in bruised or otherwise mechanically damaged fruits and vegetables than in undamaged produce.

This edited book presents recent technological developments in bruise measurement, detection, and analysis of fresh horticultural produce. First, this book introduces an overview of mechanical damage of fresh fruit and vegetable commodities. Measuring the susceptibility of produce to damage is important to determine the magnitude of damage and associated economic costs and to evaluate the efficacy and cost-effectiveness of control strategies. Given the rising demand for rapid and accurate methods of quality measurement in the horticultural produce industry, this book covers destructive and nondestructive techniques for bruise measurement. Selected applications of different nondestructive methods for various fresh produce commodities are also included in this book. As a result, the reader will experience a better understanding of how to detect the presence of bruise and eliminate damaged items on production lines, thereby significantly enhance the quality of fresh produce for sale and hence improve fresh produce economy.

With various interesting topics addressing the bruise measurement and analysis techniques for different fresh produce including the effect on produce quality and shelf life, it is expected that the book will benefit a wide array of audiences including postharvest and food scientists, technologists, industry personnel, and researchers working on different aspects of reducing the incidence and economic impacts of mechanical damage affected by postharvest handling practices. Besides, the book can serve as a readily accessible reference material for postgraduate (MSc/MEng and PhD) students and scientists to extend their knowledge in these research fields. The editors are confident that readers will find this book informative and interesting.

We are grateful to all the authors for contributing chapters to this book. We are also thankful to the staff of the editorial and production departments of Springer for their support and their efforts to bring this book to publication.

Al-Khod, Oman — Pankaj B. Pathare
Stellenbosch, South Africa — Umezuruike Linus Opara

Contents

About the Editors

Pankaj B. Pathare is an Associate Professor of Postharvest Technology at Sultan Qaboos University, Oman. He has graduated with a B.Tech degree (Agricultural Engineering) from Dr. Panjabrao Deshmukh Krishi Vidyapeeth, Akola (India), M. Engg from Maharana Pratap University of Agriculture and Technology, Udaipur (India) and PhD (Process and Chemical Engineering) from the University College Cork (Ireland). Before joining at SQU, he worked as researcher at Newcastle University, UK and Stellenbosch University, South Africa. Over the years, he has gained expertise on postharvest technology and food engineering which includes quantification of postharvest losses during transportation, mechanical damage of perishables, food agglomeration/granulation, food drying and cooling, structural design for ventilated corrugated packaging, renewable energy in food processing for Sub-Saharan Africa, and energy efficient production processes in SME food and beverage industry. The research results are well documented in 68 original scientific papers published in international peer-reviewed journals and several conference presentations. He also worked as supervisor/co-supervisor for 14 postgraduate research students. He was invited as keynote/invited speaker for more than 14 international conferences in the postharvest and food engineering area. According to the Scopus, his H-index is 20 and citation 3440. In 2021, he was recognized among world top 2% scientists, published by Stanford University, USA. He is the editorial board member for the several research journals including PLoS ONE, Frontiers in Sustainable Food Systems, AgriEngineering, Journal of Food Quality, Journal of Food Processing and Preservation, Open Agriculture, Measurement: Food, Discover Food, and Current Functional Food.

Umezuruike Linus Opara is a Distinguished Professor at Stellenbosch University, South Africa, where he is also the South African Research Chair in Postharvest Technology and the founding Director of the Africa Institute for Postharvest Technology. He graduated with degrees in Agricultural Engineering (First Class Honors) from the University of Nigeria, Nsukka, and PhD from Massey University, New Zealand. He has published over 320 peer-reviewed journal articles, book chapters, and conference papers and graduated over 70 MSc and PhD students in universities in South Africa, the Sultanate of Oman, and New Zealand. He was recognized in 2019 and 2021, respectively, as a "Highly Cited Researcher" by the Web of Science Group, which distinguishes the "world's most influential researchers of the past decade." With current H-Index of 57 (Scopus) and 64 (Google Scholar) and a leading global expert on postharvest technology of fresh produce, he is also the leading global individual researcher on pomegranates—covering the value chain from fruit development to postharvest handling, processing, and impacts on human health. Among many international, continental, and national awards for his sustained research productivity and impact, he was honored as Laureate of the African Union Kwame Nkrumah Continental Award for Life and Earth Sciences—Africa's highest research award for senior researchers, Winner of the Impact Research and Science in Africa Award for "excellence in research and building Africa's human resources capacity," and Distinguished Researcher Award by Sultan Qaboos University, Sultanate of Oman. In 2019, Stellenbosch University bestowed its higher honor on him—the Chancellor's Award, for his "sustained excellence in research." In 2022, he was honored with the Distinguished Alumnus Award of the Department of Agricultural and

Bioresources Engineering, University of Nigeria, Nsukka. He was the founding President of the Pan African Society for Agricultural Engineering, past Executive Secretary of the International Academy of Agricultural and Biosystems Engineering, and past President of the International Commission of Agricultural and Biosystems Engineering. He was the founding Editor-in-Chief (now Honorary Editor) of the International Journal of Postharvest Technology and Innovation and currently the Co-Editor-in-Chief—Food Safety and Health (Wiley), Associate Editor of Frontiers in Plant Science, Regional Editor—Agricultural Mechanization in Asia, Latin America and Africa, and serves on the Editorial Board of several international journals. His ongoing activities focus on building Africa's research capacity in engineering and biosciences for agri-food systems transformation and wealth creation.

Chapter 1
Mechanical Damage of Fresh Produce: An Overview

Umezuruike Linus Opara and Pankaj B. Pathare

Abstract Fresh produce, especially fruit and vegetables, are important components of food systems. Global trade in horticultural fresh produce is a multi-billion-dollar business which has surpassed other land-based food commodities during the past three decades. The production, handling, processing, and marketing of these commodities are sources of livelihoods and income for farmers, orchardists, packhouse operators, transport and logistics operators, marketers, and other role players along the food value chain. The high moisture content and related microstructure of fresh horticultural produce contribute mainly to their high susceptibility to mechanical damage during postharvest handling and distribution. The economic cost of mechanical damage, including bruising, can be quite substantial, thereby reducing profitability and damaging the quality reputation of agribusiness. Different types of mechanical damage affecting fresh produce include impact, compression, abrasion, splits, cracks, cuts, and puncture. By far, bruise damage by impact and compression is the most common form of mechanical damage and contributor to quality downgrading and postharvest loss of fresh fruit and vegetable commodities. Measuring the susceptibility of produce to damage is important to determine the magnitude of damage and associated economic costs and to evaluate the efficacy and cost-effectiveness of control strategies. The presence of bruise damage on fresh produce may be difficult to detect due to the severity of damage and/or the level of contrast between the characteristic color of the whole produce and the browning of the bruised tissue. Therefore, the development and application of non-destructive sensors for online and offline assessment of mechanical damage is important, and various technological advances in this regard have been reported based on

U. L. Opara (✉)
SARChI Postharvest Technology Research Laboratory, Africa Institute for Postharvest Technology, Faculty of AgriSciences, Stellenbosch University, Stellenbosch, South Africa

UNESCO International Centre for Biotechnology, Nsukka, Enugu State, Nigeria
e-mail: opara@sun.ac.za

P. B. Pathare
Department of Soils, Water & Agricultural Engineering, College of Agricultural and Marine Sciences, Sultan Qaboos University, Al-Khod, Oman

P. B. Pathare, U. L. Opara (eds.), *Mechanical Damage in Fresh Horticultural Produce*, https://doi.org/10.1007/978-981-99-7096-4_1

spectroscopy, computer vision, X-ray, and thermal imaging, among others. Various preharvest and postharvest techniques have been applied to reduce the incidence and severity of mechanical damage on fresh produce. Preharvest factors such as crop load, exposure to sunlight, irrigation amount and frequency, and crop maturity at harvest can affect the potential for damage to occur during harvesting and postharvest handling. Harvesting practices such as time during the day, gentle picking of produce, and the use of cushioning materials inside fruit containers can reduce the incidence of mechanical damage. Similarly, both the incidence and amount of damage on produce during postharvest processes can be reduced or prevented by applying postharvest treatments. Various produce simulators, such as "electronic" apples and potatoes, have been successfully applied to detect "hot spot" where mechanical damage is likely to occur along the fresh produce supply chain. Ultimately, the best strategy against mechanical damage of fresh produce is to prevent the occurrence, a priori, and where this is not possible, a combination of cost-effective measures must be applied to reduce both the occurrence and severity of damage to minimize the economic costs.

Keywords Mechanical damage · Bruising · Fruit quality · Postharvest · Nutrition loss

Now it is Autumn and the falling fruit
and the long journey towards oblivion.
The apples falling like great drops of dew,
to bruise themselves an exit from themselves.—D.H. Lawrence

1.1 Introduction

Global trade in horticultural fresh produce is a multi-billion-dollar business that has surpassed other land-based food commodities during the past three decades. The global horticulture market was valued at USD 20.4 billion in 2021 and is expected to surpass USD 56.5 billion by 2030, which has been forecast to expand at a cumulative annual growth rate (CAGR) of 9.9% during the period of 2022–2030. The growth of the market is attributed to the rising acceptance of sustainable production practices and increasing demand for fresh agricultural and horticultural products to meet the growing needs of the global population. The production, handling, processing, and marketing of these commodities are sources of livelihood and income for farmers, orchardists, packhouse operators, transport and logistics operators, marketers, and other role players along the food value chain.

Fresh produce, especially fruit and vegetables, are important components of the food system. With increasing global recognition and demand for high-quality fresh produce, there is increasing business opportunity to produce a wide range of tropical, sub-tropical, and temperate fruit and vegetables to meet demand. A high incidence of mechanical damage affects the trade and marketing of fresh produce (Kampp & Pedersen, 1990). In developing and emerging markets aiming to export, meeting the marketing standards of the importing countries is a major problem. In existing markets, severe incidences of mechanical damage can lead to product rejection and difficulties in retaining market share in both local and export markets. Mechanical damage is primarily an appearance quality problem, which can restrict the market potential of high-quality fresh produce such as fruit, vegetables, edible fungi, fresh animal meat, fresh fish, and seafood. This in turn can lead to reduced availability and consumption of fresh produce which are major sources of dietary micronutrients and phytochemicals that are essential for healthy living and mitigation against non-communicable diseases (Yahia et al., 2019; Slavin & Lloyd, 2012; Opara et al., 2009; Al-Said et al., 2009). Both epidemiological studies and clinical trials have demonstrated that diets rich in fruits and vegetables can lower the risk of non-communicable diseases such as cardiovascular disease and stroke, type II diabetes Miletus, prevent some types of cancer, and lower the risk of eye and digestive problems, and have a positive effect upon blood sugar (Hung et al., 2004; Riboli & Horel, 2003; Smith-Warner et al., 2001; Dauchet et al., 2006).

Most fresh produce, especially fruits and vegetables, contain 70–80% moisture, while some vegetables like leafy vegetables and melons contain almost 92–95% moisture, compared with field crops such as rice paddy, wheat, corn, and millet which range from 20 to 35% (Fudholi et al., 2010). The high moisture content and related microstructure of fresh horticultural produce contribute mainly to their high susceptibility to mechanical damage during postharvest handling and distribution. The economic cost of mechanical damage, including bruising, can be quite substantial, thereby reducing profitability and damaging the quality reputation of agribusiness.

Different types of mechanical damage, also referred to as physical damage (Opara et al., 2007), affecting fresh produce include impact, compression, abrasion, scuff, splits, cracks, cuts, and puncture. By far, bruise damage by impact and compression is the most common form of mechanical damage and contributor to quality downgrading and postharvest loss of fresh fruit and vegetable commodities. Measuring the susceptibility of produce to damage is important for several reasons: (a) to determine the magnitude of damage and associated losses and costs, (b) to determine the potential of damage to occur under prevailing or future postharvest handling conditions, (c) to assess the impacts of preharvest and postharvest factors contributing to damage, and (d) to evaluate the efficacy and cost-effectiveness of control strategies. The potential for produce to damage is commonly expressed as bruise susceptibility (Bollen, 2005; Opara, 2007; Hussein et al., 2017, 2019, 2020c)—which is the amount of damage (area or volume of tissue affected) divided by energy absorbed by produce during contact with the force body. Many researchers prefer to use the term "bruise resistance" (Komarnicki et al., 2016), which is the inverse of bruise susceptibility.

The presence of damage by splits, scuffs, cracks, cuts, and puncture is easy to assess visually because of the physical breach or opening of the affected underlying tissue in addition to the presence of browning discoloration in the affected area.

However, the presence of bruise damage on fresh produce may be difficult to detect due to the level of contrast between the characteristic color of the whole produce and the browning of the bruised tissue. For instance, for the same magnitude of force applied, bruise damage is easier to assess on yellowish (such as "Golden Delicious") and green (such as "Granny Smith") apple cultivars than other multi-colored cultivars such as "Fuji" and "Red Delicious." Similarly, the presence of bruise damage is easier to assess on ripe banana fruit than on mature cucumber. Therefore, the development and application of non-destructive sensors for online and offline assessment of mechanical damage is important, and various technological advances in this regard have been reported based on spectroscopy, computer vision, X-ray, and thermal imaging, among others. The incorporation of various advanced statistical, mathematical, and numerical modeling tools, including chemometrics, in bruise measurement and analysis has led to improved quantification and prediction of bruise damage, especially in early bruise detection.

For some types of fresh produce with thick rind/peel/skin such as plantains, bananas, pomegranate, and citrus, the presence of external bruise damage may not translate into internal tissue damage. However, the presence of visible damage represents a quality problem that can lead to downgrading or rejection by the buyer. Therefore, preventing and/or reducing damage to such products is equally important. Various preharvest and postharvest techniques have been applied to reduce the incidence and severity of mechanical damage on fresh produce. Preharvest factors such as crop load, exposure to sunlight, irrigation amount and frequency, and crop maturity at harvest can affect the potential for damage to occur during harvesting and postharvest handling. Harvesting practices such as time during the day, gentle picking of produce, and the use of cushioning materials inside fruit containers can reduce the incidence of mechanical damage. Similarly, both the incidence and amount of damage on produce during postharvest processes can be reduced or prevented by applying postharvest treatments. These include warming cold fruit prior to packhouse operations, incorporating various fruit deceleration techniques along the packing line, designing packaging to improve the mechanical protection of produce, training packhouse staff on improved postharvest handling practices, and improving the ergonomic design and lighting in packhouses to improve the welfare and comfort of quality control staff for optimal visual detection of damage and other defects.

Various produce simulators, such as electronic apples and potatoes, have been successfully applied to determine "hot spot" where mechanical damage is likely to occur along the fresh produce supply chain—from orchard/farm to end-user (Anderson, 1990). Equipped with accelerators for force detection and time recorder, these sensors can be added into boxes and other produce containers during harvesting and shipping or along the packaging line during packhouse operations. By downloading and analyzing the force history experienced by the electronic produce at the end of the journey, each spot and related handling practices can be better understood regarding the potential for damage of produce to occur, from which corrective actions can be taken. Ultimately, the best strategy against mechanical damage of fresh produce is to prevent the occurrence, a priori, and where this is not possible, a combination of cost-effective measures must be applied to reduce both the occurrence and severity of damage to minimize the economic costs. Notwithstanding the considerable technological advancements during the past few decades in understanding the mechanisms of mechanical damage and technologies to reduce it, adequate training of relevant personnel along the value chain remains the most basic and essential.

1.2 Terminologies

Mechanical damage is a type of tissue failure that may or not result in the rupture of the skin (rind, peel) of fresh produce due to the action of excessive external mechanical force which results in cell breakage as evidenced by a discoloration of the affected tissue. Bruise damage, which is the most common form of mechanical damage affecting fresh produce, is a type of tissue failure that does not result in the rupture of the skin of affected produce (Opara & Pathare, 2014). Bruised tissue on produce manifests a browning discoloration which generally affects the underlying tissue depending on the magnitude of mechanical force applied. Focusing on fresh fruit, Ruiz-Altisent and Moreda (2011) defined bruise as "Subcutaneous mechanical damage without rupture of the skin caused to a fruit by an impact, a quasi-static compression, or a vibration. It consists of the local degradation of the flesh tissue, combined with intracell water extravasation and browning (oxidation) of phenolic compounds from released intracell water."

Different terminologies have been used to describe the same or different types of mechanical damage (Altisent, 1991). Often, the term "mechanical damage" is commonly interchanged with mechanical injury (Knee & Miller, 2002; Lam et al., 1987), mechanical defect (Macheka et al., 2013; Nturambirwe et al., 2020), physical damage (Opara et al., 2007; Hounhouigan et al., 2014), physical defect (Nturambirwe & Opara, 2020; Ketsa et al., 1995), and physical injury (Wells & Butterfield, 1999).

In specific studies or reports, reference to each of these terminologies could also include different types of damage, injury, defect, or spoilage that may have preharvest or postharvest origin or both. For instance, in their study on the incidence of *Salmonella* on 20 different types of fresh fruits and vegetables affected by fungal rots or physical injury, Wells and Butterfield (1999) found that the forms of injury included broken, bruises, cracks, cuts, field scars, gouged, growth cracks, insect injury, pitting, and punctures. Furthermore, the authors noted that the pitting observed was a physiological disorder of peppers (Snowdon, 1990). The term "superficial defects" has also been mentioned in reference to mechanical damage of produce (Pathare & Al-Dairi, 2022); often, this connotes damage that is localized on the surface of produce. Although researchers have used these interchangeably or in combination in the same article, there are differences in meaning, and accurate terminology is important so that results can be compared, and appropriate damage reduction and prevention strategies can adequately be applied based on scientific evidence and proven industry practice.

The terms "Blackspot," "Black spot damage," or "Black spot bruise" have been widely used for mechanical damage in potato tubers caused by mechanical stress during harvest and postharvest handling operations. Such a bruise is generally difficult to detect in intact tubers without peeling because the skin remains intact and undamaged (Grudzińska & Mańkowski, 2019). Nonetheless, the presence of black spot damage can significantly reduce tuber quality and economic value.

Different terminologies have been used to describe the potential for fresh produce to suffer damage or injury when an external force is applied, including contact between the produce and a stationary or moving mechanical object, or between the produce and another produce. These terminologies include bruise susceptibility, bruise resistance, and bruise severity (Opara & Pathare, 2014), and these describe the relationship between the amount of mechanical force applied and the amount of damage suffered by the affected produce.

Considering fresh fruit, Ruiz-Altisent and Moreda (2011) defined bruise susceptibility as the "Degree of ease or difficulty by which a fruit bruises. Besides of species and cultivar, bruise susceptibility depends on the physical condition or physiological status of each individual fruit, which can be assessed, among others, based on mechanical properties. The point at which fruits start to show symptoms of bruising is called bruise threshold. When this threshold is exceeded, mechanical load-induced stress exceeds flesh tissue failure stress, and a bruise results."

Irrespective of the terminology used, bruise susceptibility, resistance or severity provides a standard index to assess the prospects for fresh produce to experience mechanical damage under specified conditions. They can, therefore, be used to assess the inherent (genetic) potential of damage to occur or to assess the effectiveness of damage-reducing technological interventions alone or in combination with preharvest and postharvest management practices. Numerically, bruise susceptibility is calculated by dividing the area or volume affected by damage by the amount of energy absorbed by the produce during contact with external force. Bruise resistance is the inverse of this relationship (Opara & Pathare, 2014; Opara, 2007). Using data on the effects of orchard management practices on the bruising of "Gala" apples, Opara (2007) recommended that two separate bruising indices be calculated: (a) specific bruise susceptibility, which expresses the potential for damage to occur per unit mass of the fruit, and (b) global bruise susceptibility, which quantifies the damage potential of the whole fruit. This was considered relevant to account for differences in fruit mass because of the well-known biological variability of fresh produce, including fruit.

"Latent" damage, also referred to as "early bruise" damage is a special case of bruise damage that presents considerable difficulty to detect because the damaged tissue has not aged sufficiently to be detectable by the naked eye. This presents a major economic problem in quality management of fresh produce because although affected produce may not be detected during normal quality control procedures, the bruised area often develops with time which may coincide with arrival at the market destination or at the consumer. Detecting early bruise or latent damage which is conducted within a few hours after occurrence is, therefore, a key research challenge. Many researchers have applied innovative sensing technologies to address this problem of detecting bruise damage during the early stages of development. Nturambirwe et al. (2021) applied machine learning classifiers, namely ensemble subspace discriminant (ESD), k-nearest neighbors (KNN), support vector machine (SVM), and linear discriminant analysis (LDA) to build models to detect early bruises and investigate the effect of time after bruising on detection performance. Overall, classifiers, detection models had a higher performance than the quantitative ones. With the highest speed in prediction and high classification performance, SVM classifier was the highest rated for detection tasks. However, ESD models had the highest classification accuracy (>85%) among quantitative models and were found to be relatively better suited for such a multiple-category classification problem. Other researchers have also reported considerable success in detecting early onset of bruise damage in mangoes (Rivera et al., 2014), peaches (Li et al., 2018), and apples (Baranowski et al., 2009, 2012; Keresztes et al., 2016; ElMasry et al., 2008).

1.3 Types and Causes of Mechanical Damage Affecting Fresh Produce

The major types of mechanical force which result in different types of damage or injury of fresh produce and where they occur along the supply chain are summarized below (Table 1.1). Overall, bruising occurs due to impact, compression, and vibration, while puncture damage occurs due to impact, compression, and puncturing. For impact, the severity of damage is primarily determined by: (1) height of fall; (2) initial velocity; (3) number of impacts; (4) type of impact surface and size; and (5) physical properties of the fruit, which may be related or not to maturity.

Table 1.1 Type of mechanical forces, types of injury they cause and mode of occurrence along the supply chain of fresh produce

Mechanical force	Type of injury	Modes of occurrence
Impact	Bruise, puncture	Produce falling onto a hard surface; dropping produce during loading into vehicle; excessive drops during loading and offloading; sudden stop or acceleration of the transport vehicle
Vibration or abrasion	Bruises, abrasion	Transport vehicles with bad shock absorbers; vehicles with small wheels; weak containers such as crates and cartons; bad roads; vibration of the transmission system
Compression	Bruises, puncture	Weak packaging; improper stacking of containers; overfilling of containers
Puncturing	Cuts, cracks, splits, puncture	Presence of nail and splinters on containers; fingernails; contact with moving objects such as forklift or other containers; hard and sharp fruit stalk

Excessive impacts leading to mechanical damage may occur during preharvest due to strong wind and hailstorms, and during harvesting and postharvest handling operations such as fruit dropping into the picking buckets or during sorting. Conversely, vibration occurs mainly during transportation and shipping (Komarnicki et al., 2016), while compression or static loading can occur during harvesting, transportation, or storage due to poorly designed bins that are overfilled and stacked, resulting in damage of produce at the bottom.

Due to logistical challenges and huge cost, comprehensive field assessment of incidence and magnitude of postharvest losses and waste along the value chain is difficult. Although the problem of losses and waste has long being recognized at both policy and research levels, including the major contributing factors such as mechanical damage, it is only during the past decade and a half that the science of postharvest loss and waste has emerged (Opara, 2006a, b). As a result, most efforts have focused on specific links in the supply chain such as farm/orchard (Opara et al., 2021a, b), packhouse (Opara et al., 2022), and retail (Munhuweyi et al., 2016).

The most detailed and comprehensive assessment of the incidence of mechanical damage on fresh produce was carried out between 1972 and 1984 on fruit passing through New York terminals in the USA (Table 1.2), which shows only data for injuries that affected more than 1% of fruit lots examined. The study reported considerable variation in bruising, including differences in batches, high incidence of abrasion on watermelons, "rolling" on pears, and bruising appearing almost unavoidable on peaches. Overall, bruise damage was by far the predominant type of mechanical damage affecting all fruit types, while puncture damage or cuts were found in about 50% of the types of fruit. Current knowledge suggests that this trend has not changed over the past half-century, with bruising still representing the major cause of damage and loss of fresh horticultural produce during postharvest handling.

1.4 Economic Importance of Mechanical Damage

In the fresh produce industry, mechanical damage, especially bruising, is the major contributor to the high incidence of postharvest losses and food waste. The presence of mechanical damage on fresh produce, such as bruises, punctures, cracks, splits, abrasion, cuts, and scuff, is a major economic problem (Valenciano Garcia, 1990), which has further negative impacts on other quality attributes as well as financial, nutritional, social, and environmental impacts. Mechanical damage also promotes water loss in affected produce, which in turn contributes to the degradation of other quality attributes and financial loss (since fresh commodities are sold on a unit weight basis).

Fresh commodities affected by mechanical damage may be downgraded and sold at lower prices, leading to financial losses. In extreme cases, whole batches of produce may be rejected, and when the cost of alternative use is higher than the economic worth, the affected produced may be dumped as waste. In a pilot study on

Table 1.2 Types and incidence of mechanical damage/injuries affecting fruit passing through New York terminal markets 1972–1984 (Cappellini & Ceponis, 1984; Ceponis et al., 1986, 1987; Wells et al., 1994)

			Percentage with injury		
Fruit	Number of lots	Type of injury	Lots	All fruit	Affected lots
Apple	4453	Bruise	75.1	5.54	7.38
		Cut/puncture	3.3	0.10	3.10
Apricot	204	Bruise	44.6	3.79	8.49
		Cut/puncture	6.9	0.23	3.36
Avocado	889	Bruise	6.4	0.40	6.23
Cherry	2455	Bruise	35.0	2.37	6.76
		Cut/puncture	1.3	0.06	4.41
Grape	8100	Crush	33.7	1.25	3.73
		Shatter	47.3	2.26	4.77
		Wet/sticky	42.3	2.33	5.51
Grapefruit	4910	Bruise	4.7	0.23	4.82
Mango	717	Bruise	8.5	0.84	9.84
Nectarine	2576	Bruise	72.7	5.31	7.31
		Cut/puncture	3.2	0.12	3.87
Orange	9104	Bruise	4.8	0.29	6.03
Papaya	209	Bruise	14.8	1.59	10.73
Peach	2610	Bruise	86.9	9.19	10.58
		Cut/puncture	1.0	0.03	3.00
Pear	4409	Bruise	17.5	0.85	4.86
		Cut/puncture	1.9	0.06	3.46
Pineapple	677	Bruise	14.2	1.36	9.57
Plum	3079	Bruise	25.7	1.38	5.38
		Cut/puncture	16.3	1.03	6.30
Strawberry	1777	Bruise	69.2	7.98	11.53
Tomato	9059	Bruise	14.0	0.66	4.73
Watermelon	894	Bruise	35.3	3.42	9.66
		Abrasion	3.2	0.31	9.48

what the consumer values in fresh fruit quality in the Sultanate of Oman (Opara et al., 2007), bruise damage were among the most common quality problems reported by consumers during purchase or use, which contributed to a high incidence of postharvest fruit losses at the household level. Among the different types of fruit investigated, the reported frequency of bruising/skin damage observed ranged from 7.83% (dates) to 16.17% (orange), 17.68% (banana), 20.94% (mango), 35.63% (apple). To assess the economic importance of quality fruit based on willingness to pay for premium quality fruit, the overall response showed a very strong willingness to pay for guaranteed good quality fruit. Over two-thirds of respondents reported their willingness to pay up to 25% extra on the unit price for guaranteed best quality fruit, and nearly a quarter of consumers were also willing to pay an extra 26–50% to assure the good quality of fruit they purchase.

1.4.1 Impacts on the Incidence of Postharvest Food Losses and Waste

High incidence of postharvest losses, particularly in fresh produce such as fruit and vegetables, has been identified as a major contributor to total global amount of about 1.3 billion tonnes of food losses and waste per annum (FAO, 2011). At the global level, 14% of food valued at an estimated USD400 billion is lost from harvest up to, but excluding retail, while an additional 17% is wasted at the retail and consumer levels. The problem of postharvest food loss and waste is widely recognized as a pressing challenge in the design and management of sustainable food systems. Together, they exacerbate food and nutrition insecurity and contribute significantly to greenhouse gas (GHG) emissions, environmental pollution, degradation of natural ecosystems, biodiversity loss, and represent a waste of resources used in food production.

Mechanical damage is a major contributor to the incidence of postharvest food losses and waste, especially fresh produce such as fruit and vegetables. Reducing the incidence and severity of mechanical damage, therefore, contributes to the SDGs, and in particular SDG 12.3—"By 2030, halve per capita global food waste at the retail and consumer levels and reduce food losses along production and supply chains, including postharvest losses."

Recent case studies by Opara et al. (2021a) assessing on-farm postharvest losses and waste of major pomegranate fruit cultivars in South Africa showed that pomegranate fruit loss ranged between 15.3 and 20.1% of the harvested fruit at the case study farm. Bruise damage alone accounted for 5.43% ("Wonderful") to 9.07% ("Hershkawitz") and 9.43% ("Acco") of losses. These amounted to an estimated R10.5 million ($618,715.34) economic loss per annum to the farmer, in addition to considerable dietary nutrient losses that are embedded in lost fruit. In a similar case study on the incidence of fruit losses and waste at the packhouse level in South Africa (Opara et al., 2021b), the authors showed that losses ranged between 6.74% and 7.69%, which was equivalent to over ZAR 29.5 million (USD 1,754,984) in revenue, in addition to the opportunity costs of resources used to produce fruit that is lost. Superficial injuries (part of mechanical damage) were found to be the second highest cause of pomegranate total fruit loss for each cultivar at the packhouse after sunburn: 23.33% for "Acco", 23.07% for "Hershkawitz", and 19.07% for "Wonderful". The proportion of fruit losses due to bruise damage ranged from 10.94% ("Wonderful") to 12.80% ("Hershkawitz") and 13.33% ("Acco").

In their assessment of the seasonal changes in fruit quality of rambutans in retail markets in Thailand, Ketsa et al. (1995) found that physical damage (also considered as part of surface defects) was the main cause of postharvest losses. These included bruising, breakage, black scars, and browning spinterns. The growing and marketing of pineapples is a major source of livelihood in Benin, West Africa. However, poor quality produce, including bruise damage, is a major handicap in export marketing to lucrative overseas markets. Due to strict export standard regulations, only about 2% of the country's production was exported to European countries in 2010, while domestic markets and other West African markets accounted for about 80% of the traded pineapple. The remaining produce (about 18%) was lost because of physical damage (Hounhouigan et al., 2014).

1.4.2 Impacts on Postharvest Physiology and Quality

The presence of bruise damage has been reported to affect a wide range of physiological and biochemical processes which impact on quality of fresh produce, leading to eventual spoilage, loss, and waste (Lee, 2005). Bruising reduces appearance quality and appeal to consumers, thereby reducing purchase and repeat buying. Physiological and biochemical process of fresh produce such as respiration and transpiration (water loss) are accelerated in damaged or injured produce, including higher rates of ethylene production and relative electrical conductivity (Maia et al., 2011; Montero et al., 2009). Together, these usually lead to mass loss, senescence, loss of nutritional value, and eventual spoilage.

The presence of bruise damage also accelerates opportunistic infection of spoilage-causing microbial organisms (Prusky, 2011; Eissa et al., 2013) and contamination of potentially hazardous microbes. Postharvest rots and decay are more prevalent in bruised or mechanically damaged fruits than in non-damaged produce. Accordingly, decay pathogens can easily enter through dead or wounded tissues and contaminate the rest of the fruit, resulting in significant losses (Pholpho et al., 2011). In pomegranates, the present of bruise damage accelerated physiological responses and reduced fruit textural properties and other quality attributes (Hussein et al., 2020a, b).

In their study of bruise damage on pear fruit, Pathare and Al-Dairi (2021a, b) found that firmness was highly reduced (92.82%) due to the increase in drop height (60 cm), storage temperature (22 °C), and storage duration (14 days); however, fruit color purity (Hue), redness (a*), and TSS were not affected by impact level. The presence of minor bruise damage on "Galaxy" apple contributed to a decrease in fruit firmness (Ergun, 2017). Similarly, bruising damage accelerated loss of firmness and the ratio of sugar to acid for Yali pears (Li et al., 2012). Bruising reduced firmness and weight loss of packaged kiwifruit after 10 days of storage (Xia et al., 2020). Decreased weight of the fruit and the unsightly shriveling lead to economic losses (Al-Dairi et al., 2021b). Impact bruising resulted in both the qualitative internal and minor external changes on fresh produce, including changes in citric acid content (Hussein et al., 2020b), soluble solid content, (Gao et al., 2021), respiration, and ethylene production (Xia et al., 2020).

Hussein et al. (2020a) demonstrated that impact bruising of pomegranates (cv. Wonderful) resulted in a twofold increase in fruit respiration rate, at least during the first 4 weeks of storage. Similarly, changes in total soluble solids (TSS) and titratable acidity (TA) were significantly ($p < 0.05$) higher at medium and high drop impact. Furthermore, high impact bruising resulted in 30% of both decay incidence and internal fruit decay of fruit after 12-week storage. Furthermore, the combination of drop impact and storage significantly ($p < 0.05$) affected peel hue angle (h°), as well as aril lightness (L*), redness (a*), chroma (C*) and h°. The radical scavenging activity and total phenolic content were higher in juice of bruised-damage fruit at medium and high impacts, presumably due to higher biochemical defense signal induced by the force applied on fruit.

Studies have shown that bruising does not only affect external fruit characteristics but also results in internal quality losses, alterations of physiological processes, and promotes postharvest rot and decay (Fatima Ali Elshiekh & Abu-Goukh, 2008; Montero et al., 2009; Scherrer-Montero et al., 2011). In addition, bruising can result in weight loss in fruit and vegetables and hence a decrease in their market value. According to Crisosto et al. (1993), 8% of weight loss in fresh fruit resulted in economic loss to the fruit growers.

Furthermore, bruise damage modifies physiological and metabolic processes, leading to faster ripening, internal browning, and quality losses (Opara & Pathare, 2014; Costa et al., 2018). For instance, minor bruise damage on "Galaxy" apple resulted in a decrease in firmness and fruit browning (Ergun, 2017). Similarly, bruising damage accelerated loss of firmness and the ratio of sugar to acid for Yali pears (Li et al., 2012). Impact bruising resulted in both the qualitative internal and minor external changes on tangerines, including losses of citric acid, soluble solid, and ascorbic acid (Montero et al., 2009).

1.4.3 Impacts on Nutrition Losses

Postharvest food losses and waste contribute to food and nutrition insecurity, which affects over 800 million people globally annually. More serious is the problem of hidden hunger—where people might have a sufficient intake of food calories but lack a sufficient intake of micronutrients and phytochemicals to maintain good health. Fruit and vegetables are the major sources of essential micronutrients and phytochemical antioxidants for a healthy diet (Yahia et al., 2017; Slavin & Lloyd, 2012), including dietary fiber (Lanza & Butrum, 1986) which has been reported to play important role in the bioaccessibility and bioavailability of fruit and vegetable antioxidants (Palafox-Carlos et al., 2011).

The quality and nutritional value of fresh produce is affected by postharvest handling practices and storage conditions, including the presence of mechanical damage which may occur. Micronutrients such as vitamin C and other essential phytochemical compounds are subject to considerable losses under adverse postharvest handling and storage conditions, which results in loss of nutritional value. While bruising damage affects the chemical composition of damaged tissue, the effect of bruise may depend on the level of damage to the tissue. Sablani et al. (2005) showed that during 16 days of storage duration at refrigerated (4 °C) and room (25 °C) conditions, respectively, the influence of bruise on vitamin C content in tomatoes was not significant ($p > 0.05$). Bruise damage was made on opposite sides of fruit by dropping a uniform round steel ball (110 g) from a height of 30 cm.

Other researchers have reported significant reductions in vitamin C content of fresh tomatoes after bruising and storage. Ismail et al. (1993) and Moretti et al. (1998) reported that vitamin C content in bruised tomato declined by about 15.5% than unbruised fruit.

1.5 Preharvest and Postharvest Factors Affecting Bruise Damage and Reduction

A wide range of preharvest, harvest, and postharvest factors affect the potential for damage to occur on fresh produce (Hussein et al., 2018, 2020a, b). Mechanical damage can occur while fruit is still hanging on the tree due to a variety of reasons: (1) forceful contact of fruit with other fruit or parts of the tree such as branches during growth which may cause abrasion, puncture, and bruising, (2) action by biological agents such as slugs, insects, birds, and mammals which puncture the skin to consume a portion of the tissue, and (3) effect of weather, such as wind and hail that can aggravate damage caused by contact of fruit with other parts of the tree, causing mechanical injury such as bruising, cleavage, slip, and buckling. For instance, Kumar et al. (2016) reported preharvest losses up to 30.4% in litchi fruit during sorting at harvest, which was contributed mainly by sunburn, cracking, bruising anthracnose, and infestation by fruit borer.

To reduce the incidence and severity of mechanical damage, several steps are necessary: (1) determine the potential of the selected produce to incur damage under simulated conditions, (2) conduct a mechanical forensic audit of the supply chain, from farm to end-user, to determine the types of damage occurring, the contributing factors, and identify the hot spots where intervention will result in higher impact in damage reduction, (3) train relevant personnel on improved techniques and procedure to reduce produce damage, (4) identify, select and cost-effective measure to reduce the incidence and severity of damage, and (5) keep accurate records for continuous process improvement. Various produce simulators, also called instrumented spheres, such as "electronic" apples and potatoes, have been successfully applied to identify hot spots where mechanical damage is likely to occur along the fresh produce supply chain.

Various preharvest and postharvest techniques have been applied to reduce the incidence and severity of mechanical damage on fresh produce. Preharvest factors such as crop load, exposure to sunlight, irrigation amount and frequency, and crop maturity at harvest can affect the potential for damage to occur during harvesting and postharvest handling. Harvesting practices such as time during the day, gentle picking of produce, and the use of force-deceleration devices such as cushioning materials inside fruit containers can reduce the incidence of mechanical damage. Similarly, both the incidence and amount of damage on produce during postharvest processes can be reduced or prevented by applying postharvest treatments such as temperature conditioning of produce. Appropriate packaging reduces bruise through its cushioning effect (Fadiji et al., 2016), while inadequate packaging can lead to significant incidence of damage, often resulting in loss of whole consignment. Reducing fruit drop height to minimum level during postharvest operations also minimizes the occurrence of mechanical damage (Shafie et al., 2017).

1.6 Conclusion and Future Prospects

A steady supply of quality, safe, and nutritious fresh produce is an important component of a healthy diet. This has spurred rising global demand and trade in fresh fruit and vegetables and other fresh commodities. Strict market standards at destination markets, both local and international, often limit trade in fresh produce when exporters are not able to meet these standards. The presence of mechanical damage including bruises, puncture, cuts, abrasion, gouge, and scuff is a major cause of downgrading and rejection of produce consignments. In extreme cases, this can lead to total loss and waste which translates to high economic loss and a reduction in food availability. Reducing the incidence and severity of mechanical damage, therefore, contributes to the SDGs, and in particular SDG 12.3—"By 2030, halve per capita global food waste at the retail and consumer levels and reduce food losses along production and supply chains, including postharvest losses."

A plethora of terminologies have been used to describe and characterize the presence of mechanical damage on produce, including mechanical injury, physical damage, physical defect, etc. A closer look at the literature reveals that in some cases, the use of a particular terminology may be misleading, especially where a physiological disorder may be described as injury. By far, bruise damage by impact and compression is the leading cause of mechanical damage leading to downgrading and loss of produce. Bruise susceptibility test is the common method for assessing the potential for damage to occur, usually by a drop test.

Numerous preharvest, harvest, and postharvest factor influence the incidence and severity of mechanical damage of fresh produce, including orchard management practices, fresh produce properties, and postharvest handling practices, among others. Knowledge of the mode of influence of these factors for a specific commodity will assist in developing and deploying appropriate cost-effective control measures. Novel instrumented sphere technologies are available to detect hot spots along the supply chains, including the application of various optical sensors combined with chemometrics and machine learning to detect the presence and severity of bruise damage.

Despite significant progress made, mechanical damage, especially bruising, remains a major problem in the fresh produce industry. Future prospects include the integration of cost-effective and rapid sensors for on-farm and online assessment, especially for the detection and prevention of damage at onset and during early stages of development. Ultimately, the best strategy against mechanical damage of fresh produce is to prevent occurrence, a priori, and where this is not possible, a combination of cost-effective measures must be applied to reduce both the occurrence and severity of damage to minimize the economic costs and the negative impacts on both food security and the environment.

Acknowledgments This work is based on the research supported in part by the National Research Foundation of South Africa (Grant Numbers: 64813). The opinions, findings, and conclusions or recommendations expressed are those of the author(s) alone, and the NRF accepts no liability whatsoever in this regard.

References

Al-Dairi, M., Pathare, P. B., & Al-Yahyai, R. (2021a). Chemical and nutritional quality changes of tomato during postharvest transportation and storage. *Journal of the Saudi Society of Agricultural Sciences, 20*, 401. https://doi.org/10.1016/j.jssas.2021.05.001

Al-Dairi, M., Pathare, P. B., & Al-Yahyai, R. (2021b). Effect of postharvest transport and storage on color and firmness quality of tomato. *Horticulturae, 7*, 163. https://doi.org/10.3390/horticulturae7070163

Al-Said, F. A., Opara, U. L., & Al-Yahyai, R. A. (2009). Physico-chemical and textural quality attributes of pomegranate cultivars (Punica granatum L.) grown in the Sultanate of Oman. *Journal of Food Engineering, 90*(1), 129–134. 169.

Altisent, M. R. (1991). Chapter 9: Damage mechanisms in the handling of fruits. In J. Matthews (Ed.), *Progress in agricultural physics and engineering* (pp. 231–257). CAB International.

Anderson, G. (1990). The development of artificial fruits and vegetables. In *Proceedings of the European workshop on impact damage of fruits and vegetables (EWIDF) Zaragoza (Spain), 27-29 March* (Vol. 90, pp. 133–139). FIMA.

Baranowski, P., Mazurek, W., Witkowska-Walczak, B., & Sławiński, C. (2009). Detection of early apple bruises using pulsed-phase thermography. *Postharvest Biology and Technology, 53*, 91–100. https://doi.org/10.1016/j.postharvbio.2009.04.006

Baranowski, P., Mazurek, W., Pastuszka-Woźniak, J., & Majewska, U. (2012). Detection of early bruises in apples using hyperspectral data and thermal imaging. *Journal of Food Engineering, 110*, 345–355. https://doi.org/10.1016/j.jfoodeng.2011.12.038

Bollen, A. (2005). Major factors causing variation in bruise susceptibility of apples (Malus domestica) grown in New Zealand. *New Zealand Journal of Crop and Horticultural Science, 33*, 201–210. https://doi.org/10.1080/01140671.2005.9514351

Cappellini, R. A., & Ceponis, M. J. (1984). Postharvest losses in fresh fruits and vegetables. In H. E. Moline (Ed.), *Postharvest pathology of fruits and vegetables: postharvest losses in perishable crops, Univ. Calif., Agric. Exp. Sta., Bull. No. 1914* (pp. 24–30).

Ceponis, M. J., Cappellini, R. A., & Lightner, G. W. (1986). Disorders in citrus shipments to the New York market, 1972-1984. *Plant Disease, 70*, 1162. https://doi.org/10.1094/PD-70-1162

Ceponis, M. J., Cappellini, R. A., Wells, J. M., & Lightner, G. W. (1987). Disorders in plum, peach, and nectarine shipments to the New York market, 1972-1985. *Plant Disease, 71*, 947–952. https://doi.org/10.1094/PD-71-0947

Costa, A. G., Braga, R. A., Jr., Boas, E. V. B., & Risso, M. (2018). Early prediction of internal bruising in potatoes by biospeckle laser technique. *African Journal of Agricultural Research, 13*(14), 691–697. https://doi.org/10.5897/AJAR2017.12959

Crisost, C. H., Garner, D., Doyle, J., & Day, K. R. (1993). Relationship between fruit respiration, bruising susceptibility, and temperature in sweet cherries. *HORTSCIENCE, 28*(2), 132–135.

Dauchet, L., Amouyel, P., Hercberg, S., & Dallongeville, J. (2006). Fruit and vegetable consumption and risk of coronary heart disease: A meta-analysis of cohort studies. *The Journal of Nutrition, 136*, 2588–2593.

Eissa, A. H. A., Albaloushi, N. S., & Azam, M. M. (2013). Vibration analysis influence during crisis transport of the quality of fresh fruit on food security. *Agricultural Engineering International: CIGR Journal, 15*, 181–190.

ElMasry, G., Wang, N., Vigneault, C., Qiao, J., & El-Sayed, A. (2008). Early detection of apple bruises on different background colors using hyperspectral imaging. *Lebensmittel-Wissenschaft & Technologie, 41*, 337–345. https://doi.org/10.1016/j.lwt.2007.02.022

Ergun, M. (2017). Physical, physiochemical and electrochemical responses of 'Galaxy' apples to mild bruising. *European Journal of Horticultural Science, 82*, 244–250. https://doi.org/10.17660/eJHS.2017/82.5.4

Fadiji, T., Coetzee, C., Pathare, P., & Opara, U. L. (2016). Susceptibility to impact damage of apples inside ventilated corrugated paperboard packages: Effects of package design. *Postharvest Biology and Technology, 111*, 286–296. https://doi.org/10.1016/j.postharvbio.2015.09.023

FAO. (2011). *Global food losses and food waste—Extent, causes and prevention*. FAO.

Fatima Ali Elshiekh, F. A., & Abu-Goukh, A. A. (2008). Effect of harvesting method on quality and storability of grape fruits. *Univ. Khartoum Journal Agricultural Science, 16*(1), 1–14. 2.

Fudholi, A., Sopian, K., Ruslan, M. H., Alghoul, M. A., & Sulaiman, M. Y. (2010). Review of solar dryers for agricultural and marine products. *Renewable and Sustainable Energy Reviews, 14*, 1–30.

Gao, M., Guo, W., Huang, X., Du, R., & Zhu, X. (2021). Effect of pressing and impacting bruises on optical properties of kiwifruit flesh. *Postharvest Biology and Technology, 172*, 111385. https://doi.org/10.1016/j.postharvbio.2020.111385

Grudzińska, M., & Mańkowski, D. (2019). Reconditioning and weather conditions affect black spot damage during storage of potato (Solanum tuberosum L.) tubers. *Journal of the Science of Food and Agriculture, 99*(11), 4913–4921. https://doi.org/10.1002/jsfa.9719

Hounhouigan, M. H., Linnemann, A. R., Ingenbleek, P. T. M., Soumanou, M. M., van Trijp, H. C. M., & van Boekel, M. A. J. S. (2014). Effect of physical damage and storage of pineapple fruits on their suitability for juice production. *Journal of Food Quality, 37*, 268–273. https://doi.org/10.1111/jfq.12094

Hung, H. C., Joshipura, K. J., Jiang, R., Hu, F. B., Hunter, D., Smith-Warner, S. A., Colditz, G. A., Rosner, B., Spiedelman, D., & Willett, W. C. (2004). Fruit and vegetable intake and risk of major chronic disease. *Journal of the National Cancer Institute, 96*, 1577–1584.

Hussein, Z., Fawole, O. A., & Opara, U. L. (2017). Investigating bruise susceptibility of pomegranate cultivars during postharvest handling. *African Journal of Rural Development, 2*, 33–39. https://doi.org/10.22004/ag.econ.263298

Hussein, Z., Fawole, O. A., & Opara, U. L. (2018). Preharvest factors influencing bruise damage of fresh fruits—A review. *Scientia Horticulturae, 229*(2018), 45–58. https://doi.org/10.1016/j.scienta.2017.10.028

Hussein, Z., Fawole, O. A., & Opara, U. L. (2019). Bruise damage susceptibility of pomegranates (Punica granatum, L.) and impact on fruit physiological response during short term storage. *Scientia Horticulturae, 246*, 664–674. https://doi.org/10.1016/j.scienta.2018.11.026

Hussein, Z., Fawole, O. A., & Opara, U. O. (2020a). Bruise damage of pomegranate during long-term cold storage: Susceptibility to bruising and changes in textural properties of fruit. *International Journal of Fruit Science, 20*, S211–S230. https://doi.org/10.1080/15538362.2019.1709602

Hussein, Z., Fawole, O. A., & Opara, U. O. (2020b). Effects of bruising and storage duration on physiological response and quality attributes of pomegranate fruit. *Scientia Horticulturae, 267*, 109306. https://doi.org/10.1016/j.scienta.2020.109306

Hussein, Z., Fawole, O. A., & Opara, U. L. (2020c). Harvest and postharvest factors affecting bruise damage of fresh fruits. *Horticultural Plant Journal, 6*(1), 1–13. https://doi.org/10.1016/j.hpj.2019.07.006

Ismail, S. M. M., Ali, H. M., & Habiba, R. A. (1993). GC-SCD and GC-MS. analysis of profenofos residues and its biochemical effects in tomatoes and tomato products. Journal of agricultural. *Food Chemistry, 41*, 610–615.

Kampp, J., & Pedersen, J. (1990). Quality of imported and domestic fruits and vegetables in the Danish retail trade with special reference to mechanical damages. In *Proceedings of the European workshop on impact damage of fruits and vegetables (EWIDF), Zaragoza (Spain), 27-29 March* (Vol. 90, pp. 11–16). FIMA.

Keresztes, J. C., Goodarzi, M., & Saeys, W. (2016). Real-time pixel based early apple bruise detection using short wave infrared hyperspectral imaging in combination with calibration and glare correction techniques. *Food Control, 66*, 215–226. https://doi.org/10.1016/j.foodcont.2016.02.007

Ketsa, S., Klaewkasetkorn, O., & Kosittrakul, M. (1995). Seasonal-changes-in-fruit-quality-of-rambutans-in-retail-markets-in-Thailand. *Tropical Science, 35*, 240–244.

Knee, M., & Miller, A. R. (2002). Chapter 7: Mechanical injury. In M. Knee (Ed.), *Fruit quality and its biological basis* (pp. 155–157). Sheffield Academic Press.

Komarnicki, P., Stopa, R., Szyjewicz, D., & Młotek, M. (2016). Evaluation of bruise resistance of pears to impact load. *Postharvest Biology and Technology, 114*, 36–44. https://doi.org/10.1016/j.postharvbio.2015.11.017

Kumar, V., Purbey, S. K., Anal, A. K. D., et al. (2016). *Crop Protection, 79*, 97–104.

Lam, P. F., Kosiyachinda, S., Lizada, M. C. C., Mendoza, D. B., Jr., Prabawati, S., & Lee, S. K. (1987). Postharvest physiology and storage of rambutan. In P. F. Lam & S. Kosiyachinda (Eds.), *Fruit development postharvest physiology and marketing in ASEAN* (pp. 39–50). ASEAN Food Handling Bureau.

Lanza, E., & Butrum, R. R. (1986). A critical review of food fiber analysis and data. *Journal of the American Dietetic Association, l86*, 732–743.

Lee, E. (2005). *Quality changes induced by mechanical stress on Roma-type tomato and potential alleviation by 1-Methylcyclopropene*. Doctoral dissertation, University of Florida, Gainesville.

Li, J., Yan, J., Cao, J., Zhao, Y., & Weibo, J. P. (2012). Preventing the wound-induced deterioration of Yali pears by chitosan coating treatments. *Food Science and Technology International, 18*, 123.

Li, J., Chen, L., & Huang, W. (2018). Detection of early bruises on peaches (Amygdalus persica L.) using hyperspectral imaging coupled with improved watershed segmentation algorithm. *Postharvest Biology and Technology, 135*, 104–113. https://doi.org/10.1016/j.postharvbio.2017.09.007

Macheka, L., Ngadze, R. T., Manditsera, F. A., Mubaiwa, J., & Musundire, R. (2013). Identifying causes of mechanical defects and critical control points in fruit supply chains: an overview of a banana supply chain. *International Journal of Postharvest Technology and Innovation, 3*(2), 109–122. https://doi.org/10.1504/IJPTI.2013.055841

Maia, V. M., Salomão, L. C. C., Siqueira, D. L., Puschman, R., Mota Filho, V. J. G., & Cecon, P. R. (2011). Physical and metabolic alterations in "Prata Anã" banana induced by mechanical damage at room temperature. *Scientia Agricola, 68*, 31–36. https://doi.org/10.1590/S0103-90162011000100005

Montero, C. R. S., Schwarz, L. L., Santos, L. C. D., Andreazza, C. S., Kechinski, C. P., & Bender, R. J. (2009). Postharvest mechanical damage affects fruit quality of 'Montenegrina' and 'Rainha' tangerines. *Pesquisa Agropecuária Brasileira, 44*, 1636–1640. https://doi.org/10.1590/S0100-204X2009001200011

Moretti, C. L., Sargent, S. A., Huber, D. J., Calbo, A. G., & Puschmann, R. (1998). Chemical composition and physical prosperities of pericarp, locule and placental tissues of tomatoes with internal bruising. *Journal of the American Society of Horticultural Sciences, 123*, 656–660.

Munhuweyi, K., Opara, U. L., & Sigge, G. (2016). Postharvest losses of cabbages from retail to consumer and the socio-economic and environmental impacts. *British Food Journal, 118*(2), 286–300. https://doi.org/10.1108/BFJ-08-2014-0280

Nturambirwe, J. F. I., & Opara, U. L. (2020). Machine learning applications to non-destructive defect detection in horticultural products. *Biosystems Engineering, 189*, 60–83. https://doi.org/10.1016/j.biosystemseng.2019.11.011

Nturambirwe, J. F. I., Nieuwoudt, H. H., Perold, W. J., & Opara, U. L. (2020). Detecting bruise damage and level of severity in apples using a contactless NIR spectrometer. *Applied Engineering in Agriculture, 36*, 257–270. https://doi.org/10.13031/aea.13218

Nturambirwe, J. F. I., Perold, W. J., & Opara, U. L. (2021). Classification learning of latent bruise damage to apples using shortwave infrared hyperspectral imaging. *Sensors (Basel), 21*(15), 4990. https://doi.org/10.3390/s21154990

Opara, L. U. (2006a). Postharvest technology for linking production to markets. *International Journal of Postharvest Technology & Innovation, 1*(2), 139–141.

Opara, L. U. (2006b). A new era in postharvest technology. *International Journal of Postharvest Technology & Innovation, 1*(1), 1–3.

Opara, U. L. (2007). Bruise susceptibilities of 'gala' apples as affected by orchard management practices and harvest date. *Postharvest Biology and Technology, 43*, 47–54. https://doi.org/10.1016/j.postharvbio.2006.08.012

Opara, U. L., & Pathare, P. B. (2014). Bruise damage measurement and analysis of fresh horticultural produce-a review. *Postharvest Biology and Technology, 91*, 9–24. https://doi.org/10.1016/j.postharvbio.2013.12.009

Opara, L. U., Al-Said, F. A., & Al-Abri, A. (2007). Assessment of what the consumer values in fresh fruit quality: Case study of Oman. *New Zealand Journal of Crop and Horticultural Science, 35*(2), 235–243. https://doi.org/10.1080/01140670709510190

Opara, U. L., Al-Ani, M. R., & Al-Shuaibi, Y. (2009). Physico-chemical properties, vitamin C content, and antimicrobial properties of pomegranate fruit (Punica granatum L.). *Food and Bioprocess Technology, 2*, 315–321.

Opara, I. K., Fawole, O. A., Kelly, C., & Opara, U. L. (2021a). Quantification of on-farm pomegranate fruit postharvest losses and waste, and implications on sustainability indicators: South African case study. *Sustainability, 13*, 5168. https://doi.org/10.3390/su13095168

Opara, I. K., Fawole, O. A., & Opara, U. L. (2021b). Postharvest losses of pomegranate fruit at the packhouse and implications for sustainability indicators. *Sustainability, 13*, 5187. https://doi.org/10.3390/su13095187

Opara, I. K., Fawole, O. A., & Opara, U. L. (2022). Pomegranate production and export during the past decade in South Africa and incidence of postharvest losses—A review. *Acta Horticulturae, 1349*, 325–332. https://doi.org/10.17660/ActaHortic.2022.1349.45

Palafox-Carlos, H., Ayala-Zavala, J. H., & Gonzalez-Aquilar, C. A. (2011). The role of dietary fiber in the bioaccessibility and bioavailability of fruit and vegetable antioxidants. *Journal of Food Science, 76*, R6.

Pathare, P. B., & Al-Dairi, M. (2021a). Bruise susceptibility and impact on quality parameters of pears during storage. *Frontiers in Sustainable Food Systems, 5*. https://doi.org/10.3389/fsufs.2021.658132

Pathare, P. B., & Al-Dairi, M. (2021b). Bruise damage and quality changes in impact-bruised, stored tomatoes. *Horticulturae, 7*, 113. https://doi.org/10.3390/horticulturae7050113

Pathare, P. B., & Al-Dairi, M. (2022). Effect of mechanical damage on the quality characteristics of banana fruits during short-term storage. *Discover Food, 2*, 4. https://doi.org/10.1007/s44187-022-00007-7

Pholpho, T., Pathaveerat, S., & P. (2011). Sirisomboon classification of longan fruit bruising using visible spectroscopy. *Journal of Food Engineering, 104*, 169–172.

Prusky, D. (2011). Reduction of the incidence of postharvest quality losses, and future prospects. *Food Security Journal, 3*, 463–474.

Riboli, E., & Horel, T. (2003). Epidemiologic evidence of the protective effect of fruit and vegetables on cancer risk. *The American Journal of Clinical Nutrition, 78*(Suppl 3), S559–S569.

Rivera, N. V., Gómez-Sanchis, J., Pérez, J. J. C., Carrasco, J. J., Millán-Giraldo, M., Lorente, D., Cubero, S., & Blasco, J. (2014). Early detection of mechanical damage in mango using NIR hyperspectral images and machine learning. *Biosystems Engineering, 122*, 91–98. https://doi.org/10.1016/j.biosystemseng.2014.03.009

Ruiz-Altisent, M., & Moreda, G. P. (2011). Fruits, mechanical properties and bruise susceptibility. In J. Gliński, J. Horabik, & J. Lipiec (Eds.), *Encyclopedia of agrophysics. Encyclopedia of earth sciences series*. Springer. https://doi.org/10.1007/978-90-481-3585-1_63

Sablani, S. S., Opara, L. U., & Al-Balushi, K. (2005). Influence of bruising and storage temperature on vitamin C content of tomato fruit. *Journal of Food, Agriculture, and Environment (Food & Health-Part A), 3-1*, 1–3.

Scherrer-Montero, C. R., dos Santos, L. C., Andreazza, C. S., Getz, B. M., & Bender, R. J. (2011). Mechanical damages increase respiratory rates of citrus fruit. *International Journal of Fruit Science, 11*(3), 256–263. https://doi.org/10.1080/15538362.2011.608297

Shafie, M. M., Rajabipour, A., & Mobli, M. (2017). Determination of bruise incidence of pomegranate fruit under drop case. *International Journal of Fruit Science, 17*, 296–309.

Slavin, J. L., & Lloyd, B. (2012). Health benefits of fruits and vegetables. *Advances in Nutrition, 3*(4), 506–516. https://doi.org/10.3945/an.112.002154

Smith-Warner, S. A., Spiegelman, D., Yaun, S. S., Adami, H. O., Beeson, W. L., van den Brandt, P. A., Folson, A. R., Fraser, G. E., Freudenheim, J. L., Goldhobohm, R. A., et al. (2001). Intake of fruits and vegetables and risk of breast cancer: A pooled analysis of cohort studies. *Journal of the American Medical Association, 285*, 769–776.

Snowdon, A. L. (1990). *A color atlas of post-harvest diseases of fruits and vegetables: Vol 2. Vegetables*. CRC.

Valenciano Garcia, J. (1990). Estimate of quality losses of fruit and vegetables in packinghouses with special reference to mechanical damages. In *Proceedings of the EWIDF, Zaragoza (Spain), 27-29 March* (Vol. 90, pp. 19–25). FIMA.

Wells, M., & Butterfield, J. E. (1999). Incidence of salmonella on fresh fruits and vegetables affected by fungal rots or physical injury. *Plant Disease, 83*(8), 722–726. https://doi.org/10.1094/PDIS.1999.83.8.72

Wells, J. M., Butterfield, J. E., & Ceponis, M. J. (1994). Diseases, physiological disorders, and injuries of plums marketed in metropolitan New York. *Plant Disease, 78*, 642–644.

Xia, M., Zhao, X., Wei, X., Guan, W., Wei, X., Xu, C., & Mao, L. (2020). Impact of packaging materials on bruise damage in kiwifruit during free drop test. *Acta Physiologiae Plantarum, 42*, 1–11. https://doi.org/10.1007/s11738-020-03081-5

Yahia, E. M., Maldonado Celis, M. E., & Svendsen, M. (2017). Chapter 1. The contribution of fruit and vegetable consumption to human health. In E. M. Yahia (Ed.), *Fruit and vegetable phytochemicals: chemistry and human health* (2nd ed., pp. 1–52). Wiley. https://doi.org/10.1002/9781119158042

Yahia, E. M., García-Solís, P., & Maldonado Celis, M. E. (2019). Chapter 2. Contribution of fruits and vegetables to human nutrition and health. In E. M. Yahia (Ed.), *Postharvest physiology and biochemistry of fruits and vegetables* (pp. 19–45). Woodhead Publishing. https://doi.org/10.1016/B978-0-12-813278-4.00002-6

Chapter 2
Factors Affecting Bruise Damage Susceptibility of Fresh Produce

Umezuruike Linus Opara and Zaharan Hussein

Abstract The acknowledgment of an increasing global demand for high quality in fruits and vegetables has grown in recent years. Nevertheless, evidence of severe problems of mechanical damage is increasing, and this is affecting the trade of fruits worldwide. Damage of fresh fruit as well as vegetables by bruising is a significant concern in the horticulture sector, with occurrences throughout preharvest, harvest, and all phases of the postharvest handling chain. This damage can cause major postharvest and economic losses, as well as decreased crop quality and potential food safety issues. The potential market for fresh high quality fresh fruits and vegetables remains constrained by the lack of quality in the majority of products that reach consumers; this is true for both local and import/export markets, resulting in a decrease in consumption of fresh fruits in favour of other non-crop commodities. Accordingly, taste/flavour, freshness, cosmetic appearance, and sanitary conditions appear to have the greatest impact on the purchasing decisions of these consumers.

For many years, scientists have been studying mechanical damage in fruits and vegetables. In recent years, research has grown significantly, and various aspects of the bruise damage issue have been addressed. These include bruise susceptibility, simulation, detection, and measurement, with both invasive and non-invasive measurements reported. Other studies have concentrated on applicable mechanical models for the contact problem, the response of biological tissues to loading, devices for detecting damage causes in machines and equipment, and methods for sensing bruises in grading and sorting.

U. L. Opara (✉)
SARChI Postharvest Technology Research Laboratory, Africa Institute for Postharvest Technology, Faculty of AgriSciences, Stellenbosch University, Stellenbosch, South Africa

UNESCO International Centre for Biotechnology, Nsukka, Enugu State, Nigeria
e-mail: opara@sun.ac.za

Z. Hussein
Department of Food Science and Technology, College of Agricultural Sciences and Technology, Centre for Innovation and Technology Transfer, Mbeya University of Science and Technology, Mbeya, Tanzania

P. B. Pathare, U. L. Opara (eds.), *Mechanical Damage in Fresh Horticultural Produce*, https://doi.org/10.1007/978-981-99-7096-4_2

Understanding the factors that influence bruising susceptibility or resistance is critical in developing prevention strategies. This chapter covers an overview of bruise damage in fresh produce, as well as its perishability, causes, and mechanisms. This chapter also discusses the development of bruises in fresh produce, as well as the various factors that have been identified as influencing the appearance and development of bruise damage in fruits. Harvest readiness, ripening, harvest time (during the day or season), time following harvest, storage condition and length, produce quality, and their respective effects on the bruise susceptibility of fresh fruits have all been discussed. This chapter covers other topics such as measuring bruise damage on fresh produce and methods for reducing/controlling bruise damage. The future research directions are also discussed.

Keywords Bruise damage · Fresh produce · Fruits and vegetables · Bruise susceptibility · Mechanical damage

2.1 Introduction

Fresh produce (fruits and vegetables) are high in micronutrients, fibre, vitamins, and phytochemicals (which have antioxidant properties) such as anthocyanins, carotenoids, polyphenols, and flavonoids. As a result, they are necessary components of the daily human diet (Opara & Al-Ani, 2010; Hussein et al., 2015, 2018). Fresh fruit and vegetables have been linked to a variety of nutritional and health benefits. There are reports indicating that eating unhealthily high-calorie foods continues to be a risk factor for a number of chronic human diseases (Hert et al., 2014; Mason-D'Croz et al., 2019). Consuming fruits and vegetables is strongly suggested as a healthy diet to battle illnesses associated with sedentary lifestyle and degenerative illnesses such as cancer, high blood pressure, cardiovascular disease, and ageing due to the strong association between fresh fruit intake and various nutritional and health advantages (Fawole et al., 2012; Hussein et al., 2015; Mphahlele et al., 2016). Therefore, increasing fruit and vegetable consumption is a critical component of the transition to healthier, more sustainable diets (Mason-D'Croz et al., 2019).

The national and international dietary guidelines for most countries recommend fruits and vegetables as a healthy diet to combat the negative health effects of sedentary lifestyles and degenerative diseases (Ramos et al., 2013; Kuriyan et al., 2014). For instance, the WHO recommended a daily fruit and vegetable consumption of 400 g per person (WHO, 2018). Increased intake of fruits and vegetables is, among other things, an important national strategy framework for public health nutrition (Kuriyan et al., 2014). As a result, dietary recommendations for most countries emphasise increased consumption of fruits and vegetables (F&Vs) as part of a broader dietary pattern to lower the risk of diet-related chronic diseases and metabolic disorders (Willett et al., 2019; Mason-D'Croz et al., 2019). This, together with the growing human populace, will lead to significant expansion in world's fruit and vegetable business into the future.

There has been a rapid increase in the production of fruits and vegetables in the last decade. The global fruit and vegetable market increased from 11.7% to $1517.3 billion in 2011. A decade later, fruits and vegetables now account for approximately 22% of global food production (Żaczek et al., 2015; Grünwald, 2021). The vegetable sector accounts for 12% of overall agricultural production, alongside fruits which account for the other 10% (Grünwald, 2021). According to recent studies, global fruit and vegetable (F&V) production will need to increase by 50–150% by 2050 to ensure sustainable and healthy diets for ten billion people (Stratton et al., 2021).

2.1.1 *Perishability and Mechanical Damage of Fresh Fruits and Vegetables*

Despite the fact that global production of F&V has increased by 50% in the last two decades, F&Vs have the highest levels of pre- and postharvest loss of any food group, estimated at 35–55% across all geographies (FAO, 2019). As a result, recent advancements in sustainable food systems place an emphasis on strategies to reduce postharvest losses and waste fresh F&Vs (Cattaneo et al., 2020; Stathers et al., 2020). Fresh fruits and vegetables (F&V) suffer from significant postharvest loss as well as retail and consumer waste (FAO, 2019; Grünwald, 2021). Perishability and loss exacerbate the gap between current and expected availability and recommended F&V intake levels (Mason-D'Croz et al., 2019), posing a significant challenge to long-term growth.

Traditional and novel food preservation techniques such as edible coatings, controlled atmosphere storage (CAS), modified atmosphere packaging (MAP), smart and intelligent packaging, and so on can both increase the duration of storage, retain nutrients, and decrease postharvest wastage (Miller & Knudson, 2014). Cold chain technology and packaging are commonly used by suppliers and producers to reduce losses along the supply chain. Furthermore, other techniques for controlling perishability of fresh fruits (during storage and distribution) such as edible coatings derived from mineral and plant compounds have gained widespread application (Herrero et al., 2020; Stratton et al., 2021).

Several obstacles limit fruit and vegetable production output, the most significant of which is postharvest loss. Due to a variety of factors, fresh produce are susceptible to losses as they travel along the postharvest chain, including improper handling, biodeterioration by microbes, insects, rodents, or birds, and mechanical damage (Kiaya, 2014; Emana et al., 2017). Other sources of significant postharvest losses include inadequate packing, insufficient storage facilities, and poor modes of transportation and the operation (driver factor). Furthermore, the F&V industry also experiences significant economic losses each year because of mechanical damage to fresh produce that occurs both before and after harvest (Hussein et al., 2018). During harvesting and postharvest handling, mechanical damage occurs when one or

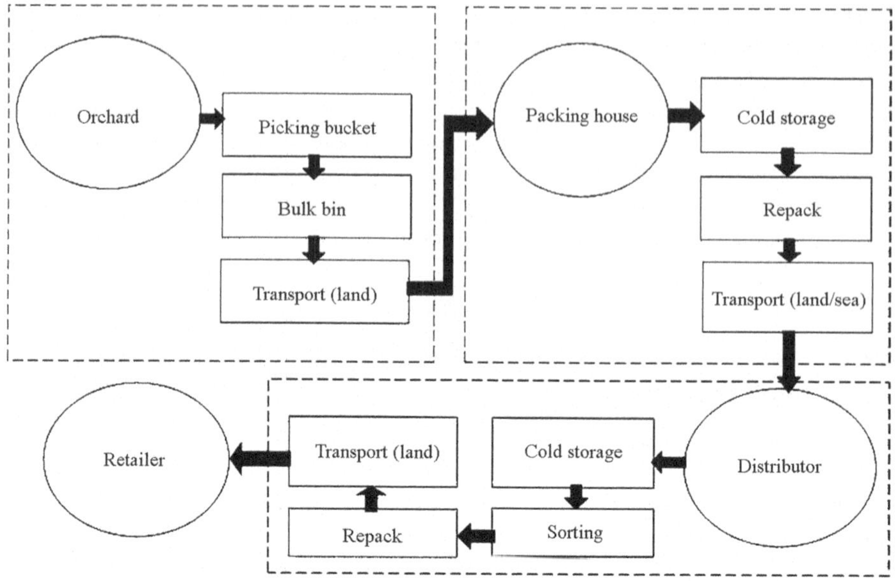

Fig. 2.1 Postharvest route of fresh F&V from the point of harvest (orchard) to retail stores. (Adapted from Hussein et al., 2020)

more loading forces occur on produce, causing the outer layer of the fresh produce pericarp to become damaged (Opara & Pathare, 2014; Al-Dairi et al., 2021).

From the orchard (at the harvest point) to the supermarket shelves, the journey of fresh fruit and vegetables is long and winding (Eissa et al., 2013; Hussein et al., 2020). That is to say, from the time they are harvested to the time they are sold in stores, fresh fruits and vegetables undergo several phases and procedures, includes harvesting, packaging, organising, storage, and ferrying (Fig. 2.1). These processes put F&V in a position to be subjected to a wide range of loading conditions, some of which can result in mechanical damage (Bugaud et al., 2014; Hussein et al., 2020; Pathare & Al-Dairi, 2022).

Bruising is the most prevalent kind of mechanical injury induced by the breakdown of subcutaneous cells when the loading intensity surpasses the failure strain of the fruit tissue (Opara & Pathare, 2014). Fruit bruises are a result of impact with another fruit or a rigid body, vibration, and compression during harvest, transportation, and handling (Eissa et al., 2008; Stropek & Gołacki, 2015). Bruised produce is characterised by undesirable blemishes (brownish to black patches) that downgrade both appearance and nutritional value of any fresh food and result in significant financial losses (Bugaud et al., 2014; Pathare & Al-Dairi, 2022). A large percentage of fresh produce quality loss is due to bruises that develop between the time of harvest and consumption, which also lowers the fruit's market value and, ultimately, lowers potential sales significantly (Ahmadi et al., 2010; Saracoglu et al., 2011).

Accordingly, low grade and low quality produce resulting from bruise damage fetch less money for both growers and packers (Eissa et al., 2008; Hussein et al., 2018).

2.2 Bruising of Fresh Produce: An Overview

Fresh produce, primarily vegetables and fruits, are subjected to a variety of postharvest procedures across the cold chain, including harvesting, organising, packing, storage, and logistics. Mild mechanical damages on fruit are difficult to identify as a result of the increased usage of mechanised activities during harvest and postharvest processing. Fruit is predisposed to various degrees of dynamic stresses due to preharvest causes, harvesting, and postharvest handling processes (Hussein et al., 2018).

Bruising is the most common type of mechanical damage resulting from such forces (Opara & Pathare, 2014). The measure of the fresh produce response to loading externally is referred to as bruise susceptibility (BS) or bruise resistance (Opara & Pathare, 2014; Ahmadi, 2012). Varieties of fresh produce are more susceptible to mechanical bruising due to variations in energy absorption capacity. Consequently, the amount of loading (dynamic or static) imparted to a produce is thought to be the key bruising component, and this loading or absorbed energy is typically described in terms of energy absorbed (Blahovec & Zidova, 2004; Blahovec, 2006). Accordingly, when harvesting and handling fresh produce, all potential dynamic loading events must be considered. These include impacts that occur when fruit fall into picking buckets or during sorting, as well as vibrations, most commonly experienced during transportation (Komarnicki et al., 2016). Static or compression loading, however, can occur during harvesting, transportation, or storage due to overfilling and stacking of poorly designed bins or crates (Lewis et al., 2007; Komarnicki et al., 2016). Table 2.1 demonstrates several loading situations and related drop heights that might put fresh fruits at risk for impacts that could lead to bruising damage throughout harvest as well as postharvest handling.

Numerous studies have demonstrated that the quantity of mechanical energy delivered to and absorbed by a product during impact, compression, or vibration, regardless of other circumstances, is a significant determining factor on the magnitude of damage that happens (Opara, 2007; Zarifneshat et al., 2010; Pathare &

Table 2.1 Potential dynamic pomegranate loading situations and associated drop heights

Loading situation	Process stage	Potential drop height (m)
Orchard	Picking bucket	0.6
Packhouse	Bulk bin	0.6–1
	Repack	0.05–0.15
Distributor	Sorting	0.05–1.15
Retailer	Putting on display	0.05–0.3

Adapted from Shafie et al. (2017)

Al-Dairi, 2021). Interestingly, other elements, including the physiological and biochemical characteristics of the product and the environment in which it produce grows, have a substantial impact on the sensitivity to bruising damage (Van Linden et al., 2006; Strehmel et al., 2010). Therefore, the best mitigation technique on bruising damage on fresh produce is without a doubt to minimise compressions or vibrations on product during mechanical handling (Li & Thomas, 2014).

Fruit factors such as turgidity and firmness have been shown to affect bruise susceptibility in fresh produce. According to García et al. (1995), the impact response analysis of apples and pears revealed that mechanical stresses in the tissues were higher in turgid fruit, making it more susceptible to bruising. The authors also demonstrated that, due to the impact of fruit turgidity, fruits collected early in the day were less prone to bruising than those collected later in the day. It is pertinent to consider that when turgor pressure increases, cell wall flexibility diminishes and fruit tissue becomes more brittle, making it more prone to bruise injury (Perkins et al., 2017).

Yurtlu and Erdoğan (2005) investigated the bruising susceptibility of pears and apples with regard to storage time and discovered that mechanical properties such as modulus of flexibility changed depending on harvest date, preservation time, and conditions of storage, influencing their susceptibility to bruising. The authors found that because fresh produce cells are biological materials and living tissues, they are susceptible to environmental effects such as humidity, temperature, oxygen, and energy absorption throughout development and storage. As a result, mechanical qualities including modulus of elasticity vary depending on harvest date, storage period, and storage circumstances.

Some fruits are more susceptible to bruising than others, and this is due to the most visible physical attributes, and the histological and physiological traits (Ruiz-Altisent, 1991). In this regard, fresh produce are divided into 'rigid' (hard) and 'liquid' (plastic or soft) fruits (Ruiz-Altisent, 1991). This classification, however, may be somehow inaccurate, in part because many fruits change from one texture to another due to ripening or the influence of storage environmental conditions. Table 2.2 shows a simple classification of fresh fruits and vegetables based on their susceptibility to impact bruise damage.

Despite other factors relating to environmental conditions and the magnitude of impact applied to produce, the severity of bruise damage to the fresh produce is primarily related to the following: (1) height of fall; (2) initial velocity; (3) number of impacts; (4) type of impact surface and size; and (5) physical properties of the produce (Ruiz-Altisent, 1991). Since it is well known that impact energy is directly proportional to the mass of the dropping object, fruit mass plays a crucial role in bruise damage susceptibility. This could mean that the smaller the fruit, the safer it will be to handle, and the lesser the impact damage it will receive compared to larger fruit at the same drop height. This emphasises the need to conduct appropriate research on bruise damage susceptibility for each variety of fruit in order to accurately describe and simulate their bruise damage susceptibility behaviour.

Table 2.2 Fresh fruits and vegetables classification based on impact bruise damage susceptibility

Produce category	Fresh produce physical and physiological characteristics	Susceptibility to impact bruising in the[a] scale of 1–4
1. Rigid produce (apples, pears, peaches, nectarines, apricots, avocados, mangoes, papaya, kiwi fruits, potatoes, etc.)	• Produce whose strength is based on a mostly rigid structure, surrounded by a thin elastic skin • Resistance to bruising for these commodities is based on the produce flesh's histological and physiological traits	1
2. Liquid produce (plums, tomatoes, grapes, cherries, and berries)	• Produce composed of a liquid or 'soft' mass enclosed in a mostly elastic skin	2
3. Thick-skinned produce (melons, watermelons, pomegranates, and bananas)	• Although these produces are very resistant to impacts, skin rupture issues may occur	3
4. Fibrous produce (pineapples, cucumber)	• React in a very different manner to impacts, not widely studied	4

[a](1 = highly susceptible; 4 least susceptible). Table modified from Ruiz-Altisent (1991)

2.3 Factors Contributing to Fresh Produce Bruising

2.3.1 *Effect of Maturity and Ripening at Harvest*

The lack of a recognised maturity index for certain commodities is one of the key reasons of bruising damage occurrence at the harvest stage (Kiaya, 2014). As a result, for many fruits, the maturation stage is one of the most critical elements determining bruise damage susceptibility (García et al., 1995; Lee, 2005). The degree of ripeness influences mechanical qualities, physiological processes, and consequently mechanical susceptibility. Fruit undergoes a number of physiological changes as it grows, which may impact its sensitivity to bruising injury. Reports indicate that mature fruit is more sensitive to bruise damage than young fruit (García et al., 1995; Martinez-Romero et al., 2004; Canete et al., 2015). Furthermore, harvest ripeness may affect produce susceptibility to water loss as well as mechanical damage (Van Linden et al., 2006; Canete et al., 2015).

Response to mechanical stress on freshy harvested produce is influenced by the maturity phase at harvest. Canete et al. (2015) observed an increase in the bruise size of 'Algerie' loquat fruit as the maturation stage increased. Fruit firmness, however, has proved beneficial as a benchmark for classifying fruit into various phases of maturity or for sorting overripe and damaged fruit from wholesome ones (Tabatabaekoloor, 2013). Firmness is an indicator of cell wall strength in fruit tissue and a method of determining fruit ripeness. Firmness decreases as ripening progresses, whereas bruise susceptibility increases as firmness decreases (White et al., 2009; Perkins et al., 2017).

Fig. 2.2 Bruising in 'Hass' avocados subjected to a 'very slight' 5 N thumb compression at various stages of ripeness. The arrow points the direction of ripening stages. (Modified from Perkins et al., 2017)

In apples and pears, excessive pressure of turgor in the flesh/tissue has been associated with higher bruise sensitivity (García et al., 1995). These authors reported a link between fruit firmness, turgor, ripening process, and bruise susceptibility, and the bruise model obtained demonstrated that fruit turgidity and hardness changes occur throughout ripening, such that bruise damage reduced with declining fruit turgor in apples while fruit firmness increased. Firmness is a gauge for cell wall resilience in fruit tissue that may be used to predict maturity. Firmness reduces with ripening in avocados, for example, and is described as hard, rubbery, springy, softening, firm-ripe, soft-ripe, overripe, and extremely overripe (White et al., 2009; Perkins et al., 2017). According to Perkins et al. (2017), 'Hass' avocados began to bruise at various stages of ripeness after a 'very slight' thumb compression of 5 N. Overripe (least firm) avocado had the largest bruise size and softened or firm-ripe fruit had the smallest. This shows that hard 'green or firm' avocado fruits are more resistant to bruising than ripe or overripe avocado fruits (Fig. 2.2).

Fruit ripening has also been shown to affect bruising damage susceptibility in other fruits that include apples, tomatoes (Van Zeebroeck et al., 2007a), and loquat (Canete et al., 2015). According to the authors, the degree of ripeness in fruits impacts the frequency of bruising. Canete et al. (2015) reported an increase of 56 percent in bruise area from the beginning of maturity to the end of ripening. Van Linden and De Baerdemaeker (2005) observed similar effects in tomatoes, noting greater bruise damage as maturity progressed. The ripening process, according to Studman (1997), is connected with a decline in cell wall strength. In this sense, the cells of a riper fruit are less able to sustain external loading, making the fruit more susceptible to bruise damage (Studman, 1997; Van Linden et al., 2006). To prevent bruising and retain the quality attributes of fruit, it is necessary to understand the various phases of maturity and ripening and their respective effect on bruise susceptibility before harvest. Ripe or overripe fruits are more susceptible to bruising than unripe ones.

2.3.2 *Effect of Harvest Time (During the Day or Season)*

The time of day when fresh fruit is collected influences its bruise susceptibility. It has been found that, independent of fruit cultivar, fruit are more likely to experience higher bruise damage during the early hours than in the late afternoon when exposed to the same impact energy. Abbott et al. (2009) studied the influence of harvest time on the bruising susceptibility of 'Cripps Pink' and 'Granny Smith' apples and discovered that fruit harvested later in the day was less sensitive to bruising than fruit harvested earlier in the day. The authors hypothesised that the observed reduction in bruise susceptibility as the day continued was due to changes in cell turgidity. Furthermore, as the temperature rises from morning to afternoon, fruit senescence, shrivelling, and wilting increase, contributing to reduced impact or compression energy absorption at harvest (García et al., 1995; Abbott et al., 2009).

According to other reported research, delayed harvest worsens bruise damage, while others found a reduction in bruise susceptibility. For apples, fruit collected near the conclusion of the commercial harvest period (late harvest) are more prone to bruising than fruit harvested at the start of the season (early harvest) (García et al., 1995; Ericsson & Tahir, 1996; Gunes et al., 2002; Opara, 2007). It was concluded that early-harvested apples are less susceptible to bruising due to higher pulp firmness, which reduces the absorption of impact energy compared to less mature fruits. According to Opara (2007), the increase in bruise susceptibility of 'Gala' apples from early-, mid-, to late-season harvest was attributed to the differences in fruit size that could be affected by advancing maturity and delaying in harvesting coupled with the differences in the fruit's physico-textural attributes at harvest. Similarly, Gunes et al. (2002) reported higher cranberry fruit losses in the late harvest lot than in the early harvest lot. They argued that the late-harvested cranberries were riper and thus more susceptible to bruising during harvesting and handling than earlier harvested berries. Overall, these findings emphasise the importance of crop harvesting at the appropriate time and the right maturity stage.

2.3.3 *Effect of Time After Harvest*

Evaluating the impact of postharvest time lapse on bruise susceptibility of fresh fruit is crucial since it may serve as a valuable tool for fruit packers in determining the ideal time for fruit packing and/or storage. Fresh vegetables and fruits are often conveyed from the farm to a packaging plant or distribution hub after harvest Given the perishable nature of these crops, it is necessary that the time between harvest and these subsequent operations is not extended. Martinez-Romero et al. (2004)

identified that an increase in the time elapsed (days) after harvest is attributed to the decline in fruit turgidity of young tissue which consequently improves their resistance to bruise damage. Further reports revealed that prolonged time between harvest and transportation of fruit for further operations leaves the produce with field heat for a longer time, which subsequently leads to faster senescence and reduced fruit turgidity (Mowatt, 1997; Bollen, 2005).

Apart from mechanical damage, there is a high risk of biological deterioration if the produce is kept for too long before reaching the intended market, processing, or packhouse (Kiaya, 2014). This effect may be associated with the rate of respiration, the rate of ethylene synthesis and action, the rate of compositional changes (related with colour, texture, flavour, and nutritional value), water stress, sprouting and roots, physiological diseases, and pathological breakdown. If the produce must be kept in the orchard for an extended period of time for any reason, it is critical to keep in place the proper conditions of the environment, including as humidity levels, temperature, air velocity, and atmospheric composition (concentration of oxygen, carbon dioxide, and ethylene), and the sanitation procedures.

2.3.4 *Effect of Storage Condition and Duration*

Several research investigations have found storage time as an important factor in reducing bruising susceptibility for several types of fresh food, including apples (Pang, 1993; García et al., 1995; Timm et al., 1998); peaches (Hung & Prussia, 1989; Vursavus & Ozguven, 2003); Pears (Yurtlu & Erdoğan, 2005); plums (Lippert & Blanke, 2004); and pomegranates (Shafie et al., 2015; Hussein et al., 2019a, b). The decrease in sensitivity to bruising has been linked to a decrease in fruit turgidity during storage (García et al., 1995). Garcia et al. also reported that turgid fruit deforms less than flaccid fruit at the same impact energy. This has been related to the greater absorption of mechanical forces in turgid fruit, which results in increased bruise susceptibility (Timm et al., 1998).

Fresh food that has been stored for an extended period of time may be more resistant to mechanical impact. The 5% increase in bruise size after 120 days of pomegranate fruit storage reported by Ekrami-Rad et al. (2011) was attributed to physiological and structural changes in fruit during cold storage, typically loss of cell wall integrity due to pectin substance breakdown, resulting in an increase in soluble pectin and a decrease in fruit firmness. The effect of storage time on fruit bruising might be influenced by the environment parameters such as temperature and humidity, which also have a significant impact on physiological changes in the fruit during storage.

One of the most significant environmental factors influencing the bruising of diverse fresh products is temperature. The temperature of the fruit flesh influences tissue flexibility and, as a result, bruise susceptibility (Studman, 1997; Hertog et al., 2004). Temperature changes in the surroundings cause the water volume inside the fruit to increase or decrease, providing a turgor-like effect (Hertog et al., 2004).

Product temperature may also improve tissue bruising resistance by influencing cell hydration, resulting in enhanced turgor (Lee, 2005). Because of higher cell tension, greater fruit turgor may enhance tissue stiffness and modulus of elasticity as a result, a rise in cell internal pressure tends to minimise the extra force necessary to fail the preloaded tissue (Bajema & Hyde, 1998), meaning that tissue susceptibility to bruise injury increases. Several research on the preservation of fresh produce have discovered that temperature has an influence on the sensitivity to bruise damage. An increase in handling temperature during pear storage enhanced tissue failure strain and consequent bruise injury (Baritelle & Hyde, 2001). Similarly, Chun and Huber (1998) discovered that low temperatures diminish tomato sensitivity to bruise injury. The scientists hypothesised that when storage temperature rises, so does the rate of softening in fresh fruit, as does polygalacturonase activity, an enzyme responsible for enhanced bruise sensitivity in larger quantities (Chiesa et al., 1998; Chun & Huber, 1998).

In many ways, storage factors such as humidity and temperature impact the quality of harvested products. As a result, the majority of storage methods committed to increasing the shelf life of fresh food rely on proper temperature management and a favourable humidity balance. The use of controlled atmospheric storage (CAS) in conjunction with suitable temperature control has the ability to regulate features such as physico-mechanical properties of fruit during storage, which are strongly related to the susceptibility of many types of fresh produce to bruise damage. According to Eckhoff et al. (2009), both CAS and ultra-low oxygen (ULO) storage at 2 °C preserved the bruise sensitivity of 'Braeburn' and 'Jonagold' apples in comparison to the normal environment. The beneficial effect of CAS/ULO storage was related to reduced pulp strength degradation, which preserves the fruit's resistance to the formation of pressure sores.

The influence of storage humidity on bruise damage in fresh vegetables is poorly studied. Nonetheless, a few published studies have discovered that humidity affects fresh vegetable bruising during storage. There was a disparity in bruise damage susceptibility between 'Golden Supreme' and 'Golden Delicious' apples kept in dry and humid air, according to García et al. (1995). The authors discovered that apples held at ambient temperature and dry conditions (20–25 °C; 35–49% RH) for 16 hours had lower bruising than those stored at greater humidity (100% RH) for 'Golden Supreme' and 'Golden Delicious', respectively. In comparison to CAS, normal atmospheric storage has low humidity levels, which influences water loss from stored fruit (Eckhoff et al., 2009). Increased water loss during storage may lower fruit bruise sensitivity (Kupferman, 2006). However, during storage, the circulation of water vapour between stored food and its surrounding environment must be maintained until the water activity in the product and the atmosphere reaches equilibrium (Atanda et al., 2011).

According to studies on sweet cherries (Stow et al., 2004), bananas (Bugaud et al., 2014), strawberries (Ferreira et al., 2009), and cherries, the fruit susceptibility to bruise damage is favourable at low-temperature storage (Crisosto et al., 1993). Despite these contradicting findings, understanding how fruits respond to compression or impact loading at different temperatures might be utilised to reduce both the severity and frequency of fruit bruising during harvest and postharvest storage. This information might also be used to generate guidelines for the best time of day for fruit harvesting, postharvest handling duties, and recommended storage period.

2.3.5 Effect of Produce Properties

The properties of fresh produce such as fruit size, mass, and shape or morphology play a significant role in bruise susceptibility during mechanical impact (Blahovec et al., 2003; Blahovec & Paprstein, 2005). As a result, physical and mechanical properties of fresh produce such as compressibility, impact energy absorption, and rolling resistance on impact surface may all be affected by these properties. For instance, if we consider the mass of the produce, fruit from the same or different cultivars may have different susceptibilities to bruising due to differences in fruit mass (Blahovec et al., 2003; Ahmadi, 2012).

Differences in strength of tissue between smaller and more extensive fruit, for example, partly explain the effect of produce size on bruise damage susceptibility, where the latter is thought to be weaker tissue that is more susceptible to bruise damage upon impact than the former (Van Zeebroeck et al., 2007b; Ahmadi et al., 2010). The difference in tissue structure between small and large apples, according to Johnson and Dover (1990), explains their noticeable disparities in bruise susceptibility, since big apples contain larger cells with thinner cell walls than smaller apples. As a result, the former absorbed more total energy and, as a result, suffered more bruise damage upon impact.

Other fruit qualities that have a substantial influence on the susceptibility of fresh produce to bruise damage include physico-morphological parameters, radius of curvature, fruit length to maximum fruit diameter ratio, and acoustic stiffness. Under the same loading circumstances, Menesatti and Paglia (2001) discovered that apple fruit had the most rigorous bruise susceptibility among the four examined fruit species of drupe (peach and apricot) and pomes (apple and pear), followed by pear, peach, and apricot. The found variations in bruise damage susceptibility were linked to cultivar differences in fruit physico-morphological attributes such as fruit mass, volume, and fruit equatorial diameters (i.e., the radius of curvature). Similar findings were reported by Aliasgarian et al. (2013), who investigated the effect of physical parameters (linear dimensions such as length and diameter, volume, geometric mean diameter, as well as fruit mass and sphericity) on strawberry fruit bruising.

Aliasgarian et al. (2013) evaluated the influence of physical parameters (linear dimensions i.e., length and diameter, volume, geometric mean diameter, and fruit mass and sphericity) on bruising of strawberry fruit cultivars. Based on their findings, the authors reported that the 'Gaviota' cultivar was more susceptible to bruise damage than the 'Selva' cultivar, and it was speculated that fruit size was the most important element contributing to the observed variations in sensitivity to damage.

The radius of curvature is an essential produce index that relates to the form and morphology and has been found to have a significant impact on the bruise sensitivity of certain fruits (Van Zeebroeck et al., 2007b; Zarifneshat et al., 2010; Shafie et al., 2017). The influence of radius of curvature on absorbed energy has been well reported in the literature for kiwifruit, apple, apricot, peach, and tomato. Produce with a low radius of curvature (at the contact region) suffers more bruising damage than produce with a larger radius of curvature. The radius of curvature of fresh produce impacts its susceptibility to bruise damage in two ways: (1) during heavy impact, the radius of curvature of a smaller fruit raises peak stress but lowers contact area. This may imply that sufficient tissue in contact (i.e., a large radius of curvature) is more important than inducing a high peak stress for the fruit to sustain more bruise damage at high impact, and (2) at little effect, the fruit may sustain more bruise damage if a greater peak anxiety is used rather than its greater area of contact during impact (Van Zeebroeck et al., 2007b; Zarifneshat et al., 2010).

Acoustic stiffness, however, has an indirect effect on bruising due to its influence on the fruit's contact time with the impact surface during impact (Van Linden et al., 2006). Produce stiffness is a complicated textural trait that fluctuates with fruit ripening (Van Zeebroeck et al., 2007c; Zarifneshat et al., 2010). Ahmadi (2012) discovered that kiwifruit with lower acoustic stiffness (softer) had greater bruise damage due to more absorbed energy upon impact than fruit with higher acoustic stiffness (stiff). It has been reported that fruit with higher stiffness will have shorter contact time with the impact surface, which may lower the risk of bruising in tomatoes (Van Linden et al., 2006; Van Zeebroeck et al., 2007c).

However, there are conflicting findings regarding how acoustic stiffness affects the bruising of tomatoes and apples (Van Zeebroeck et al., 2007b, c). Van Zeebroeck et al. (2007b) found that the 'Jonagold' apple cultivar suffered more bruise damage when the acoustic stiffness was higher. It has been established that higher peak stress during impact is caused by greater acoustic stiffness (Van Zeebroeck et al., 2007c). Additionally, Van Zeebroeck et al. (2007b) clarified the apparent contradiction by failure stress, observing that while stiffer unripe tomatoes have higher peak stress, this does not inevitably result in increased bruise damage (or greater absorption energy) because the failure stress is also superior for the unripe tomatoes.

2.4 Measurement of Bruise Damage on Fresh Produce

Bruise damage is a kind of subcutaneous tissue failure that does not result in skin rupture, and the colouring of wounded tissues shows the location of the damaged site (Opara & Pathare, 2014). Fruit flesh bruising is regularly documented and quantified in a variety of methods:

1. The quantity of damaged fruits in a particular fruit production lot is referred to as the bruise incidence (BI). BI can also be reported as a proportion of the total number of afflicted fruits in the sample (Opara & Pathare, 2014; Hussein et al.,

2019a, b). Bruise incidence data, rather than the degree to which fresh produce is damaged, indicates the number of fruits that sustained visible and quantifiable bruises at a certain impact strength, also known as the likelihood of bruise occurrence (Jarimopas et al., 2007).

2. Bruise severity refers to the size of a bruise or the extent to which bruising occurs. It is commonly measured as the area, volume, or depth of affected flesh in individual produce. The key parameters used to characterise bruise size are the diameter and depth. The intensity of the bruise can also be stated as a percentage of the total area or volume of fruit flesh. The degree of a bruise is determined by the amount of mechanical energy exerted and absorbed by injured produce during collision, compression, or vibration (Opara, 2007; Zarifneshat et al., 2010).
3. The varying degree of darkness or browning of a bruise is measured by bruise intensity. It can be scored visually (for example, from light brown to black) or quantified using a pre-calibrated digital colour metre. For the latter, three colour coordinate values (for example, CIE L, a, b) that identify a specific colour in a three-dimensional colour space of all conceivable hues must be captured (Fairchild, 2013).
4. Bruise susceptibility (BS) is the ease with which fresh produce suffers bruising when subjected to a given damaging pressure. It is calculated as the quantity of injured flesh per unit of absorbed impact or compression energy, or as the ratio of bruise volume to absorbed impact or compression energy (e.g., cm^2/J or cm^3/J).
5. Bruise resistance (BR) is often used synonymously with the term 'bruise susceptibility' and represents a potential for bruise damage to occur. However, a more accurate definition of BR is the amount of mechanical energy required to cause a unit amount of bruised tissue (e.g., J/cm^2 or J/cm^3).

Direct impact or compression forces to the outermost layer of fresh produce might cause superficial and/or the inner bruising (Li & Thomas, 2014; Opara & Pathare, 2014). External bruising is often identified by the existence of a skin rupture and/or the development of browning on the produce's exocarp surface. Internal bruising, however, entails either damage to fruit tissues underneath the exocarp or tissues that are not in touch with the exocarp (Li & Thomas, 2014). Internal bruising, often known as 'latent injury', is characterised by concealed damage due to the following characteristics: (1) readily ignored and difficult to quantify and (2) incurred at one stage of a postharvest system but not obvious until a later stage in the handling chain (Shewfelt, 1986).

There are three principal techniques for evaluating the severity of bruise damage in fresh produce: (1) manually taking measurements of bruised tissue dimensions to determine the area, depth, and/or amount of damaged tissue (Bollen et al., 2001; Opara, 2007; Opara et al., 2007), (2) visualisation of bruised tissue and afterwards the utilisation of image analysis software to quantify the size of damage (Menesatti & Paglia, 2001), and (3) employing nondestructive analysis (Xing et al., 2006; Dang et al., 2012). Due to lower research costs, destructive methods are more prevalent but are time-consuming. Conversely, the nondestructive method requires the utilisation of specialised machinery, which is expensive but results in improved test efficiency.

Despite differing viewpoints on the accuracy of both procedures, there has been a surge in interest in nondestructive treatment in the last several years (Shafie et al., 2015; He et al., 2022).

Manual measurement of bruised tissue dimensions is always very tricky and is associated with human errors. Bruises on fresh produce are often hidden beneath the surface of the produce skin, making them difficult to detect and quantify manually (Opara & Pathare, 2014). In addition, many factors influence the reliability of bruise detection methods, including the age of the bruise, the extent of the bruise (be it exterior or interior), the depth of the bruise, the kind of product and variety, and the fruit's preharvest and postharvest circumstances (Lu, 2003; Opara & Pathare, 2014).

Traditional methods of estimating or measuring internal bruising involve extrapolating the shape of the damage, which is not visible to the naked eye (Li & Thomas, 2014). Internal bruising is assumed to be either spherical (Ahmadi et al., 2010; Ahmadi, 2012), an elliptical cone (Shafie et al., 2015), or ellipsoidal (Lü & Tang, 2012; Kitthawee et al., 2011). Slicing across the core of the affected region (brown colour, indicated area) is used to cut out the affected area to quantify bruised area (either internal or external). The presence of injured tissues that were clearly identifiable from other unbruised regions of the same fruit is connected with the bruise damage of the fruit sliced through the impact location. After measuring the diameter, breadth, and depth of the bruised tissues with digital callipers, the bruise volume (BV) or bruise area (BA) of an internal bruising may then be calculated (Ahmadi et al., 2010; Kitthawee et al., 2011; Shafie et al., 2015).

Researchers have used impact or drop impact test (Fu et al., 2020; Pathare & Al-Dairi, 2021, 2022; Pathare et al., 2022); drop or free fall test (Hussein et al., 2019a, b, 2020); or compression tests (Ferreira et al., 2008; Kitthawee et al., 2011; Polat et al., 2012; Opara & Fadiji, 2018) to obtain controlled bruise damage on a wide range of fresh horticultural produce.

Pendulum impactor is commonly used for impact tests and involves swinging an individual produce in pendulum arrangement from different heights onto a hard surface (Polat et al., 2012; Fu et al., 2023). During testing, individual produce is tied to the end of an impacting arm of a known length, and the arm is manually pulled to a starting point and then released to strike an impact or surface freely arranged on an elastic pallet. Following impact, the fruit is trapped by a catching box outfitted with polyurethane foam or by any other method (Polat et al., 2012). In the test sequence, different fruits are utilised for each location and height. A mechanical release mechanism is placed at a desired drop height during impact testing, and the fruit is attached to it. In order to designate the position of the impact region on the fruit surface, the metal impact surface is commonly inked or coated with white chalk powder (Polat et al., 2012; Opara & Pathare, 2014).

However, assessing the rebound height of the fruit after the initial impact has always been the most difficult aspect of utilising the pendulum impact test. The rebound height is important in determining the real impact energy received by the tissue and causing bruising damage. Opara (2007) utilised a conventional video camera as an objective way to determine rebound height during a pendulum test for testing bruise susceptibility. Other writers have utilised a rebound height scale on a pendulum frame to calculate the volume of energy absorbed by the produce during impact by reading the rebound height from the drop after impact (Polat et al., 2012).

Drop test method is categorised into two approaches (1) dropping a steel ball (free fall) of a known mass through PVC hollow pipe into every individual fruit from pre-determined drop heights (Opara, 2007; Pathare & Al-Dairi, 2021, 2022; Pathare et al., 2022). During the test, the ball is usually fixed by hand after each drop to prevent multiple impacts into the fruit; (2) dropping an individual fruit at pre-determined drop heights against a rigid surface/floor along with a graduated wooden ruler. To avoid multiple impacts (rebound height) which may interfere in calculating the precise impact energy received by product to cause damage, the product is held by hand after the initial rebound to avoid further impact (Hussein et al., 2019a, 2020; Opara & Pathare, 2014; Opara et al., 2007).

Compression test uses quasi-static compression loading procedure, where either of the two approaches could be employed; (1) using spherical indenters of known mass and diameter, each indenter, where the compression is applied at several contact points on the fruit surface (Kitthawee et al., 2011), or (2) employing a static compression test, where a test device can be set to compress a fresh produce at a pre-set compression force and speed. Compression can be applied on produce which results in various deformations (measured in mm), and the force of compression used for each deformation amount is recorded (Polat et al., 2012). Mechanical properties at any point (e.g., bio-yield point) can be measured using the force-deformation curve. The amount of energy needed to compress the fruit is estimated from the area under the force curve versus deformation.

In both laboratory methods used obtain controlled bruise damage, simulation of bruise damage is usually followed by holding the bruised damaged fruit for 24–72 h (depending on type of produce) for bruise development (Kitthawee et al., 2011; Hussein et al., 2019a, b).

2.5 Methods to Reduce and Control Bruise Damage

Bruising is one of the most significant constraints in producing high-quality fresh fruit. Furthermore, bruising is a significant barrier restricting the mechanisation and automation of soft fruit and vegetable picking, sorting, and transportation (Polat et al., 2012; Opara & Pathare, 2014). As a result, reducing bruising damage and other mechanical damage has drawn the interest of researchers and producers (Pathare & Al-Dairi, 2022). The following techniques should be implemented to reduce/control bruise damage magnitude and incidences of fresh product at harvest, throughout transit, and future postharvest handling procedures.

2.5.1 *Reduce Drop Height and Compression Loading*

Developing techniques for decreasing or even avoiding the problem of bruising requires insight into how fresh fruit bruises under various impact or compression circumstances (Fu et al., 2020, 2023). Application of knowledge obtained through the science-based laboratory analysis for reproducing bruise damage is warranted to both growers, packers, and sellers of fresh produce (Opara & Pathare, 2014; Fu et al., 2023). Given that bruising is caused by mechanical force in close contact with fresh food, avoiding impacts or compression might be a significant tactic to avoid or reduce damage. As high incidences of postharvest handling damages are linked to mechanised harvesting, transport, and packhouse operations, harvesters and fruit packers must limit produce exposure to impact levels (including low 0.1 to high 1 m drop impacts) during harvest as well as postharvest operations and processes (Shafie et al., 2015, 2017; Hussein et al., 2019a). This might be accomplished by limiting the exposure of fresh produce to impact or compression damage events that induce bruising, such as falling or squeezing.

2.5.2 *Redesign the Architecture of the Produce Handling Lines*

A better knowledge of the process of fruit bruising will aid in the development of harvesting technologies that can decrease fruit damage (Fu et al., 2020). Fruit acceleration and deceleration must be properly regulated throughout the handling chain to prevent impact bruising damage (Opara & Pathare, 2014). Reducing speed of produce along handling systems and avoiding produce-to-produce contact during sorting and grading could be one of the important features to modify in the existing produce handling lines. The adoption of innovations such as instrumented spheres allows for real-time monitoring and assessment of packing lines in order to identify crucial control points in order to limit the occurrence of mechanical damage to fresh horticulture. Additionally, the use of cushioning materials such as padding hard surfaces in the handling lines, in fruit harvesters, or harvesting and handling buckets and bins could also reduce energy absorption at impact or compression.

2.5.3 *Provide Adequate Training on Appropriate Harvest and Postharvest Handling of Fresh Produce*

Minimising impact damage during postharvest handling is the first recommended action, followed by storage management, particularly when grading, sorting, and packaging (Opara & Pathare, 2014). Therefore, it is possible to reduce the risk of bruise damage by training workers to: (a) handle the crop gently, (b) harvest crop at the right maturity and whenever possible to avoid wetness on produce, (c) handling produce only as much as is necessary, (d) installing padding or cushioning materials inside bulk bins, and (e) avoiding over- or under-packing of containers.

2.5.4 *Modulate Postharvest Physical and Mechanical Properties of Fresh Produce*

Most fresh horticultural produce become susceptible to bruising due to poor storage and handling conditions, such as temperature and humidity. Accordingly, temperature and humidity conditions can influence change in physical and mechanical properties during handling or storage. Therefore, it is crucial to manage these conditions properly during postharvest handling, particularly sorting, grading, and packing (Hussein et al., 2019b). Approaches such as air storage techniques that rely on air modification (e.g., controlled atmosphere storage) could be suitable in improving resistance of produce to bruise damage.

To increase the resistance to impact bruising, heat treatment can be used in conjunction with controlled environment storage (Tahir et al., 2009; Li et al., 2016). For instance, pre-heating fresh food reduces the impact pressure and creates a cushioning effect by melting skin wax and causing structural changes in the fruit (Roy et al., 1994).

2.6 Conclusions and Future Prospects

Brusing and other forms of mechanical damage contribute to postharvest food losses and waste and thereby contribute to food and nutrition insecurity and reduce the profitability of agri-food business. The prospect for bruising continues to be a significant problem preventing the mechanisation of fresh produce harvesting and postharvest handling. The need of suitable postharvest technology is highlighted by the overwhelmingly high demand for high-quality fruits and vegetables. To maintain freshness quality attributes, increase the shelf life of fresh produce, and lower postharvest losses, mechanical damage reduction strategies must be put in place. The most frequent causes of bruising during postharvest handling operations are excessive compression and impact forces.

Economic modelling indicates that future horticultural produce supply will be insufficient to meet recommended levels in many countries. If there is also the problem of bruise damage and subsequent postharvest losses, the situation becomes even worse. This will necessitate a surge in investments and interventions aimed at increasing the production of horticultural produce, developing technologies and methods to reduce bruise damage, and addressing other sources of postharvest losses.

There are numerous techniques and approaches available that, if implemented, would enable the reduction or elimination of bruise damage in the horticultural industry, both for smallholders and large-scale producers; these include, but are not limited to (a) minimising unnecessary exposure of fresh produce to dropping, compression, and quizzing; (b) redesigning the architecture of produce handling lines, harvesting buckets, and handling bins to allow them to absorb some of the energy during impact or compression; (c) providing adequate training to growers, packers, and sellers (distributors) on proper harvest and postharvest handling of fresh produce; and (d) pre-treatment of fresh produce to modulate their postharvest physical and mechanical properties in favour of improving mechanical damage resistance.

Acknowledgments This work is based on the research supported in part by the National Research Foundation of South Africa (Grant Numbers: 64813). The opinions, findings, and conclusions or recommendations expressed are those of the author(s) alone, and the NRF accepts no liability whatsoever in this regard.

References

Abbott, B., Holford, P., & Golding, J. B. (2009). Comparison of 'Cripps Pink' apple bruising. *Acta Horticulturae, 880*, 223–230.

Ahmadi, E. (2012). Bruise susceptibilities of kiwifruit as affected by impact and fruit properties. *Research in Agricultural Engineering, 58*, 107–113.

Ahmadi, E., Ghasemzadeh, H. R., Sadeghi, M., Moghadam, M., & Zarifneshat, S. (2010). The effect of impact and fruit properties on the bruising peach. *Journal of Food Engineering, 97*, 110–117.

Al-Dairi, M., Pathare, P. B., & Al-Yahyai, R. (2021). Quality changes kinetic of tomato during postharvest transportation and storage. *Journal of Food Process Engineering, 44*, e13808.

Aliasgarian, S., Ghassemzadeh, H. R., Moghaddam, M., & Ghaffari, H. (2013). Mechanical damage of strawberry during harvest and postharvest operations. *World Applied Sciences Journal, 22*, 969–974.

Atanda, S. A., Pessu, P. O., Agoda, S., Isong, I. U., & Ikotun, I. (2011). The concepts and problems of post-harvest food losses in perishable crops. *African Journal of Food Science, 5*(11), 603–613.

Bajema, R. W., & Hyde, G. (1998). Instrumented pendulum for impact characterization of whole fruit and vegetable specimens. *Transactions of ASAE, 41*, 1399–1405.

Baritelle, A. L., & Hyde, G. M. (2001). Commodity conditioning to reduce impact bruising. *Postharvest Biology and Technology, 21*, 331–339.

Blahovec, J. (2006). Shape of bruise spots in impacted potatoes. *Postharvest Biology and Technology, 39*, 278–284.

Blahovec, J., & Paprstein, F. (2005). Susceptibility of pear varieties to bruising. *Postharvest Biology and Technology, 38*, 231–238.

Blahovec, J., & Zidova, J. (2004). Potato bruises spot sensitivity dependence on modes of cultivation. *Research in Agricultural Engineering, 50*, 89–95.

Blahovec, J., Mares, V., & Paprstein, F. (2003). Static low-level pear bruising in a group of varieties. *Scientia Agriculturae Bohemica, 34*, 140–145.

Bollen, A. F. (2005). Major factors causing variation in bruise susceptibility of apples (*Malus domestica*) grown in New Zealand. *New Zealand Journal of Crop and Horticultural Science, 33*, 201–210.

Bollen, A. F., Cox, N. R., Dela Rue, B. T., & Painter, D. J. (2001). A descriptor for damage susceptibility of a population of produce. *Journal of Agricultural Engineering Research, 78*, 391–395.

Bugaud, C., Ocrisse, G., Salmon, F., & Rinaldo, D. (2014). Bruise susceptibility of banana peel in relation to genotype and post-climacteric storage conditions. *Postharvest Biology and Technology, 87*, 113–119.

Canete, M. L., Hueso, J. J., Pinillos, V., & Cuevas, J. (2015). Ripening degree at harvest affects bruising susceptibility and fruit sensorial traits of loquat (*Eriobotrya japonica* Lindl.). *Scientia Horticulturae, 187*, 102–107.

Cattaneo, A., Sánchez, M. V., Torero, M., & Vos, R. (2020). Reducing food loss and waste: Five challenges for policy and research. *Food Policy, 98*, 101974.

Chiesa, A., Diaz, L., Cascone, O., Panak, K., Camperi, S., Frezza, D., & Fraguas, A. (1998). Texture changes on normal and long shelf-life tomato (*Lycopersicon esculentum* mill.) fruit ripening. *Acta Horticulturae, 464*, 488.

Chun, J. P., & Huber, D. J. (1998). Polygalacturonase-mediated solubilization and depolymerization of pectic polymers in tomato fruit cell walls. Regulation by pH and ionic conditions. *Plant Physiology, 117*, 1293–1299.

Crisosto, C. H., Garner, D., Doyle, J., & Day, K. R. (1993). Relationship between fruit respiration, bruising susceptibility, and temperature in sweet cherries. *HortScience, 28*(2), 132–135.

Dang, H. Q., Kim, I., Cho, B. K., & Kim, M. S. (2012). Detection of bruise damage of pear using hyperspectral imagery. In *2012 12th international conference on control, Automation and Systems (ICCAS)* (pp. 1258–1260). IEEE.

Eckhoff, L., Dierend, W., & Köpcke, D. (2009). Influence of foliar fertilization and storage method on bruise sensitivity of apples. *Erwerbs-Obstbau, 51*(4), 133–144.

Eissa, A. H. A., Albaloushi, N. S., & Azam, M. M. (2008). Vibration analysis influence during crisis transport of the quality of fresh fruit on food security. *Agricultural Engineering International: CIGR Journal, 15*, 181–190.

Eissa, A. H. A., Albaloushi, N. S., & Azam, M. M. (2013). Vibration analysis influence during crisis transport of the quality of fresh fruit on food security. *Agricultural Engineering International: CIGR Journal, 15*, 181–190.

Ekrami-Rad, N., Khazaei, J., & Khoshtaghaza, M. (2011). Selected mechanical properties of pomegranate peel and fruit. *International Journal of Food Properties, 14*, 570–582.

Emana, B., Afari-Sefa, V., Nenguwo, N., Ayana, A., Kebede, D., & Mohammed, H. (2017). Characterization of pre-and postharvest losses of tomato supply chain in Ethiopia. *Agriculture & Food Security, 6*(1), 1–11.

Ericsson, N. A., & Tahir, I. I. (1996). Studies on apple bruising: II. The effects of fruit characteristics, harvest date and precooling on bruise susceptibility of three apple cultivars. *Acta Agriculturae Scandinavica Section B Soil and Plant Science, 46*(4), 214–217.

Fairchild, M. (2013). *Color appearance models* (3rd ed.). Wiley.

FAO. (2019). *The state of food and agriculture: Moving forward on food loss and waste reduction*. Food and Agriculture Organization.

Fawole, O. A., Opara, U. L., & Theron, K. I. (2012). Chemical and phytochemical properties and antioxidant activities of three pomegranate cultivars grown in South Africa. *Food and Bioprocess Technology, 5*, 425–444.

Ferreira, M. D., Sargent, S. A., Brecht, J. K., & Chandler, C. K. (2008). Strawberry fruit resistance to simulated handling. *Scientia Agricola, 65*, 490–495.

Ferreira, M. D., Sargent, S. A., Brecht, J. K., & Chandler, C. K. (2009). Strawberry bruising sensitivity depends on the type of force applied, cooling method, and pulp temperature. *HortScience, 44*(7), 1953–1956.

Fu, H., Karkee, M., He, L., Duan, J., Li, J., & Zhang, Q. (2020). Bruise patterns of fresh market apples caused by fruit-to-fruit impact. *Agronomy, 10*(1), 59.

Fu, H., Du, W., Yang, J., Wang, W., Wu, Z., & Yang, Z. (2023). Bruise measurement of fresh market apples caused by repeated impacts using a pendulum method. *Postharvest Biology and Technology, 195*, 112143.

García, J. L., Ruiz-Altisent, M., & Barreiro, P. (1995). Factors influencing mechanical properties and bruise susceptibility of apples and pears. *Journal of Agricultural Engineering Research, 61*(1), 11–17.

Grünwald, N. J. (2021). Entering the international year of fruits and vegetables: Tradeoffs between food production and the environment. *CABI Agriculture and Bioscience, 2*, 1–3.

Gunes, G., Liu, R. H., & Watkins, C. B. (2002). Controlled-atmosphere effects on postharvest quality and antioxidant activity of cranberry fruits. *Journal of Agricultural and Food Chemistry, 50*, 5932–5938.

He, Y., Xiao, Q., Bai, X., Zhou, L., Liu, F., & Zhang, C. (2022). Recent progress of nondestructive techniques for fruits damage inspection: A review. *Critical Reviews in Food Science and Nutrition, 62*(20), 5476–5494.

Herrero, M., Thornton, P. K., Mason-D'Croz, D., Palmer, J., Benton, T. G., Bodirsky, B. L., & West, P. C. (2020). Innovation can accelerate the transition towards a sustainable food system. *Nature Food, 1*(5), 266–272.

Hert, K. A., Fisk, P. S., II., Rhee, Y. S., & Brunt, A. R. (2014). Decreased consumption of sugar-sweetened beverages improved selected biomarkers of chronic disease risk among US adults: 1999 to 2010. *Nutrition Research, 34*, 58–65.

Hertog, M. L. A. T. M., Ben-Arie, R., Roth, E., & Nicolai, B. M. (2004). Humidity and temperature effects on invasive and non-invasive firmness measures. *Postharvest Biology and Technology, 33*, 79–91.

Hung, Y. C., & Prussia, S. E. (1989). Effect of maturity and storage time on the bruise susceptibility of peaches (CV. Red globe). *Transactions of the American Society of Agricultural Engineers, 32*(4), 1377–1382.

Hussein, Z., Caleb, O. J., & Opara, U. L. (2015). Perforation-mediated modified atmosphere packaging of fresh and minimally processed produce—A review. *Food Processing and Shelf Life, 6*, 7–20.

Hussein, Z., Fawole, O. A., & Opara, U. L. (2018). Preharvest factors influencing bruise damage of fresh fruits–a review. *Scientia Horticulturae, 229*, 45–58.

Hussein, Z., Fawole, O. A., & Opara, U. L. (2019a). Bruise damage susceptibility of pomegranates (*Punica granatum*, L.) and impact on fruit physiological response during short term storage. *Scientia Horticulturae, 246*, 664–674.

Hussein, Z., Fawole, O. A., & Opara, U. O. (2019b). Bruise damage of pomegranate during long-term cold storage: Susceptibility to bruising and changes in textural properties of fruit. *International Journal of Fruit Science, 20*(sup2), S211–S230.

Hussein, Z., Fawole, O. A., & Opara, U. L. (2020). Harvest and postharvest factors affecting bruise damage of fresh fruits. *Horticultural Plant Journal, 6*(1), 1–13.

Jarimopas, B., Singh, I. S. P., Sayasoonthorn, S., & Singh, J. (2007). Comparison of package cushioning materials to protect postharvest impact damage to apples. *Packaging Technology and Science, 20*, 315–324.

Johnson, D. S., & Dover, C. J. (1990). Factors influencing the bruise susceptibility of Bramley's seedling apples. In *Proceedings of the European workshop on impact damage of fruits and vegetables, Zaragoza, Spain* (pp. 87–93).

Kiaya, V. (2014). Post-harvest losses and strategies to reduce them. Technical paper on postharvest losses. *Action Contre la Faim (ACF), 25*, 1–25.

Kitthawee, U., Pathaveerat, S., Srirungruang, T., & Slaughter, D. (2011). Mechanical bruising of young coconut. *Biosystems Engineering, 109*, 211–219.

Komarnicki, P., Stopa, R., Szyjewicz, D., & Mlotek, M. (2016). Evaluation of bruise resistance of pears to impact load. *Postharvest Biology and Technology, 114*, 36–44.

Kupferman, E. (2006). Minimizing bruising in apples. In *Postharvest information network*. Washington State University, Tree Fruit Research and Extension Center.

Kuriyan, R., Griffiths, J. K., Finkelstein, J. L., Thomas, T., Raj, T., Bosch, R. J., Kurpad, A. V., & Duggan, C. (2014). Innovations in nutrition education and global health: The Bangalore Boston nutrition collaborative. *BMC Medical Education, 14*, 5.

Lee, E. (2005). *Quality changes induced by mechanical stress on Roma-type tomato and potential alleviation by 1-methylcyclopropene*. Master's Thesis. University of Florida, USA.

Lewis, R., Yoxall, A., Canty, L. A., & Romo, E. R. (2007). Development of engineering design tools to help reduce apple bruising. *Journal of Food Engineering, 83*, 356–365.

Li, Z., & Thomas, C. (2014). Quantitative evaluation of mechanical damage to fresh fruits. *Trends in Food Science and Technology, 35*, 138–150.

Li, J., Yan, J., Ritenour, M. A., Wang, J., Cao, J., & Jiang, W. (2016). Effects of 1-methylcyclopropene on the physiological response of Yali pears to bruise damage. *Scientia Horticulturae, 200*, 137–142.

Lippert, F., & Blanke, M. M. (2004). Effect of mechanical harvest and timing of 1-MCP application on respiration and fruit quality of European plums Prunus domestica L. *Postharvest Biology and Technology, 34*, 305–311.

Lu, R. (2003). Detection of bruises on apples using near-infrared hyperspectral imaging. *Transactions of ASAE, 46*, 523–530.

Lü, Q., & Tang, M. (2012). Detection of hidden bruise on kiwi fruit using hyperspectral imaging and parallelepiped classification. *Procedia Environmental Sciences, 12*, 1172–1179.

Martinez-Romero, D., Serrano, M., Carbonell, A., Castillo, S., Riquelme, F., & Valero, D. (2004). Mechanical damage during fruit postharvest handling: Technical and physiological implications. In R. Dris & S. M. Jain (Eds.), *Production practices and quality assessment of food crops* (pp. 233–252). Springer.

Mason-D'Croz, D., Bogard, J. R., Sulser, T. B., Cenacchi, N., Dunston, S., Herrero, M., & Wiebe, K. (2019). Gaps between fruit and vegetable production, demand, and recommended consumption at global and national levels: An integrated modelling study. *The Lancet Planetary Health, 3*(7), 318–329.

Menesatti, P., & Paglia, G. (2001). Development of a drop damage index of fruit resistance to damage. *Journal of Agricultural Engineering Research, 80*, 53–64.

Miller, S. R., & Knudson, W. A. (2014). Nutrition and cost comparisons of select canned, frozen, and fresh fruits and vegetables. *American Journal of Lifestyle Medicine, 8*, 430–437.

Mowatt, C. M. (1997). *Factors influencing the susceptibility of apples to bruising*. Unpublished PhD thesis. Massey University, Palmerston North, New Zealand.

Mphahlele, R. R., Fawole, O. A., Makunga, N. P., & Opara, U. L. (2016). Effect of drying on the bioactive compounds antioxidant, antibacterial and antityrosinase activities of pomegranate peel. *BMC Complementary and Alternative Medicine, 16*, 1–12.

Opara, L. U. (2007). Bruise susceptibilities of 'gala' apples as affected by orchard management practices and harvest date. *Postharvest Biology and Technology, 43*, 47–54.

Opara, U. L., & Al-Ani, M. R. (2010). Antioxidant contents of pre-packed fresh-cut versus whole fruit and vegetables. *British Food Journal, 112*(8), 797–810.

Opara, U. L., & Fadiji, T. (2018). Compression damage susceptibility of apple fruit packed inside ventilated corrugated paperboard package. *Scientia Horticulturae, 227*, 154–161.

Opara, U. L., & Pathare, P. B. (2014). Bruise damage measurement and analysis of fresh horticultural produce—A review. *Postharvest Biology and Technology, 91*, 9–24.

Opara, L. U., Al-Ghafri, A., Agzoun, H., Al-Issai, J., & Al-Jabri, F. (2007). Design and development of a new device for measuring susceptibility to impact damage of fresh produce. *New Zealand Journal of Crop and Horticultural Science, 35*(2), 245–251.

Pang, D. W. (1993). *Prediction and quantification of apple bruising*. PhD Thesis. Massey University, New Zealand.

Pathare, P. B., & Al-Dairi, M. (2021). Bruise susceptibility and impact on quality parameters of pears during storage. *Frontiers in Sustainable Food Systems, 5*, 658132.

Pathare, P. B., & Al-Dairi, M. (2022). Effect of mechanical damage on the quality characteristics of banana fruits during short-term storage. *Discover Food, 2*(1), 1–13.

Pathare, P. B., Al-Dairi, M., Al-Yahyai, R., & Al-Mahdouri, A. (2022). Simulated handling to investigate the effect of mechanical damage on stored pomegranate fruit. *PRO, 10*(12), 2695.

Perkins, M. L., Mazhar, M., Joyce, D. C., Ainsworth, N., Coates, L., & Hofman, P. (2017). Factors affecting avocado flesh bruising susceptibility. *Talking Avocados, 28*(2).

Polat, R., Aktas, T., & Ikinci, A. (2012). Selected mechanical properties and bruise susceptibility of nectarine fruit. *International Journal of Food Properties, 15*(6), 1369–1380.

Ramos, B., Miller, F. A., Brandão, T. R. S., Teixeira, P., & Silva, C. L. M. (2013). Fresh fruits and vegetables—An overview on applied methodologies to improve its quality and safety. *Innovative Food Science and Emerging Technologies, 20*, 1–15.

Roy, S., Conway, W. S., Watada, A. E., Sams, C. E., Erbe, E. F., & Wergin, W. P. (1994). Heat treatment affects epicuticular wax structure and postharvest calcium uptake in 'Golden delicious' apples. *HortScience, 29*, 1056–1058.

Ruiz-Altisent, M. (1991). Damage mechanisms in the handling of fruits. In *Progress in agricultural physics and engineering* (pp. 231–257). C.A.B. International.

Saracoglu, T., Ucer, N., & Ozarslan, C. (2011). Engineering properties and susceptibility to bruising damage of table olive (*Olea Europaea*) fruit. *International Journal of Agriculture and Biology, 13*, 801–805.

Shafie, M. M., Rajabipour, A., Castro-Garcia, S., Jiménez-Jiménez, F., & Mobli, H. (2015). Effect of fruit properties on pomegranate bruising. *International Journal of Food Properties, 18*, 1837–1846.

Shafie, M., Rajabipour, A., & Mobli, H. (2017). Determination of bruise incidence of pomegranate fruit under drop case. *International Journal of Fruit Science, 17*(3), 296–309.

Shewfelt, R. L. (1986). Postharvest treatment for extending the shelf life of fruits and vegetables. *Food Technology, 40*, 70–89.

Stathers, T., Holcroft, D., Kitinoja, L., Mvumi, B. M., English, A., Omotilewa, O., & Torero, M. (2020). A scoping review of interventions for crop postharvest loss reduction in sub-Saharan Africa and South Asia. *Nature Sustainability, 3*(10), 821–835.

Stow, J. R., Jameson, J., & Senner, K. (2004). Storage of cherries: The effects of rate of cooling, store atmosphere and store temperature on storage and shelf life. *The Journal of Horticultural Science and Biotechnology, 79*(6), 941–946.

Stratton, A. E., Finley, J. W., Gustafson, D. I., Mitcham, E. J., Myers, S. S., Naylor, R. L., & Palm, C. A. (2021). Mitigating sustainability tradeoffs as global fruit and vegetable systems expand to meet dietary recommendations. *Environmental Research Letters, 16*(5), 055010.

Strehmel, N., Praeger, U., Konig, C., Fehrle, I., Erban, A., Geyer, M., Kopka, J., & Dongen, J. (2010). Time course effects on primary metabolism of potato (*Solanum tuberosum*) tuber tissue after mechanical impact. *Postharvest Biology and Technology, 56*, 109–116.

Stropek, Z., & Gołacki, K. (2015). A new method for measuring impact related bruises in fruits. *Postharvest Biology and Technology, 110*, 131–139.

Studman, C. (1997). Factors affecting bruise susceptibility of fruit. In G. Jeronimidis & J. F. V. Vincent (Eds.), *Proceedings of the 2nd International Conference of Plant Biomechanics* (pp. 273–281). Centre for Biomimetics, University of Reading.

Tabatabaekoloor, R. (2013). Engineering properties and bruise susceptibility of peach fruits (*Prunus persica*). *Agricultural Engineering International: CIGR Journal, 15*(4), 244–252.

Tahir, I. I., Johansson, E., & Olsson, M. E. (2009). Improvement of apple quality and storability by a combination of heat treatment and controlled atmosphere storage. *HortScience, 44*(6), 1648–1654.

Timm, E. J., Bollen, A. F., Dela Rue, B. T., & Woodhead, I. M. (1998). Apple damage and compressive forces in bulk bins during orchard transport. *Applied Engineering in Agriculture, 14*(2), 165–172.

Van Linden, V., & De Baerdemaeker, J. (2005). The phenomenon of tomato bruising: Where biomechanics and biochemistry meet. *Acta Horticulturae, 682*, 925–932.

Van Linden, V., Scheerlinck, N., Desmet, M., & De Baerdemaeker, J. (2006). Factors that affect tomato bruise development as a result of mechanical impact. *Postharvest Biology and Technology, 42*, 260–270.

Van Zeebroeck, M., Van Linden, V., Ramon, H., De Baerdemaeker, J., Nicola, B. M., & Tijskens, E. (2007a). Impact damage of apples during transport and handling—A review. *Postharvest Biology and Technology, 45*, 157–167.

Van Zeebroeck, M., Van Linden, V., Darius, P., De Ketelaere, B., Ramon, H., & Tijskens, E. (2007b). The effect of fruit factors on the bruise susceptibility of apples. *Postharvest Biology and Technology, 46*, 10–19.

Van Zeebroeck, M., Darius, P., De Ketelaere, B., Ramon, H., & Tijskens, E. (2007c). The effect of fruit properties on the bruise susceptibility of tomatoes. *Postharvest Biology and Technology, 45*, 168–175.

Vursavus, K., & Ozguven, F. (2003). Determining the strength properties of the Dixired peach variety. *Turkish Journal of Agriculture and Forestry, 27*(3), 155–160.

White, A., Woolf, A., Hofman, P. J., & Arpaia, M. L. (2009). *The international avocado quality manual*. UC Davis Press.

WHO. (2018). *Healthy diet*. Retrieved January 16, 2023, from https://www.who.int/news-room/factsheets/detail/healthy-diet.

Willett, W., Rockström, J., Loken, B., Springmann, M., Lang, T., Vermeulen, S., Murray, C. J., et al. (2019). Food in the Anthropocene: The EAT–lancet commission on healthy diets from sustainable food systems. *The Lancet, 393*(10170), 447–492.

Xing, J., Bravo, C., Moshou, D., Ramon, H., & De Baerdemaeker, J. (2006). Bruise detection on 'Golden delicious' apples by Vis/NIR spectroscopy. *Computers and Electronics in Agriculture, 52*, 11–20.

Yurtlu, Y. B., & Erdoğan, D. (2005). Effect of storage time on some mechanical properties and bruise susceptibility of pears and apples. *Turkish Journal of Agriculture and Forestry, 29*(6), 469–482.

Żaczek, M., Weber-Dąbrowska, B., & Górski, A. (2015). Phages in the global fruit and vegetable industry. *Journal of Applied Microbiology, 118*(3), 537–556.

Zarifneshat, S., Ghassemzadeh, H. R., Sadeghi, M., Abbaspour-Fard, M. H., Ahmadi, E., Javadi, A., & Shervani-Tabar, M. T. (2010). Effect of impact level and fruit properties on Golden delicious apple bruising. *American Journal of Agricultural and Biological Sciences, 5*, 114–121.

Chapter 3
Imaging Techniques for Fresh Produce Damage detection

Naveen Kumar Mahanti, Pankaj B. Pathare, Upendar Konga, and Jithender Bhukya

Abstract Fresh fruits and vegetables are susceptible to mechanical damage and physiological problems induced by static and dynamic forces acting on them during postharvest handling. Detecting these mechanical damages is critical to quality control in fresh produce processing and handling. The detection of bruises is primarily based on manual inspection, which is time-consuming and error-prone, particularly for early bruises. As a result, industries are looking for fast, non-destructive, and precise technology for fresh produce inspection along the postharvest value chain. Non-destructive imaging techniques have the ability to identify mechanical damage in real-time while overcoming the drawbacks of traditional methods which are tediousness, destructiveness, and time consumption. The application of imaging techniques including biospeckle imaging, computer vision, hyperspectral imaging, thermal imaging, and other imaging techniques for real-time and automated bruise detection is proposed in this chapter.

Keywords Mechanical damage · Imaging techniques · Biospeckle imaging · Computer vision · Hyperspectral imaging · Thermal imaging · Bruise detection

N. K. Mahanti
Post Harvest Technology Research Station, Dr YSR Horticultural University, Venkataramannagudem, West Godavari, Andhra Pradesh, India

P. B. Pathare (✉)
Department of Soils, Water & Agricultural Engineering, College of Agricultural and Marine Sciences, Sultan Qaboos University, Al-Khod, Oman
e-mail: pankaj@squ.edu.om

U. Konga · J. Bhukya
School of Agricultural and Bio-Engineering, Centurion University of Technology and Management, Paralakhemundi, Odisha, India

P. B. Pathare, U. L. Opara (eds.), *Mechanical Damage in Fresh Horticultural Produce*, https://doi.org/10.1007/978-981-99-7096-4_3

3.1 Introduction

Fruits and vegetables contain vitamins, minerals, phytochemicals, antioxidants, and dietary fibre (Hussein et al., 2015; Pandiselvam et al., 2019). People need to eat fruits and vegetables daily to get all the necessary nutrients. By 2050, the world population is expected to be nine billion (FAO). The demand for food grains, fruits, vegetables, ready-to-eat (RTE) and minimally processed foods will increased drastically. Mechanical damage contributes to postharvest losses in fruits and vegetables. Mechanical damage to fruits and vegetables may occur due to static or dynamic forces acting on them during harvesting, handling, transportation, and storage (Pathare & Al-Dairi, 2021). Due to their pulpy composition, fruits and vegetables are susceptible to mechanical injury (Li et al., 2013). The result of mechanical damage on fruits and vegetables includes crushing, rupture, and bruising. It also leads to plant tissue destruction or superficial rupture and plastic deformation (Polat et al., 2012).

This mechanical damage significantly increases metabolic and physiological processes such as ethylene production, microbial spoilage, transpiration, respiration, and conductivity (Kumar et al., 2016). Conversely, the damaged part may act as an entryway for external microorganisms and pathogens; these infect the healthy portion of fruit and vegetables and result in significant losses (Pholpho et al., 2011). Apart from the mechanical damage, fruit and vegetables become defective due to tissues' non-pathogenic and non-mechanical breakdown. According to Sandhu and Gill (2013), these physiological disorders are brought on by abiotic stress during growth and development, which is brought on by an excess or deficiency of chemicals, temperature, nutritional deficiencies, and RH. The physiological disorder influences various areas of fruits and vegetables; some affect the skin, and others influence the flesh or core. The physiological or mechanical damage of fruits and vegetables within the supply chain leads to their rejection because of their low quality and short shelf life. It also causes more postharvest losses, which means farmers and retailers lose money (Hussein et al., 2015). Recently, consumers have been more concerned about product quality; therefore, product quality decides consumer acceptance and market value. Therefore, detecting mechanical or physiological damage in fruits and vegetables during the initial period is important to minimize postharvest losses.

Manual detection of damaged fruits and vegetables in the processing industry continuous process lines requires skilled personnel and is both time and cost intensive. Also, the defect cannot be seen with the naked eye at the time of imitation, and it is hard to check each fruit or vegetable on a fast-moving belt in a process line. Consumers are demanding high-quality produce. Therefore, the fruit and vegetable processing industries are searching for alternative techniques to inspect every fruit and vegetable rapidly and invasively with greater accuracy (Yildiz et al., 2019). Over the last few decades, researchers have been exploring various technologies and developing numerous machine learning algorithms to simultaneously detect damage and assess quality parameters in fruits and vegetables.

The non-destructive techniques rely on food's mechanical, chemical, physical, and structural properties. With advancements in sensing technologies, there are several types of non-destructive techniques available to assess the damage of fruits and vegetables. The non-destructive sensing techniques are biospeckle (Pandiselvam et al., 2020), hyperspectral imaging (HSI) (Xiong et al., 2018), machine vision (Bhargava & Bansal, 2021), magnetic resonance imaging (MRI) (Kamal et al., 2019), nuclear magnetic resonance (NMR) (Dar et al., 2020), X-ray imaging (Azadbakht et al., 2019), and thermal imaging (Baranowski et al., 2008). NMR, X-ray imaging, and MRI techniques are suitable for detecting internal damage, whereas biospeckle, thermal imaging, hyperspectral imaging, and machine vision are employed to detect surface damage.

3.2 Biospeckle Imaging Technique

The biospeckle technique is an advanced optical technology that is non-invasive, cost-effective, rapid, and simple and can detect damage in fruits and vegetables (Gao & Rao, 2019). Biospeckle system comprises laser lamp (to illuminate), camera (to record images), and a computer along with the frame grabber (to record a set of images with a constant time lag) (Fig. 3.1) (Pandiselvam et al., 2020). When the laser light falls on the biological sample some portion scatters and some interacts with tissue components and forms a granular-like structure which is known as 'speckle' that are recorded by the camera. When the laser light hits a substance, it shows different kinds of non-biological or biological dynamic processes (Ansari & Nirala, 2013). If the sample does not show any biological activity, then the resulting

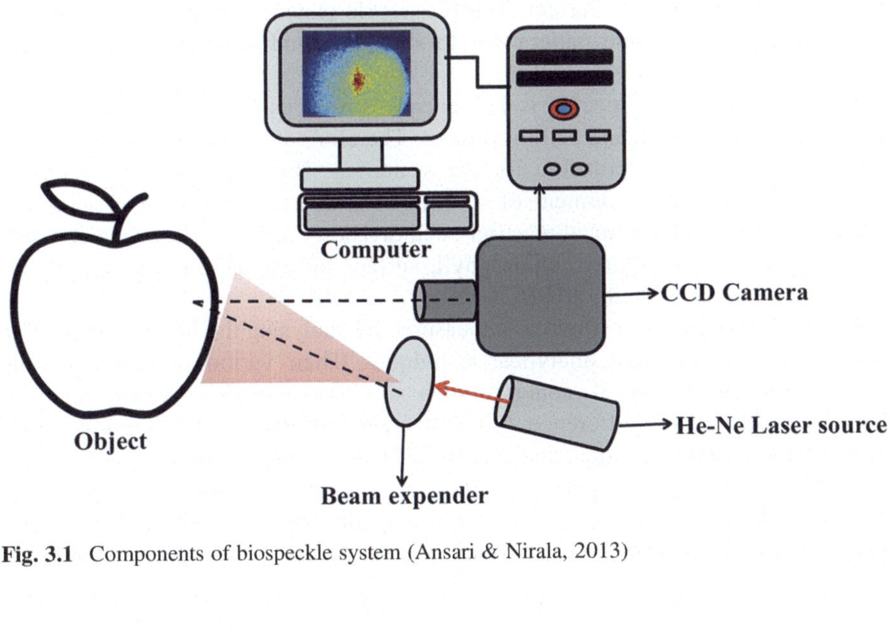

Fig. 3.1 Components of biospeckle system (Ansari & Nirala, 2013)

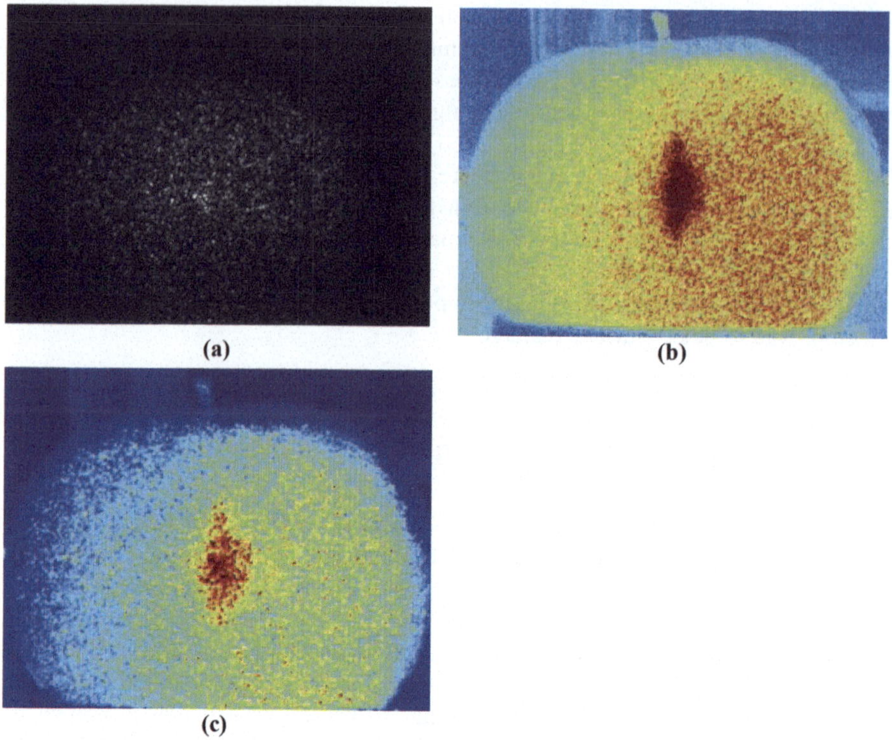

Fig. 3.2 Biospeckle images of apples subjected to mechanical damage: (**a**) raw biospeckle image, (**b**) analysed by Fujji method, and (**c**) analysed by LASCA method (Zdunek et al., 2014)

images are consistent or invariant. The biospeckle pattern comprises two patterns such as static and variable pattern due to stationary and moving particles in the biological sample (Ansari et al., 2012).

The formation of biospeckle is influenced by various factors such as cellular structure, age of the cell, and cell movement. Therefore, it is difficult to identify the component which is responsible for the scattering. Biospeckle activity (BA) is defined as the extent of moment of speckle, it is due to the Brownian motions, organelle moment, biochemical reactions, cell divisions, and cytoplasmic streaming. It changes with water, starch, chlorophyll, surface properties, and the age of the produce (Pandiselvam et al., 2020).

The BA activity of the sample was measured by analysing the biospeckle images with the help of different analytical techniques. Their various algorithms were reported for the graphical analysis such as laser speckle contrast analysis (LASCA), generalized difference (GD), average difference (AD)/Fujji method, laser speckle temporal contrast analysis (LASTCA), temporal difference (TD), and weighted generalized difference (WGD). In general, the graphical analysis is adopted to heterogenous sample where the spatial variation within the sample needs to distinguish. The activity maps of grey levels or pseudocolour will be generated as an outcome of the graphical analysis. In the case of grey levels images the white colour signifies the high activity and black colour signifies the low activity whereas in pseudocolour images red and blue indicate the high and low activity, respectively. The biospeckle images of apple subjected to mechanical damage are illustrated in Fig. 3.2. A clear separation of sound and damaged area (dark spot at the centre) was observed even when the external fruit damage was invisible (Fig. 3.2b). The spatial biospeckle activity of apple after applying Fujji and LASCA method is illustrated in Fig. 3.2c. A distinct separation in biospeckle activity can be observed between the damaged (red) and sound areas (yellow) (Zdunek et al., 2014).

Find the mean of the region of interest (ROI) on the graph to get the value in numbers. However, there are better ways to compare things since they can only be between 0 and 255 grey levels. The numerical methods are beneficial for quantifying activity levels in time. The level or magnitude of change can be derived from the graphical results. Changes in speckle patterns can be shown in different ways, such as the inertia moment (IM), the absolute value difference (AVD), and autocorrelation. Frequency analysis moves a signal from the time domain to the frequency domain. This frequency analysis is done through filtering in frequency bands, continuous wavelet transform (CWT), and discrete wavelet transform (DWT).

Yan et al. (2017) adopted Fujii's, GD, and LASTCA algorithms to detect bruises in apples. They also observed that the BA in intact areas was higher than in damaged areas. They observed that the GD method had a greater standard deviation than LSTCA algorithm; therefore, the LSTCA algorithm could detect areas that were damaged with minimum error.

In the recent study, Wu et al. (2020) used laser backscattering image technique in conjunction with machine learning techniques such as conventional neural network (CNN), the support vector machine (SVM), the particle swarm optimization (PSO), and the back propagation neural networks (BPNN) to discriminate the defect apples from fresh apples. They found that the classification accuracy of CNN model (92.5%) was higher than the other classification techniques. In another study the authors explored the feasibility of biospeckle technology (He-Ne laser 632.8 nm and diode laser 635 nm) for Bull's eye rot detection in apples during the storage at refrigerated condition (Adamiak et al., 2012). They observed that the BA increased with increase in fungal disease and decreased with storage period. They stated that the change in BA during the fruit ripening was due to biochemical reactions. The chlorophyll had negative effect on BA because of the reduction in light absorption as a result of it the penetration depth of light increases. However, the starch degradation had negative effect on BA due to decrease in number of light scattering particles. During fruit ripening process the chlorophyll and starch content of fruits alter with time eventually these affect the BA. They observed that the BA increased in between 40 and 100 days because of the increase in ethylene concentration and transpiration as a result of fungal growth. A lot of biological substances like enzymes, toxins, and growth regulators released during the fungal infestation affect the physiological process or structural integrity of fruit and vegetable cells. The BA of apples decreased in between the 100th and 140th storage period, it is due to the release of pectin enzyme that leads to the degradation of pectin and the fruit tissue looses it coherence and the cells die eventually. Ansari and Nirala (2013) observed that the biospeckle activity (BA) of pears, tomatoes, and apples declined with ageing and they correlated the BA with the fruits respiration rate.

Similarly, the chilling injury in bananas was explored using the biospeckle technique. The experimental results reported LDA techniques along with backscattering parameters to classify damaged and sound bananas; the lowest misclassification accuracy of early chilled samples was 1.33% (Hashim et al., 2013). The laser penetration depth was limited to the tissue of localized skin; therefore, it is not suitable for the internal defects detection. In case of apple tissue, laser penetration depth was 7–10 mm, whereas 2 mm in apple peel. Apart from the application limitations there are some technical hurdles in terms of hardware and software. Due to the huge demand for electronics in the international market, substandard and non-standardized materials are available in market. Experimental errors result from usage of non-standardized and substandard power source and laser diodes. The BA depends on the light intensity also so the intensity of light must be considered while conducting the experiment.

3.3 X-Ray CT Imaging

X-ray computed tomography (CT) imaging is superior to other imaging methods for detecting subsurface damage since it is quick, sophisticated, and does not destroy the object being imaged. It has potential applications in the fruit and vegetable processing industries for detecting internal defects of vegetables and fruits

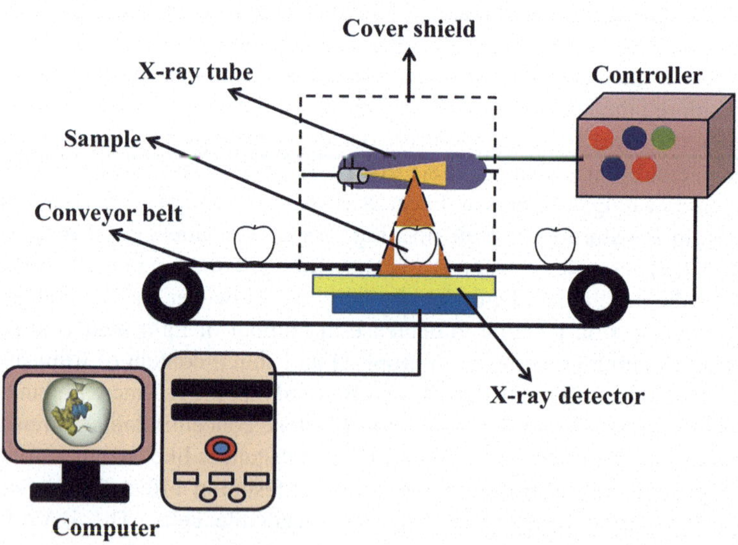

Fig. 3.3 X-ray CT imaging (Kotwaliwale et al., 2007)

(Azadbakht et al., 2019). The X-rays fall in the magnetic spectrum within 0.01–10 nm wavelength range and 0.12–120 keV energy range (Donis-Gonzalez et al., 2014). Cai et al. (2014) used soft X-rays with wavelengths ranging from 0.1 to 10 nm and energies ranging from 0.12 to 12 keV for food commodity inspection. The X-ray imaging system's components are illustrated in Fig. 3.3. Radioactive material or X-ray tubes are used to generate X-rays. The X-rays will be generated in X-ray tubes when the high-energy electrons hit the targeted atoms (Tungsten). The rays produced by X-ray tubes are polychromatic, whereas rays produced by radioactive material are monochromatic (Mathanker et al., 2013). X-ray tubes have the flexibility to alter the intensities of X-rays, and these are used to generate X-rays in radiography of agricultural commodities (Kotwaliwale et al., 2007). The most important things about X-rays are their energy and their current. The energy shows how well the X-rays can penetrate, and the current shows how many photons are made.

Kotwaliwale et al. (2007) stated that X-rays have the unique property of travelling straight lines without deflecting in an electric or magnetic field. While X-ray photons pass through material, they get absorbed by the material and lose their total energy to electrons. The amount of energy lost to electrons depends on the energy of incident photons and the density of the material (Schoeman et al., 2016). The detector in the system used to capture the variation in X-ray transmission occurred due to the variation in material density in the form of visual contrast (Kotwaliwale et al., 2007). The power rating determines the image quality. Hence, accurate control of current and energy is necessary to get good quality images. In traditional X-ray imaging, the object is only seen from one side, usually the top. This makes it hard to understand the results. In X-ray CT imaging, images of the object are taken from different angles and put together to make a three-dimensional image. The voxel is the basic building block of a 3D image, and the CT number, which is also known as the mean attenuation coefficient, represents each voxel. The variation in densities of objects was responsible for the variation in CT number. The computed tomography number is greater for higher density objects than lower density objects. The CT number is 0 for water and -1000 for air (Du et al., 2019).

The pre-processing and analysis of acquired images are necessary to acquire the required information and to visualize the data. Before proceeding to image analysis unwanted noise was removed by image pre-processing. The noise was removed by using filters such as Gaussian or median filter. Image segmentation is necessary for phase differentiation within the material as per the region of interest (ROI), it is usually carried out using thresholding technique. The morphological operations like erosion, watershed, and dilution were used for the clear segmentation and the small pixels (artefacts) were removed by cleaning. For the detection of damaged tissue various pattern recognition and machine vision techniques are available. Conversely, for quantitative analysis the data such as particle size, density, sample thickness, and particle size could be found out from the X-ray CT images. Van Dael et al. (2019) adopted multisensor system to detect the browning in apples. Initially they developed shape and density models with the help of CT data obtained from the apples having different levels of defects. They have developed an internal gradient distribution adjusted method in order to avoid intrusion caused by the density gradients. With this model the browning prediction accuracy has been increased from 73 to 86% and the number of false positive rates lowered from 14 to 5%. They concluded that this system can be used to detect every sort of bruise, but specific identification and segmentation techniques are to be developed.

In a recent study, Azadbakht et al. (2019) used an X-ray imaging technique for detecting the intensity of bruises in pears during transportation. The pressure deformation device and pendulum were used to simulate the static and moving loads that happened during transportation. They affirmed that the bruises that occurred were not visible at lower loads.

During storage and transport, the porosity of fruits and vegetables influences the moment of water and metabolic gases. Tanaka et al. (2018a, b) employed X-ray CT technology to study the changes in the internal structure of cucumbers during storage. They found a strong positive (0.885) correlation between grey scale values (GS) and density; conversely, a strong negative (0.869) correlation was observed between GS and the porosity of cucumber. However, a poor correlation was observed GS with moisture content and elastic modulus of cucumber. They noticed that the grey scale values of cucumber significantly decreased with storage period in endocarp and mesocarp tissues. However, no significant changes were observed in the placenta tissue. In another study, X-ray CT technology to map the porosity in pears and apples was employed by Nugraha et al. (2019). A strong correlation ($R^2 = 0.99$) between porosity and grey scale values was obtained. It was found that the grey scale correlation models developed are suitable for non-destructive mapping of porosity in various vegetables and fruits. It can also be used for online/inline quality inspection. In another study, Diels et al. (2017) employed X-ray technique and multi-level Otsu's algorithm to measure the amount of damage to apple fruits that had been bruised. They reported that detecting bruises using an X-ray CT scan was impossible during the initial stage due to the inadequate moment of cell fluids from injured place to adjacent tissues. X-ray CT imaging was employed by Jarolmasjed et al. (2016) to find the bitter pits in apples. They noticed that the bitter pit developed at the centre of the fruit without any surface symptoms. During storage period, bitter pits increased within the fruit and extended to the surface. The experimental results revealed that the X-ray CT technique could detect the defects in fruits at early stages so that the economic losses could be decreased by avoiding packaging and transporting defective fruits and vegetables.

X-ray imaging techniques have not yet been widely used in the food and vegetable processing industries because of the high-voltage power source and high initial cost needed to make X-rays. For online inspection, 2D images are acquired using the X-ray CT technique, it is possible for detecting the presence or absence of a defect with 2D images, but the bruise volume cannot be assessed (Lu & Lu, 2017). Conversely, Yang et al. (2011) and Mathanker et al. (2013) mentioned that the X-ray CT method is unsuitable for an online application because it needs a lot of computing power. The limited use of this imaging technique in the processing industries of fruit and vegetable was due to the misconception of the X-ray ionizing effect. The X-rays used for food inspection were soft X-rays with very low energy. These are minimally harmful to operators than X-rays employed in medical field. Taking appropriate care while designing the system can eliminate most safety issues.

During the controlled atmospheric storage, unfavourable storage conditions may lead to the development of browning symptoms in apple fruit. Browning disorder in 'Braeburn' apples was characterized by using X-ray imaging technique (Herremans et al., 2013). The authors used high resolution X-ray CT imaging system for inspecting the internal air network, individual cells, and 3D distribution of pores. A PLS-DA model was developed with the help of morphometric parameters, they achieved 97% classification accuracy between healthy and disordered tissue.

X-ray imaging technique is suitable to detect the presence and development of internal defects of the fruit, this is suitable to quantify the level of detection by calculating the detection area or volume. Moreover, the X-ray imaging technique has certain limitations that restrict commercial usage. The cost of equipment is prime concern; it is quite high as compared to other techniques such as spectroscopy and HSI. Since the image acquisition time is high, it limits the online application, and it is difficult to acquire the images in multiple angles in real-world application for the defect volume calculation. In a safety point of view, the leakage of X-rays in the work site poses health hazards to people.

3.4 Hyperspectral Imaging (HSI)

Hyperspectral imaging (HSI) is the combined application of image (for spatial information) and spectroscopic (for spectral information) technologies. The hyperspectral image is three-dimensional cube and comprises two spatial and one spectral dimension. It is also known as a spectral cube, hyper-cube, data cube, spectral volume, or data volume (Feng & Sun, 2012). The spectral signature of any particular commodity is due to the fundamental vibrations and molecular overtones owing to the O-H, N-H, and C-H groups bending and stretching. These are the fundamental tools for any commodity's quantitative and qualitative assessment. The pixels with the same chemical compositions yield the same spectral properties (Baiano, 2017). Therefore, it can be used to develop chemical distribution maps to investigate the spatial distribution of different quality attributes to assess the commodity's internal and external quality. In general, the HSI deals at the molecular level; therefore, a minor deviation in a sample can be discriminated against easily. HSI is a cost-effective, non-invasive, and time-intensive technique that can be used for real-time inspection of various food commodities (Liu et al., 2017). HSI is governed by the interaction between biological materials, light and food samples. The push broom hyperspectral imaging system, along with its working components, is depicted in Fig. 3.4. A HSI system comprises an illumination source, light dispersion device, image acquisition, and hardware system and computer (Wu & Sun, 2013).

The light exerted by the illumination source interacts with the biological materials, and a fraction of it is absorbed, reflected, or transmitted by raising or lowering

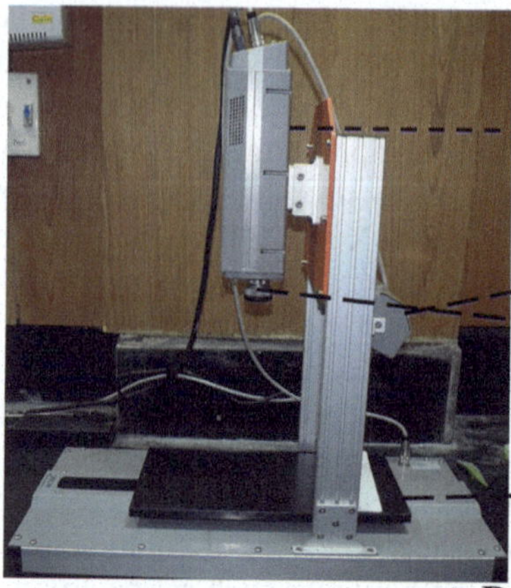

Fig. 3.4 Push broom hyperspectral imaging system

the molecules energy levels. The wavelength and strength of this light fraction depend on commodity's physiochemical properties (Fu & Ying, 2016; ElMasry & Sun, 2010; Ravikanth et al., 2017). The spectrograph turns the detected energy, sensitive to the physical and chemical state of the product, into spectra. With the help of HSI and the Beer-Lambert law, the chemical parts of the product were analysed. This rule states that the sample absorbance is strongly correlated (positively) with concentration of chemical constituents present in sample (Sun, 2010). The HSI system works in absorbance, reflectance, fluorescence, or transmittance mode. Multispectral spectral imaging is a branch of HSI where the images are acquired at particular wavelengths instead of across the spectrum.

3.4.1 *Pre-processing and Analysis of Hyperspectral Image*

Hyper-spectral image consists of millions of data points stored in a pixel form; they are highly correlated with adjacent pixels. Handling of this massive data and extraction of relevant information can be obtained with the help of appropriate chemometric techniques. The steps involved during the HSI processing and their role are tabulated in Table 3.1.

3.4.2 *Application*

The presence of defects decreases quality and price of the fruits and vegetables. The early detection and separation of defected fruits and vegetables can avoid the spoilage of whole batch. The traditional computer vision technique has been

Table 3.1 The steps involved during HSI image processing

Spatial pre-processing	• This process denotes the choosing region of interest (ROI) by eliminating the regions which are not covered by the object • The elimination of unnecessary portion decreases the size of image and avoids the background noise • The unnecessary regions are removed by creating a mask
Identifying and handling dead pixels and spikes	• The dead pixels are those pixels whose intensity values are zero or missing. These are formed due to the faulty detector • The dead pixels in the image were abolished by interpolation with the neighbouring pixels • The spike is defined as a sudden rise or fall in spectrum, these are formed due to the improper behaviour of electric system and surrounding environment • In general the spikes were identified by manual supervision; it is a time-intensive process and requires human attention
Spectral pre-processing	• The main object of pre-processing of spectral data is to avoid the adverse effect on the spectral data associated with light scattering, size of particles, the morphology of surface roughness, product surface, and instrumental noise • To enhance the difference among the spectral data (Amigo et al., 2015) • MSC, SNV, Savitzky-Golay (SG) smoothing, and derivative are commonly used pre-processing techniques
Multivariate analysis	• The captured HSI consists of useful information, it can be used to extract spatial and spectral information • Quantitative and qualitative information related to the product can be obtained from collected data • The exploratory analysis or supervised (PLS-DA, ANN, SVM, KNN) and or unsupervised (PCA and k-means) classification machine learning techniques are suitable for qualitative analysis

implemented commercially for sorting purpose, the computer vision technique works based on the colour, texture, size, and shape variation. But the application of computer vision technique in fruit and vegetable industry for sorting is still a challenging task, because the early defects cannot cause any surface textural or colour changes, high variance of defect types, and existence of stem/calyx concavities. Due to the absence of spectral information the traditional imaging technique is not effective for identification of defects which have similar colour and texture as sound fruits and vegetables. Hyperspectral and multi-spectral imaging are powerful techniques that can detect the surface defects and the defects which are not visible and they can be able to differentiate the defects those have similar colour and texture. Chilling injury is a frequent problem that occurs during low-temperature storage and transit. It is a major problem for the fruit and vegetable industry since it not only affects the safety/quality grade of the produce, but it may also result in large economic losses. Liu et al. (2006) developed protocols for detection of chilling damage in cucumber using VIS-NIR (447–951 nm) hyperspectral imaging system in conjunction with suitable chemometric techniques. The results revealed that using

PCA model within the spectral region of 733–848 nm or band ratio algorithm (811/756) can detect chilling injury with more than 90% accuracy. They revealed that identification of chilling injury during first 2 days of post-chilling room temperature storage was difficult due to the insignificant manifestation of chilling induced symptoms.

In the recent study, the invisible damage in Rojo Brillante permission fruit at different stages was detected by using the VIS-NIR (450–1040 nm) hyper-spectral imaging in conjunction with chemometrics (Munera et al., 2021). They observed that the reflectance values of intact fruit were higher than the damaged fruit within the spectral range of 450–660 nm and the opposite trend was observed after 670 nm. It may be due to the development of browning in the flesh of damaged fruit, it can change the colour of skin in this area. The damaged portion in fruit was identified using PCA and they achieved 90.8% accuracy. The intact and damaged fruits were classified using PLS-DA technique and they achieved 100%, 97.4%, 100%, and 100% classification accuracy for 0, 1, 2, and 3 days, respectively.

Fu and Wang (2021) were adopted fluorescence HSI (350–1100 nm) technique to detect early bruises in pears. They distinguished the fruits as per the bruised time (immediate, 15 min, 24 h, 48 h, 72 h) and bruise levels (sound, minor, severe) using machine learning techniques such as support vector machine (SVM) and random forest (RF). They achieved 93.33% classification accuracy when the fruits were classified as per the bruise level after 15 minutes of bruised. Conversely, 99.33% classification accuracy was obtained when the samples were classified into intact and bruised.

In the recent study the bruise in tomatoes caused due to falling was detected by using VIS-NIR HIS system (400–1000 nm). The effect of drop height, fruit size, and time of detection from the injury on spectral signature was explored (Sun et al., 2021). The reflectance values of bruised tomatoes were lower than healthy tomatoes and the wavelength 810 nm had a significant role for bruise detection. They found that the size of fruit had significant effect on the spectral values and the time of detection had lowest effect. The fruits were classified into damaged and intact using PLS-DA technique with a classification accuracy of 90.93%.

Gowen et al. (2008) adopted HSI (400–1000 nm) along with PCA to detect the damage in mushroom. They applied PCA separately on hypercube and data set consists of average spectrum obtained from sound and bruised tissue. They noticed that the application of PCA on spectral data performed better than the application on hypercube.

Similarly, Zhu and Li (2019) used VIS-NIR hyperspectral imaging system to discriminate bruised and fresh apples. They observed that spectral signature of both sound and damaged apples was similar. The featured wavelengths (421.9, 481.8, 491.8, 603.3, 634.1, 639.2, 644.4, 649.5659.9, 696.1, and 945.2 nm) representing the bruise detection were selected using minimum redundancy maximum relevance (mRMR) algorithm. They observed that the maximum classification accuracy (95.97%) was observed in extreme learning machine (ELM) algorithm used along with whole spectral data and 92.9% classification accuracy was obtained with the featured wavelengths and minimum noise fraction (MNF) algorithm.

Xu et al. (2018, 2019) adopted HSI system which has wavelength range of 900–1700 nm to evaluate quantitatively and qualitatively the mechanical damage in mango and apple. The mechanical damage was created in the equatorial region of fruit by drop test. They observed that the spectral absorption decreased with increase in the height of drop and it was lower than the sound fruits. The mango fruits were classified as per the level of damage using discriminant analysis (DA), they achieved 77.8% classification accuracy. The PLSR models developed between the damaged area of apple and absorbed energy, undamaged firmness, contact load, and spectra yield good prediction accuracies ($R^2 = 0.53–0.89$).

Another study (Rivera et al., 2014) used NIR-HSI with machine learning methods for detecting mechanical damage in mango. It was found that the undamaged mango had greater reflectance values than the damaged mango. To distinguish between damaged and healthy fruits, the spectral bands 700–780, 890–900, and 1070–1080 nm were chosen as relevant bands. The 700–780 nm spectral bands showed the greatest variation in spectral data. The k-NN model's classification accuracy using total spectral data and chosen wave bands was 95% and 91%, respectively; it outperformed other classification techniques such as extreme learning machine (ELM), Naive Bayes (NBC), decision trees (DT), and linear discriminant analysis (LDA).

Xiong et al. (2018) used VIS-HSI (300–1000 nm) in reflectance mode to detect the micro-damage in litchi fruit. The classification accuracy was found to be 94.10% with the PLS-DA technique. The featured wavelengths 694, 725, and 798 nm were finalized, representing the damage of litchi fruit. The list of recent research works on various fruits and vegetables to detect the damage using HSI is tabulated in Table 3.2.

The penetration depth of light is restricted to a few millimetres of fruit surfaces (Ravikanth et al., 2017); therefore, it is not suggested for detecting internal damage of fruits. Conversely, HSI cameras or sensors are expensive compared to normal ones. Acquisition time of hyperspectral images was high; therefore, it is unsuitable for real-time (on the go) application (Pu et al., 2015). The fruits and vegetables with glossy surfaces, such as tomatoes and apples, create bright spots during image acquisition. The application of HSI for spherical objects develops shining spots due to light reflection and dark edges of an object due to Lambert's cosine law. The computational speed and accuracy of prediction and classification models were affected by the morphology of the commodity and the other parts of the commodity, such as stem and calyx, etc.

3.5 Thermal Imaging (TI)

Thermal imaging (TI) technique is a non-invasive and non-contacting type. TI measures the thermal radiation emitted by objects with temperatures above −273.15 °C instead of reflected light. A thermal imaging system camera transfers invisible electromagnetic radiation into visible data (thermograms) (Chelladurai

Table 3.2 Recent application of HSI along with different chemometric techniques to detect the fruit damage

Product/ application	Wavelength (mode of imaging), nm	Chemometric techniques	Accuracy	References
Apple/fruit quality	400–1100 (R)	PLS, MR, ANN	$R^2 = 0.93$	Hasanzadeh et al. (2022)
Apple/bruise	400–1000(R)	ELM, PLS-DA, CART	81.67–97.82%	Zhu and Li (2019)
Apples/ infested fruits	400–1000 (R)	DT	96%	Ekramirad and Eyvani (2017)
Apple/bruise detection	400–1000 (R)	PCA, MNF, SIMCA, LDA, SVM	93%	Baranowski et al. (2012)
Apple/impact damage	900–1700(A)	PLS	$R^2 = 0.93$	Xu et al. (2019)
Apple/ bruising		PLS-DA	90%	Luo et al. (2012)
Cucumber fruit/chilling injury	500–1000 (relative intensity)	NB, SVM, and KNN	76.2–100%	Cen et al. (2016)
Jujube/ bruising		CARS-PLS-DA	91.1%	Yuan et al. (2021)
Cucumber fruit/insect damage	450–740 (R) 740–1000 (T)	PLS-DA	82–93%	Lu and Ariana (2013)
Blueberry/ mechanical damage	–	ResNet/ResNeXt	F_1:0.90, 0.89	Wang et al. (2018)
Citrus fruit/ decay	325–1100 (R)	Multispectral algorithm	98.6%	Li et al. (2016)
Jujube fruit/ insect damage	900–1700 (R)	PCA	93.1%	Liu et al. (2015)
Kiwi fruit/ bruise	408–1117 (R)	PCA	85.5%	Lu and Tang (2012)
Mushroom/ mechanical damage	–	SVM, ANN	92.48–94.19%	Rojas-Moraleda et al. (2016)
Litchi/micro-damage	300–1000 (R)	PLS-DA LS-SVM	93.95% 95%	Xiong et al. (2018)
Mushroom/ Browning	–	PLS-DA	69.4–94.5%	Mollazade (2017)
Orange/defect	400–1000 (R)	Banda ratio + PCA	92.9–95.4%	Li et al. (2011)
Pear/physical damage	950–1605 (R)	A classification algorithm based on *F*-values of ANOVA	92%	Lee et al. (2014)

(continued)

Table 3.2 (continued)

Product/ application	Wavelength (mode of imaging), nm	Chemometric techniques	Accuracy	References
Peaches/ defects	400–1000 (R)	–	93.3%	Zhang et al. (2015)

PLS partial least square, *MR* multiple linear regression, *ELM* extreme learning machine, *DA* discriminant analysis, *DT* decision tree, *SD* stepwise discrimination, *MDC* Mahalanobis distance classification, *ND* normalized difference, *SAM* spectral angle mapper, *SVM* support vector machines, *ASD* asymmetric second difference; *MNF* minimum noise fraction, *SSD* symmetric second difference, *LDA* linear discriminant analysis, *LLR* linear logistic regression, *RBF* radical basis function, *NBC* Naıve Bayes, *BNN* back propagation neural networks, *FURIA* fuzzy unordered rule induction algorithm, *ANN* artificial neural network, *PCA* principal component analysis, *SVM* support vector machine, *RF* random forests, *k-NN* k-nearest neighbour, *SMO* sequential minimal optimization algorithm

et al., 2012). TI has gained more attention in the agriculture and food industries. The thermograms are generated owing to the distinction in thermal diffusivity; as a consequence of it sound and damaged tissue have different temperatures. This can be used as an important variable for classification (Zeng et al., 2020). The physico-chemical properties of any material influence its diffusivity (Du et al., 2020). A TI system is generally categorized into active and passive. In a passive TI system, no external heating source is required. It deals the temperature difference between the surrounding environment and the object. However, in the case of an active TI system, an external heating source is used to heat the object. Active thermography can detect the defects initiated underneath surface and the defect depth based on temperature distribution and the heat pulse characteristics with respect to time (He et al., 2021). Various thermography techniques are available, such as vibro, lock in, pulsed-phase, and pulse thermography (Gowen et al., 2010).

Earlier advancement of thermal camera sensors was cooled by cryogens but nowadays due to technical advancement the cameras are operating at room temperature. There is no need of illumination source for thermal imaging system as like other imaging techniques; therefore, the problems with uneven lighting and scattering can be rid off, however it requires constant surrounding temperature (Lu & Lu, 2017). TI can be used for online/inline application due to their non-contact, non-invasive, and quick measurement potential.

3.5.1 Thermal Imaging (TI) System

Thermal imaging system consists of IR detectors, image acquisition system, and signal processing (Fig. 3.5). The detectors absorb and convert the infrared energy exerted by the object into electrical signal, the signal processing unit transforms the electrical impulses into thermal images. The sensitivity, resolution of intensity, and scanning speed define the performance of thermal imaging system. The cameras

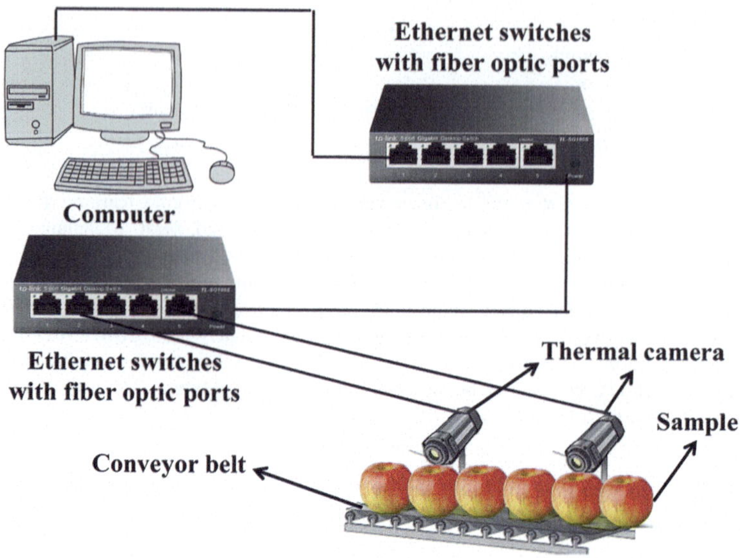

Fig. 3.5 Thermal image system for fruit damage detection (Vadivambal & Jayas, 2011)

which have high scanning speed and wide range of sensing temperature were fit to study thermal changes. However, the cameras which have high spatial and intensity resolution were suitable for objects with temperature gradient (Vadivambal & Jayas, 2011).

3.5.2 Thermal Imaging (TI) Technique Application for Fruit Damage Detection

The active or passive thermal imaging cameras can be used to distinguish the sound and bruised fruits (Fig. 3.6). The difference in rate of heating and cooling between the fresh and bruised tissues can be used as a tool to identify the bruises of the fruits in the early stage (Vadivambal & Jayas, 2011).

In a recent study, Zeng et al. (2020) adopted TI (8–14 μm) in conjunction with a hot air system to differentiate healthy and bruised pears. They observed that the TI alone is not effective in identifying invisible bruises. So, to classify, they used GLCM and deep learning techniques and got a classification accuracy of 99.25%. In another study, pulsed thermography was used to detect bruises in blueberries (Kuzy et al., 2018). The time-domain and frequency-domain features were used with different classification techniques (LDS, SVM, k-NN, random forest, and logistic regression) to classify fresh and bruised fruits. The maximum classification accuracy (89.5%) was achieved with the LDA technique.

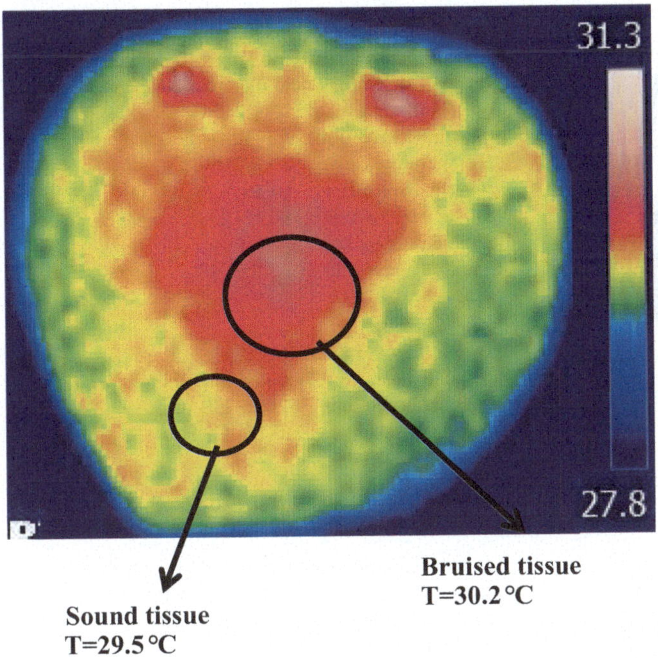

Fig. 3.6 Thermal image of apple subjected to mechanical damage (Varith et al., 2001)

Yogesh et al. (2018) used segmentation techniques like SURF, OTSU, k-mean, watershed, and texture along with TI to find the defective apple parts. They reported that the results from texture-based segmentation were better than those from other methods. The thermal behaviour of fruits varies with the surrounding environment. They concluded that the TI technique could detect the fruit's internal defects.

The chilling injury in guava during the storage at different temperatures (5–20 °C) was identified by using the TI technique (Gonçalves et al., 2015). The maximum temperature difference between sound and damaged portion was observed at 5 °C storage temperature as compared to other storage temperatures. A temperature difference of 0.5–3 °C was observed between mechanically damaged or cold injured tissues and sound tissues.

Most of the research has been done on fruits that were damaged mechanically in a lab. This is unlike damage that occurs in the actual world. The physiochemical properties of wounded fruits change over time. Consequently, the fruits thermal and mechanical properties should be considered when detecting business. In contrast, lock-in-thermography and photothermal wave models were used to quantify business in pears during the early stage in a separate study (Kim et al., 2014). The lock-in correlation technique was used to regulate thermal signals acquired at different frequencies (0.05–0.2 Hz) and the phase information in conjunction with photothermal wave model was used to detect the depth of damage. They took into account the thermal and mechanical qualities of fruits when detecting damage, as

Table 3.3 List of some commercially available thermal imaging cameras and their specifications

	Camera model: FLIR E5-XT Type of detector: Focal plane array (FPA), uncooled microbolometer Spectral range (μm): 7.5–13 Temperature range: –20 to 250 °C Thermal sensitivity: <0.10 °C/<100 mK Image size (pixels): 160 × 120 Frame rate: 9 Hz Field of view (FOV): 45° × 34° Accuracy: ±2 °C
	Camera model: FLIR E4 Type of detector: Uncooled microbolometer Temperature range: −20 to 250 °C Thermal sensitivity: <0.15 °C Image size (pixels): 80 × 60 Frame rate: 9 Hz Field of view (FOV): 45° × 34°
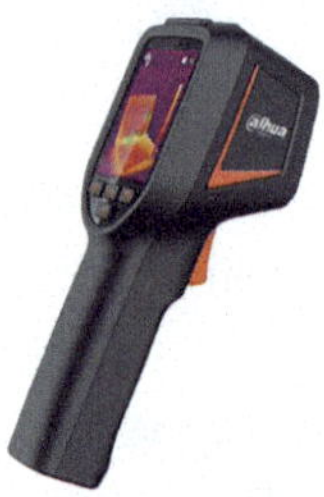	Camera model: Dahua (DH-TPC-HT2201) Type of detector: Uncooled VOx FPA Spectral range (μm): 8–14 μm Temperature range: 30–45 °C Thermal sensitivity: ≤50 mK at F1.0 Image size (pixels): 320 × 240 Field of view (FOV): 50.6° × 37.8 Accuracy: ±0.5 °C Pixel size: 12 μm
	Camera model: Thermo GEAR G120EX Type of detector: Uncooled focal plane Array (microbolometer) Spectral range (μm): 8 to 14 Temperature range: −20 °C to 60 °C Image size (pixels): 320 × 240 Field of view (FOV): 32 × 24 Accuracy: ±2 °C or ±2% of Reading Frame rate: 60 Hz

well as faults when determining the real thermal route in the fruit and quantitatively estimating bruise depth.

TI system has unique advantage as compared to other detection techniques in terms of its non-invasive, non-destructive, rapid, low cost, and non-contact nature. Table 3.3 presents the list of commercially available thermal imaging cameras. This technology can even identify small bruise areas during the initial stage and suitable

Table 3.4 Difference among various imaging techniques

	Biospeckle	X-ray imaging	HSI	TI
Cost	Low	High	High	Low
Early detection of surface damage	√	×	√	√
Detection of internal disorder	×	√	×	×
Sensitivity to internal disorder	√	√	×	×
Illumination/energy source	√	√	√	×
Real-time detection	√	×	×	√
Potential safety hazard	×	√	×	×
Complexity	Low	High	High	Low
Sensitivity to external environment	Low	Low	High	High
Image acquisition time	Low	High	High	Low

to detect bruises in dark coloured fruits also. Therefore, it is gaining more attention by fruit and vegetable industries for the detection of damages. However, the application of this technology was limited to the laboratory due to the measurement sensitivity and application of active thermal imaging during in/online sorting at commercial level is not possible. Further technological development and research is necessary in this area to bring this technology at commercial level. Table 3.4 presents the difference among biospeckle, thermal imaging, hyperspectral imaging, and X-ray imaging techniques employed to detect surface damage.

3.6 Conclusion

Manual detection of damaged fruits and vegetables in the processing industry continuous process lines requires skilled personnel and is a time-and-cost-intensive. The non-destructive techniques rely on food's mechanical, chemical, physical, and structural properties. NMR, X-ray imaging, and MRI techniques are suitable for detecting internal damage, whereas biospeckle, thermal imaging, hyperspectral imaging, and machine vision are employed to detect surface damage. Hyperspectral imaging is capable of detection of surface damage but they cannot detect inside bruise. The combination of above discussed imaging techniques will make path to inside bruise detection of fruits. Deep learning techniques are strong machine learning methods that have demonstrated their superiority on a variety of applications; nevertheless, it also has a number of drawbacks. When a situation becomes more complex, it is typically necessary to collect additional data. Supervised models necessitate an extensive labelled training dataset, which is a time-consuming and therefore costly procedure, especially for hyperspectral data.

References

Adamiak, A., Zdunek, A., Kurenda, A., & Rutkowski, K. (2012). Application of the biospeckle method for monitoring bull's eye rot development and quality changes of apples subjected to various storage methods—Preliminary studies. *Sensors, 12*(3), 3215–3227.

Amigo, J. M., Babamoradi, H., & Elcoroaristizabal, S. (2015). Hyperspectral image analysis. A tutorial. *Analytica Chimica Acta, 896*, 34–51.

Ansari, M. Z., & Nirala, A. K. (2013). Biospeckle activity measurement of Indian fruits using the methods of cross-correlation and inertia moments. *Optik, 124*(15), 2180–2186.

Ansari, M. Z., Minz, P. D., & Nirala, A. K. (2012). Fruit quality evaluation using biospeckle techniques. In *In 2012 1st international conference on recent advances in information technology (RAIT)* (pp. 873–876). IEEE.

Azadbakht, M., VahediTorshizi, M., & Mahmoodi, M. J. (2019). The use of CT scan imaging technique to determine pear bruise level due to external loads. *Food Science & Nutrition, 7*(1), 273–280.

Baiano, A. (2017). Applications of hyperspectral imaging for quality assessment of liquid based and semi-liquid food products: A review. *Journal of Food Engineering, 214*, 10–15.

Baranowski, P., Lipecki, J., Mazurek, W., & Walczak, R. T. (2008). Detection of watercore in 'Gloster' apples using thermography. *Postharvest Biology and Technology, 47*, 358–366.

Baranowski, P., Mazurek, W., Wozniak, J., & Majewska, U. (2012). Detection of early bruises in apples using hyperspectral data and thermal imaging. *Journal of Food Engineering, 110*(3), 345–355.

Bhargava, A., & Bansal, A. (2021). Fruits and vegetables quality evaluation using computer vision: A review. *Journal of King Saud University, Computer and Information Sciences, 33*(3), 243–257.

Cai, J., Yuan, L., Liu, B., & Sun, L. (2014). Nondestructive gender identification of silkworm cocoons using X-ray imaging with multivariate data analysis. *Analytical Methods, 6*(18), 7224–7233.

Cen, H., Lu, R., Zhu, Q., & Mendoza, F. (2016). Nondestructive detection of chilling injury in cucumber fruit using hyperspectral imaging with feature selection and supervised classification. *Postharvest Biology and Technology, 111*, 352–361.

Chelladurai, V., Kaliramesh, S. J., & Digvir, S. J. (2012). *Detection of Callosobruchus Maculatus (F.) infestation in mung bean (Vigna radiata) using thermal imaging technique*. NABEC/CSBE 12-121.

Dar, A. H., Makroo, H. A., Shah, S., & Khan, S. (2020). Nuclear magnetic resonance (NMR) spectroscopy for quality determination of fruits and vegetables. In *Sensor-based quality assessment systems for fruits and vegetables* (pp. 199–230). Apple Academic Press.

Diels, E., van Dael, M., Keresztes, J., Vanmaercke, S., Verboven, P., Nicolai, B., Saeys, W., Ramon, H., & Smeets, B. (2017). Assessment of bruise volumes in apples using X-ray computed tomography. *Postharvest Biology and Technology, 128*, 24–32.

Donis-Gonzalez, I. R., Guyer, D. E., Pease, A., & Barthel, F. (2014). Internal characterisation of fresh agricultural products using traditional and ultrafast electron beam X-ray computed tomography imaging. *Biosystems Engineering, 117*, 104–113.

Du, Z., Hu, Y., Ali Buttar, N., & Mahmood, A. (2019). X-ray computed tomography for quality inspection of agricultural products: A review. *Food Science & Nutrition, 7*(10), 3146–3160.

Du, Z., Zeng, X., Li, X., Ding, X., Cao, J., & Jiang, W. (2020). Recent advances in imaging techniques for bruise detection in fruits and vegetables. *Trends in Food Science & Technology, 99*, 133–141.

Ekramirad, N., & Eyvani, A. (2017). Non-destructive detection of codling moth (Cydiapomonella L) damage in apple fruit using hyperspectral imaging method. *Iranian Journal of Biosystems Engineering, 48*(2), 241–249.

ElMasry, G., & Sun, D. W. (2010). Principles of hyperspectral imaging technology. In *Hyperspectral imaging for food quality analysis and control* (pp. 3–43). Academic Press.

Feng, Y. Z., & Sun, D. W. (2012). Application of hyperspectral imaging in food safety inspection and control: A review. *Critical Reviews in Food Science and Nutrition, 52*(11), 1039–1058.

Fu, X., & Wang, M. (2021). Detection of early bruises on pears using fluorescence hyperspectral imaging technique. *Food Analytical Methods, 15*, 115. https://doi.org/10.1007/s12161-021-02092-3

Fu, X., & Ying, Y. (2016). Food safety evaluation based on near infrared spectroscopy and imaging: A review. *Critical Reviews in Food Science and Nutrition, 56*(11), 1913–1924.

Gao, Y., & Rao, X. (2019). Blackspot bruise in potatoes: Susceptibility and biospeckle activity response analysis. *Journal of Food Measurement and Characterization, 13*(1), 444–453.

Gonçalves, B. J., Giarola, T. M., Pereira, D. F., Vilas Boas, E. V., & de Resende, J. V. (2015). Using infrared thermography to evaluate the injuries of cold-stored guava. *Journal of Food Science and Technology, 53*(2), 1063–1070.

Gowen, A. A., O'Donnell, C. P., Taghizadeh, M., Cullen, P. J., Frias, J. M., & Downey, G. (2008). Hyperspectral imaging combined with principal component analysis for bruise damage detection on white mushrooms (Agaricus bisporus). *Journal of Chemometrics, 22*, 259–267.

Gowen, A. A., Tiwari, B. K., Cullen, P. J., McDonnell, K., & O'Donnell, C. P. (2010). Applications of thermal imaging in food quality and safety assessment. *Trends in Food Science & Technology, 21*(4), 190–200.

Hasanzadeh, B., Abbaspour-Gilandeh, Y., Soltani-Nazarloo, A., Hernández-Hernández, M., Gallardo-Bernal, I., & Hernández-Hernández, J. L. (2022). Non-destructive detection of fruit quality parameters using hyperspectral imaging, multiple regression analysis and artificial intelligence. *Horticulturae, 8*(7), 598.

Hashim, N., Pflanz, M., Regen, C., Janius, R. B., Rahman, R. A., Osman, A., Shitan, M., & Zude, M. (2013). An approach for monitoring the chilling injury appearance in bananas by means of backscattering imaging. *Journal of Food Engineering, 116*(1), 28–36. https://doi.org/10.1016/j.jfoodeng.2012.11.018

He, Y., Deng, B., Wang, H., Cheng, L., Zhou, K., Cai, S., & Ciampa, F. (2021). Infrared machine vision and infrared thermography with deep learning: A review. *Infrared Physics & Technology, 116*, 103754.

Herremans, E., Verboven, P., Bongaers, E., Estrade, P., Verlinden, B., Wevers, M., Hertog, M., & Nicolai, B. (2013). Characterisation of 'Braeburn' browning disorder by means of X-ray micro-CT. *Postharvest Biology and Technology, 75*, 114–124.

Hussein, Z., Caleb, O. J., & Opara, U. L. (2015). Perforation-mediated modified atmosphere packaging of fresh and minimally processed produce—A review. *Food Packaging and Shelf Life, 6*, 7–20.

Jarolmasjed, S., Espinoza, C. Z., Sankaran, S., & Khot, L. R. (2016). Postharvest bitter pit detection and progression evaluation in 'Honeycrisp' apples using computed tomography images. *Postharvest Biology and Technology, 118*, 35–42.

Kamal, T., Cheng, S., Khan, I. A., Nawab, K., Zhang, T., Song, Y., Wang, S., Nadeem, M., Riaz, M., Khan, M. A. U., & Zhu, B. W. (2019). Potential uses of LF-NMR and MRI in the study of water dynamics and quality measurement of fruits and vegetables. *Journal of Food Processing and Preservation, 43*(11), e14202.

Kim, G., Kim, G. H., Park, J., Kim, D. Y., & Cho, B. K. (2014). Application of infrared lock-in thermography for the quantitative evaluation of bruises on pears. *Infrared Physics & Technology, 63*, 133–139.

Kotwaliwale, N., Weckler, P. R., Brusewitz, G. H., Kranzler, G. A., & Maness, N. O. (2007). Non-destructive quality determination of pecans using soft X-rays. *Postharvest Biology and Technology, 45*(3), 372–380.

Kumar, V., Purbey, S. K., & Anal, A. K. D. (2016). Losses in litchi at various stages of supply chain and changes in fruit quality parameters. *Crop Protection, 79*, 97–104.

Kuzy, J., Jiang, Y., & Li, C. (2018). Blueberry bruise detection by pulsed thermographic imaging. *Postharvest Biology and Technology, 136*, 166–177.

Lee, W.-H., Kim, M. S., Lee, H., Delwiche, S. R., Bae, H., Kim, D.-Y., & Cho, B.-K. (2014). Hyperspectral near-infrared imaging for the detection of physical damages of pear. *Journal of Food Engineering, 130*, 1–7.

Li, J., Rao, X., & Ying, Y. (2011). Detection of common defects on oranges using hyperspectral reflectance imaging. *Computers and Electronics in Agriculture, 78*(1), 38–48.

Li, Z., Yang, H., Li, P., Liu, J., Wang, J., & Xu, Y. (2013). Fruit biomechanics based on anatomy: A review. *International Agrophysics, 27*, 97–106.

Li, J., Huang, W., Tian, X., Wang, C., Fan, S., & Zhao, C. (2016). Fast detection and visualization of early decay in citrus using Vis-NIR hyperspectral imaging. *Computers and Electronics in Agriculture, 127*, 582–592.

Liu, Y., Chen, Y. R., Wang, C. Y., Chan, D. E., & Kim, M. S. (2006). Development of hyperspectral imaging technique for the detection of chilling injury in cucumbers; spectral and image analysis. *Applied Engineering in Agriculture, 22*, 101–111.

Liu, G., He, J., Wang, S., Luo, Y., Wang, W., Wu, L., Si, Z., & He, X. (2015). Application of near-infrared hyperspectral imaging for detection of external insect infestations on jujube fruit. *International Journal of Food Properties, 19*(1), 41–52.

Liu, Y., Pu, H., & Sun, D. W. (2017). Hyperspectral imaging technique for evaluating food quality and safety during various processes: A review of recent applications. *Trends in Food Science and Technology, 69*, 25–35.

Lu, R., & Ariana, D. P. (2013). Detection of fruit fly infestation in pickling cucumbers using a hyperspectral reflectance/transmittance imaging system. *Postharvest Biology and Technology, 81*, 44–50.

Lu, Y., & Lu, R. (2017). Non-destructive defect detection of apples by spectroscopic and imaging technologies: A review. *Transactions of the ASABE, 60*(5), 1765–1790.

Lu, Q., & Tang, M. (2012). Detection of hidden bruise on kiwi fruit using hyperspectral imaging and parallelepiped classification. *Procedia Environmental Sciences, 12*(part B), 1172–1179.

Luo, X., Takahashi, T., Kyo, K., & Zhang, S. H. (2012). Wavelength selection in Vis/NIR spectra for detection of bruises on apples by ROC analysis. *Journal of Food Engineering, 109*(3), 457–466. https://doi.org/10.1016/j.jfoodeng.2011.10.035

Mathanker, S. K., Weckler, P. R., & Bowser, T. J. (2013). X-ray applications in food and agriculture: A review. *Transactions of the ASABE, 56*(3), 1227–1239.

Mollazade, K. (2017). Non-destructive identifying level of browning development in button mushroom (Agaricus bisporus) using hyperspectral imaging associated with chemometrics. *Food Analytical Methods, 10*(8), 2743–2754. https://doi.org/10.1007/s12161-017-0845-y

Munera, S., Rodríguez-Ortega, A., Aleixos, N., Cubero, S., Gómez-Sanchis, J., & Blasco, J. (2021). Detection of invisible damages in 'Rojo Brillante' Persimmon fruit at different stages using hyperspectral imaging and chemometrics. *Food, 10*(9), 2170.

Nugraha, B., Verboven, P., Janssen, S., Wang, Z., & Nicolaï, B. M. (2019). Non-destructive porosity mapping of fruit and vegetables using X-ray CT. *Postharvest Biology and Technology, 150*, 80–88.

Pandiselvam, R., Subhashini, S., BanuuPriya, E. P., Kothakota, A., Ramesh, S. V., & Shahir, S. (2019). Ozone based food preservation: A promising green technology for enhanced food safety. *Ozone: Science & Engineering, 41*(1), 17–34.

Pandiselvam, R., Mayookha, V. P., Kothakota, A., Ramesh, S. V., Thirumdas, R., & Juvvi, P. (2020). Biospeckle laser technique—a novel non-destructive approach for food quality and safety detection. *Trends in Food Science & Technology, 97*, 1–13.

Pathare, P. B., & Al-Dairi, M. (2021). Bruise susceptibility and impact on quality parameters of pears during storage. *Frontiers in Sustainable Food Systems, 5*, 658132.

Pholpho, T., Pathaveerat, S., & Sirisomboon, P. (2011). Classification of longan fruit bruising using visible spectroscopy. *Journal of Food Engineering, 104*(1), 169–172.

Polat, R., Aktas, T., & Ikinci, A. (2012). Selected mechanical properties and bruise susceptibility of nectarine fruit. *International Journal of Food Properties, 15*(6), 1369–1380.

Pu, Y. Y., Feng, Y. Z., & Sun, D. W. (2015). Recent progress of hyperspectral imaging on quality and safety inspection of fruits and vegetables: A review. *Comprehensive Reviews in Food Science and Food Safety, 14*(2), 176–188.

Ravikanth, L., Jayas, D. S., White, N. D. G., Fields, P. G., & Sun, D. W. (2017). Extraction of spectral information from hyperspectral data and application of hyperspectral imaging for food and agricultural products. *Food and Bioprocess Technology, 10*(1), 1–33.

Rivera, N. V., Gomez-Sanchis, J., Chanona-Perez, J., Carrasco, J. J., Millan-Giralolo, M., Lorente, D., Ubero, S., & Blasco, J. (2014). Early detection of mechanical damage in mango using NIR hyperspectral images and machine learning. *Biosystems Engineering, 122*, 91–98.

Rojas-Moraleda, R., Valous, N. A., Gowen, A., Esquerre, C., Hartel, S., Salinas, L., & O'Donnell, C. (2016). A frame-based ANN for classification of hyperspectral images: Assessment of mechanical damage in mushrooms. *Neural Computing & Applications, 28*(S1), 969–981. https://doi.org/10.1007/s00521-016-2376-7

Sandhu, S., & Gill, B. S. (2013). *Physiological disorders of fruit crops* (p. 189). New India Publishing Agency.

Schoeman, L., Williams, P., du Plessis, A., & Manley, M. (2016). X-ray micro-computed tomography (μCT) for non-destructive characterisation of food microstructure. *Trends in Food Science & Technology, 47*, 10–24.

Sun, D. W. (Ed.). (2010). *Hyperspectral imaging for food quality analysis and control*. Elsevier.

Sun, Y., Pessane, I., Pan, L., & Wang, X. (2021). Hyperspectral characteristics of bruised tomatoes as affected by drop height and fruit size. *LWT, 141*, 110863.

Tanaka, F., Imamura, K., Tanaka, F., & Uchino, T. (2018a). Determination of thermal diffusivity of persimmon flesh tissue using three-dimensional structure model based on X-ray computed tomography. *Journal of Food Engineering, 221*, 151–157.

Tanaka, F., Nashiro, K., Obatake, W., Tanaka, F., & Uchino, T. (2018b). Observation and analysis of internal structure of cucumber fruit during storage using X-ray computed tomography. *Engineering in Agriculture, Environment and Food, 11*(2), 51–56.

Vadivambal, R., & Jayas, D. S. (2011). Applications of thermal imaging in agriculture and food industry—A review. *Food and Bioprocess Technology, 4*(2), 186–199.

Van Dael, M., Verboven, P., Zanella, A., Sijbers, J., & Nicolai, B. (2019). Combination of shape and X-ray inspection for apple internal quality control: In silico analysis of the methodology based on X-ray computed tomography. *Postharvest Biology and Technology, 148*, 218–227.

Varith, J., Hyde, G. M., Baritelle, A. L., Fellman, J. K., & Sattabongkot, T. (2001). *Thermal image bruise detection*. ASAE Paper No. 016031.

Wang, Z., Hu, M., & Zhai, G. (2018). Application of deep learning architectures for accurate and rapid detection of internal mechanical damage of blueberry using hyperspectral transmittance data. *Sensors, 18*(4), 1126.

Wu, D., & Sun, D. W. (2013). Application of visible and near infrared hyperspectral imaging for non-invasively measuring distribution of water-holding capacity in salmon flesh. *Talanta, 116*, 266–276.

Wu, A., Zhu, J., & Ren, T. (2020). Detection of apple defect using laser-induced light backscattering imaging and convolutional neural network. *Computers and Electrical Engineering, 81*, 106454.

Xiong, J., Lin, R., Bu, R., Liu, Z., Yang, Z., & Yu, L. (2018). A micro-damage detection method of litchi fruit using hyperspectral imaging technology. *Sensors, 18*(3), 700.

Xu, D., Wang, H., Ji, H., Zhang, X., Wang, Y., Zhang, Z., & Zheng, H. (2018). Hyperspectral imaging for evaluating impact damage to mango according to changes in quality attributes. *Sensors, 18*(11), 3920.

Xu, D., Wang, H., Ji, H., Zhang, X., Cerbu, C., Hu, E., & Dong, F. (2019). Quantitative evaluation of impact damage to apple by hyperspectral imaging and mechanical parameters. *Food Analytical Methods, 12*(2), 371–380.

Yan, L., Liu, J., & Men, S. (2017). The biospeckle method for early damage detection of fruits. *Modern Physics Letters B, 31*(19–21), 1740034.

Yang, W., Xu, X., Duan, L., Luo, Q. M., Chen, S. B., Zeng, S. Q., & Liu, Q. (2011). High-throughput measurement of rice tillers using a conveyor equipped with x-ray computed tomography. *Review of Scientific Instruments, 82*, 761–793.

Yildiz, F., Özdemir, A. T., & Uluışık, S. (2019). Evaluation performance of ultrasonic testing on fruit quality determination. *Journal of Food Quality, 2019*, 1.

Yogesh, Dubey, A. K., & Arora, R. R. (2018). A comparative approach of segmentation methods using thermal images of apple. In *Proceedings of the 7th International Conference on Reliability, Infocom Technologies and Optimization (Trends and Future Directions) (ICRITO).*

Yuan, R. R., Liu, G. S., He, J. G., Wan, G. L., Fan, N. Y., Li, Y., & Sun, Y. R. (2021). Classification of Lingwu long jujube internal bruise over time based on visible near-infrared hyperspectral imaging combined with partial least squares-discriminant analysis. *Computers and Electronics in Agriculture, 182*, 106043. https://doi.org/10.1016/j.compag.2021.106043

Zdunek, A., Adamiak, A., Pieczywek, P. M., & Kurenda, A. (2014). The biospeckle method for the investigation of agricultural crops: A review. *Optics and Lasers in Engineering, 52*, 276–285.

Zeng, X., Miao, Y., Ubaid, S., Gao, X., & Zhuang, S. (2020). Detection and classification of bruises of pears based on thermal images. *Postharvest Biology and Technology, 161*, 111090.

Zhang, W., Pan, L., Tu, S., Zhan, G., & Tu, K. (2015). Non-destructive internal quality assessment of eggs using a synthesis of hyperspectral imaging and multivariate analysis. *Journal of Food Engineering, 157*, 41–48.

Zhu, X., & Li, G. (2019). Rapid detection and visualization of slight bruise on apples using hyperspectral imaging. *International Journal of Food Properties, 22*(1), 1709–1719.

Chapter 4
Hyperspectral Imaging Techniques for Quality Assessment in Fresh Horticultural Produce and Prospects for Measurement of Mechanical Damage

M. C. Alamar, N. Aleixos, J. M. Amigo, D. Barbin, and J. Blasco

Abstract Traditional techniques for quality assessment of fresh agricultural products are onerous and costly. Imaging systems based on the visible wavelength range have been widely used for automated processing because they are relatively inexpensive and fast and designed to represent the human eye to detect external features such as size, shape, colour, or the presence of damages and diseases. However, traditional cameras cannot identify damage that is internal or non-visible to the human eye. Hyperspectral imaging combines conventional two-dimensional imaging systems with spectroscopy to simultaneously obtain spatial and spectral information from an object. This technology is being progressively applied for postharvest inspection of fruits and vegetables to develop predictive models for estimating internal quality or identifying key wavelengths related to specific problems. Nevertheless, to obtain accurate and trustable detection, the use of this technology requires adequate equipment, robust calibrations, and powerful statistical tools to create prediction or classification models that are consistent over time. This chapter reviews the application, advantages, and requirements of visible and

---

M. C. Alamar
Plant Science Laboratory, Cranfield University, Bedfordshire, UK

N. Aleixos
Departamento de Ingeniería Gráfica, Universitat Politècnica de València, Valencia, Spain

J. M. Amigo
Department of Analytical Chemistry, University of the Basque Country, Leioa, Spain

IKERBASQUE, Basque Foundation for Science, Bilbao, Spain

D. Barbin
Department of Food Engineering, University of Campinas (UNICAMP), Cidade Universitária, Campinas-SP, Brazil

J. Blasco (✉)
Centro de Agroingeniería, Instituto Valenciano de Investigaciones Agrarias (IVIA), Moncada (Valencia), Spain

P. B. Pathare, U. L. Opara (eds.), *Mechanical Damage in Fresh Horticultural Produce*, https://doi.org/10.1007/978-981-99-7096-4_4

near-infrared hyperspectral imaging for mechanical/bruise damage and quality assessment of fresh horticultural produce.

Keywords Automated inspection · Computer vision · Chemometrics · In-line inspection · Food quality

4.1 Introduction

Quality inspection of fresh produce is essential to ensure excellent product quality and prevent biological, chemical, and physical risks in the food supply chain. Quality can be defined by focusing on the appearance, safety, and shelf-life of the product or the consumers' needs and preferences (Shewfelt, 1999). Emphasis on the visual attributes of the product can be performed visually or by creating computer vision systems that provide practical non-destructive means to facilitate product standardisation, ensure product homogeneity, and maximise profit by sorting fruit into commercial categories. These categories are associated with an appearance that the consumer associates with a good product, like the size, shape, colour, absence of defects, etc. (Pathare et al., 2013). However, inspection based on this definition of quality does not consider aspects related to the organoleptic characteristics of the product. In the current context of an increasingly demanding and informed consumer concerned with sustainability, good cultivation practices, and traditional flavours, one should not only think about homogeneous fruits with a perfect appearance. The focus should also be on the quality attributes that the consumer demands, that is, those factors perceived as organoleptic quality such as taste (sugar content, acid content, etc.), texture, colour (external and internal), or aroma (Kyriacou & Rouphael, 2018). For this reason, it is essential to investigate systems capable of detecting damage and invisible properties, determining their composition or predicting shelf-life.

In this regard, spectroscopy, due to its capacity to obtain information from inside some products, has been widely used to evaluate the internal quality and composition of fresh fruits and vegetables (Cortés et al., 2019a; Walsh et al., 2020). This technology is rapid and relatively inexpensive but can present problems when assessing small samples, such as grains, or a lack of accuracy in heterogeneous samples since only information about a single point can be collected. Moreover, the adjustments and arrangements needed for in-line inspections are highly complex due to the different sizes and shapes of the product and the movement of the samples when they are transported at high speed (Cortés et al., 2019b). An alternative to this technology is hyperspectral imaging (HSI) (Lorente et al., 2012), which combines the advantages of spectroscopy with conventional two-dimensional digital imagery. A red, green, and blue image (RGB) is composed of three monochromatic images in these bands.

Similarly, a hyperspectral image comprises many monochrome images captured in consecutive bands. Thus, it allows spatial and spectral information to be obtained

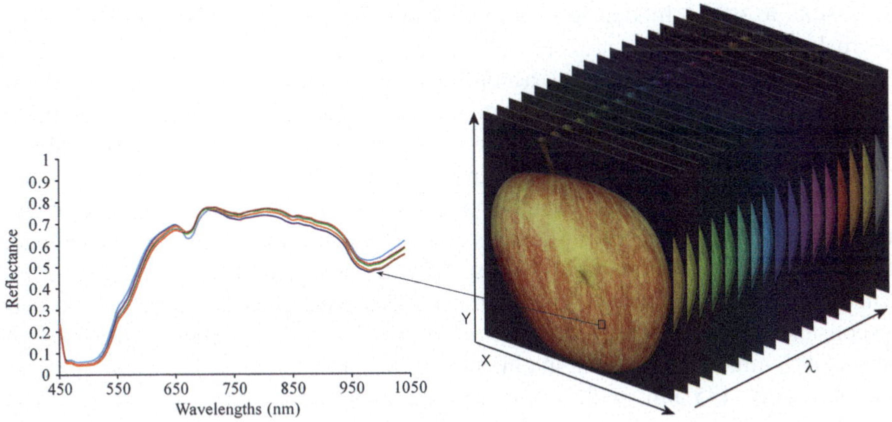

Fig. 4.1 Scheme of a hyperspectral image showing the spatial (x, y) and spectral (λ) coordinates. Each pixel in the image contains the spectrum of this pixel

simultaneously, providing an individual spectrum for each image pixel (Fig. 4.1). These spectra contain information on the physicochemical composition of the sample, and their analysis is used to characterise and visualise the composition or properties of different regions of interest (ROI) in the image.

Hyperspectral imaging has been extensively employed in recent years to predict internal attributes of fresh horticultural produce (Munera et al., 2019a; Ma et al., 2021; Riccioli et al., 2021; Kasampalis et al., 2021) to detect non-visible damage (Fazari et al., 2021) or to determine cultivars or origin (Munera et al., 2018). This work explores the latest applications of hyperspectral imaging to inspect fresh agricultural products, as well as to analyse the advantages and disadvantages of its practical application and delve into the main problems to be overcome to carry out an effective transfer to industry.

4.2 Basic Principles

When a ray of light is emitted towards a piece of fresh fruit or a vegetable, part of the light is reflected by the surface, the sample absorbs another part, some light part passes through the fruit, and the rest is scattered by the internal structure of the fruit and exits near the point of entry. Hyperspectral cameras capture the radiation reflected or scattered by the fruit to determine its state. This radiation is influenced by the composition and the internal properties of the products (spectral signature). Therefore, fruits with different concentrations or compounds have different spectral signatures, which allows the estimation of their internal attributes. Furthermore, a blow or disease can also affect the internal structure, altering the spectral signature. For this reason, the analysis of the spectrum obtained at each point of the fruit makes it possible to detect hidden lesions or to know the content and distribution of its internal constituents.

However, there are differences according to the spectral range studied. The colour of the fruit is related to the development of fresh products. Hence, it can be helpful to determine the time of harvest or the estimation of postharvest ripening. The chlorophyll content in the skin provides the green colour of the fruit. In general, when the fruit ripens, the green colour decreases due to a drop in the chlorophyll content and an increase in the synthesis of pigments associated with the colours yellow, orange, and red. Carotenes and xanthophylls absorb light in wavelengths from 420 to 503 nm with three maximum peaks. The pigment β-carotene, present in many fruits, is responsible for the yellow to orange-red colour, with a maximum absorption of around 475 nm. Xanthophylls, lutein, and violaxanthin show an absorption peak at approximately 435 nm, with strong absorption between 350 and 500 nm. The absorbance in the region of 500–550 nm is caused by anthocyanins, being strongly influenced by the pH value (Qin et al., 2010). For this reason, the determination of the colour distribution has been measured using a visible hyperspectral image to look for alterations caused by invisible damage or to determine the content of these pigments to estimate the maturity or firmness of the fruit (Liu et al., 2021).

The near-infrared (NIR) region is typically used for the evaluation of internal properties such as soluble solids content (SSC), titratable acidity (TA), or relative humidity content (RH), which are properties related to changes that occur during the ripening process. The starches are transformed into sugars, the tannins become insoluble, and the acidity is corrected due to the degradation of the organic acids. This causes, as the fruit ripens, the flavour to change, and it becomes more attractive to the consumer. In addition, the fruits acquire a softer texture due to the degradation of cell walls and factors such as the hydrolysis of starches and pectins (Seymour et al., 2013).

The camera receives both specular and diffusely reflected light from the sample. Still, specular reflections do not contain any information about the internal attributes of the sample and should be disregarded. Spectral reflectance is typically measured as the ratio of reflected energy to incident energy as a function of wavelength. In the case of fresh food products, reflectance varies with wavelength depending on the biochemical composition of the product since energy is dispersed or absorbed in different degrees and wavelengths for each compound. The valleys of the spectral curves indicate the wavelength ranges for which the incident energy is absorbed, which differ depending on the composition. The general shape of the spectral curve and the position and intensity of the absorption bands can be used to identify and discriminate the composition or structure. Another way to capture images is in transmission mode. In this case, the camera is positioned to capture the images of the product illuminated from behind and, therefore, to capture the radiation that passes through the sample. This radiation can be altered by the presence of stones or seeds that can block the incident light, which makes them appear in the image and thus, their presence is detected. The radiation captured by the camera is influenced by the composition or the internal structure of the product. Therefore, it is an effective technique to measure the internal properties in some cases. Munera et al.

(2019b) used this technique to detect split pit and quality in nectarines. However, it also presents some problems, such as the need for a large amount of light that can only be applied to products that allow light to pass through.

4.3 Equipment, Image Capture, and Calibration

Digital images are formed by the spatial information of a scene captured in given wavelengths of the electromagnetic spectrum. The difference among RGB, multispectral and hyperspectral images are thus related to the number of wavelengths included. RGB images are formed by three images in the visible range, corresponding to the bands associated with the colours red (R), green (G), and blue (B). Similarly, spectral images are composed of a group of monochrome images at different wavelengths, but the main differences are the number of spectral bands, which is commonly larger, and the wavelength range, usually within the visible and near-infrared (Vis-NIR). According to the number of wavelengths that comprises the images, spectral imaging devices are categorised according to their *modus operandi*: multispectral, hyperspectral, and ultraspectral imaging systems, where multispectral imaging systems have a smaller amount of spectral bands when compared to ultraspectral imaging and hyperspectral imaging systems.

Hyperspectral systems obtain images in a large number of adjacent bands, forming a spectrum for each pixel in the image (Ariana & Lu, 2008a, b). This particularity allows the application of hyperspectral imaging for the detailed assessment of spectral information and the spatial characteristics of a sample. In comparison, multispectral imaging systems provide images with a smaller number of spectral bands, widely and sometimes irregularly spaced, thus providing a set of images at distinctive wavelengths but not a comprehensive spectrum for each pixel or a given sample. This unique aspect of hyperspectral imaging systems provides a wide range of information that allows the analysis of extrinsic attributes of a sample, like shape, size, etc., and also spectral information in the wavelength range comprised by the equipment according to the sensor used.

Multispectral imaging systems have particular applications considering the specific wavelengths designed in the equipment. Hyperspectral imaging systems are commonly used for the preliminary investigation of effective wavelengths, which can later be used to create multispectral imaging systems. Recently it has been proposed that the more general terms "imaging spectroscopy" or "spectral imaging" should be used to mention this technique, without the prefixes "hyper-", "super-," and "ultra-", to prevent confusion (Polder & Gowen, 2021).

4.3.1 Image Acquisition System

Hyperspectral image acquisition devices are designed in four different imaging approaches, where images are acquired one point at a time (point-scan), one line at a time, comprising the entire spectral range (line-scan), one spatial image at a time at each wavelength (wavelength scan), and one spatial image at a time at each wavelength (snapshot) (Wang et al., 2017).

In a wavelength scan, the sample is kept still for image acquisition one wavelength after another, while in snapshot mode, the image is acquired in the full spectrum (Kim et al., 2001; Lawrence et al., 2003). In line-scan (also called 'pushbroom') and point-scan (also called 'whiskbroom') image acquisition, the sample is conveyed under the field of view of the camera, or the camera is moved over the sample, since the image is acquired in the full spectral range for each point or line in the spatial dimension. These methods thus demand the scanning of all the spatial features of a sample to acquire the full image.

This aspect makes line-scan imaging the most suitable method for most applications, as it can be adapted to processing lines by moving the sample under the camera using a conveyor belt or platform. It can also be adapted to be used in field applications, where the camera can move along a platform for image acquisition of a large scene.

4.3.2 Equipment in a Hyperspectral Imaging System

An HSI system typically comprises the following equipment: a charge-coupled device (CCD) camera, a device to allow wavelengths to be captured separately, and an illumination system. Some devices only let the wavelengths of interest pass through to obtain the hyper or multispectral images, such as those based on liquid crystal tunable filters (LCTF). Other sensors, like spectrographs, separate a ray of light into its different wavelengths. The light that reaches the camera from the sample, is dispersed in the wavelength range when it passes through the spectrograph, thus going across the camera lens and being acquired by the camera and recorded on a computer. LCTFs make it possible to obtain images with a higher spatial resolution but require the sample to remain static while the images are captured (Gómez-Sanchis et al., 2014). In contrast, imaging spectrographs are the most widely used because they have higher spectral resolution and require the sample to move to capture images using the line-scanning technique, which facilitates the development of in-line applications. Figure 4.2 shows a system based on LCTF (left) and an imaging spectrograph (right).

A hyperspectral image comprises a spatial image (x versus y) at each wavelength (λ) of the range encompassed by the system, thus forming a hypercube composed of numerous monochromatic images in successive wavelengths across the spectral range of the system. Hence, it comprises a spectrum for each pixel in the image. In

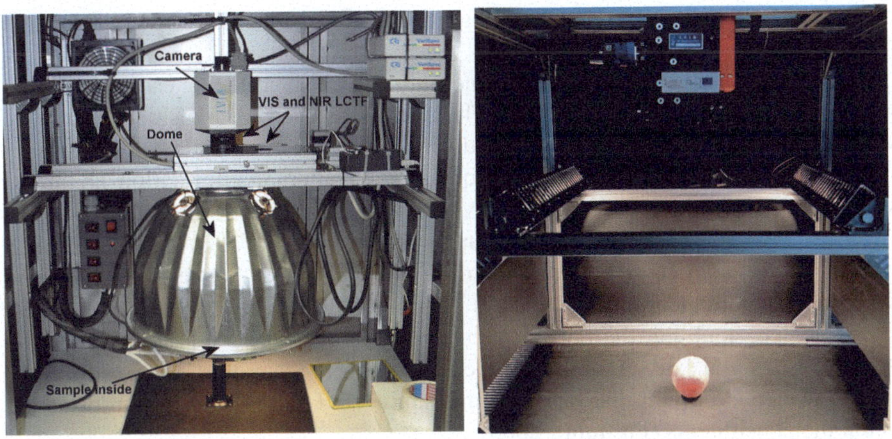

Fig. 4.2 (Left) Snapshot hyperspectral imaging system based on LCTF; (Right) line-scan hyperspectral imaging system based on imaging spectrograph

a pushbroom system, one of the spatial dimensions is fixed by the opening of the camera lens, and the other dimension is controlled by moving the sample in front of the camera lens or moving the camera across the scene.

The spectral range is determined by the type of sensor used in the system, varying from about 200 nm, in the ultraviolet range, to wavelengths up to 2500 nm, in the NIR range. Hyperspectral imaging systems in the Vis-NIR range (380–800 nm or 400–1000 nm) and NIR range (900–1700 nm or 900–2500 nm) are the more commonly used for biological samples, such as in agricultural and food processing applications. The system illuminates the sample within the whole field of view (Driver, 2009), and the spectrograph is responsible for limiting the amount of light that reaches the camera within the wavelength range of the system.

Two types of cameras are usually used in hyperspectral imaging systems: CCDs, which have been successfully used for many years, and complementary metal-oxide semiconductor (CMOS) sensors, which have some advantages that indicate their use for hyperspectral applications in the future (Goetz, 2000; Jerram et al., 2017). These advantages are that CMOS does not require a frame transfer, like CCD, and that spectral lines can be read several times within a frame, thereby providing a significant expansion in dynamic range (Jerram et al., 2017).

4.3.3 Image Acquisition and Calibration

There are different sensing modes according to the interaction of the light with the sample analysed: reflectance, transmittance, transflectance, or interactance. In the reflectance mode, the sample is illuminated, and the light not absorbed by the sample (thus being reflected) is captured by the detector, which must stand in a position to prevent specular reflection. Hence, the light source and the camera stand in a position facing the same sample surface.

In transmittance mode, the camera and light source stand facing each other, with the sample positioned between them. This method is used for translucent samples, since the light is transmitted partially or diffusely through the sample before reaching the camera lens, thus allowing the acquisition of internal information about the sample (Walsh et al., 2020).

When optical bundle probes are employed, a specific method to perform a transmittance acquisition is transflectance. The difference between this and the standard transmittance method consists of passing the radiation beam through the sample twice, doubling the optical path. In the interactance mode, the light beam penetrates to some extent into the sample and returns to the sensor. As a result, it has a higher interaction with the inside of the sample and contains further information about the constituents (Pasquini, 2003).

Considering the comparison with maximum and minimum illumination conditions, spectral images must be calibrated using dark and white reference images. A dark reference image is typically obtained with the camera lens closed to acquire an image of the noise caused by thermal effects (dark current) that is added to the signal generated. Therefore, the subsequently acquired images are adjusted by considering the dark current of the camera. The white image consists of imaging a reference substance that has uniform reflectance in the wavelength range of the system, so the spectral differences observed in the spectrum of the white reference are due to the equipment used.

4.4 Data Analysis Methods

With such an amount of spectral and spatial information obtained per sample, powerful mathematical and statistical methods are needed to extract critical analytical information in both the spectral and the spatial dimensions. Regardless of the sample, the aim is usually the point to decide the analytical strategy to follow: Detection (that implies segmentation or classification strategies) or quantitation (that implies regression or curve resolution/spectral unmixing strategies) (Amigo, 2019). The spatial shape of the sample plays a vital role since the reflectance of the light is severely affected by the shape, the roughness, or the anomalies that the surface of horticultural products might have. Therefore, a combined action between spatial/spectral preprocessing and the machine learning strategy is needed to optimise the final solution.

4.4.1 Structure and Preprocessing

Hyperspectral images usually are visualised as large datacubes containing three directions, two spatial and one spectral, as shown in Fig. 4.1. For each pixel, a spectrum is obtained in the wavelength range that has been measured. Depending on the type of image, the spectrum collected can be visualised as a continuum spectrum or a series of spectral bands with no spectral continuity. As pointed out before, this is a common source of confusion in the literature since the terms "hyperspectral" and "multispectral" have been continuously interchanged. The distinction between hyperspectral and multispectral must be made considering the spectral resolution of the final datacube. Hyperspectral images are characterised by a continuous spectrum measured for each pixel. Moreover, the wavelengths usually are equispaced, giving a sense of continuity in the radiance/reflectance/absorbance values obtained between consecutive wavelengths. Conversely, multispectral images (a.k.a. multi-band images) capture individual images at specific wavenumbers or wavelengths (Amigo & Grassi, 2020).

The spectral and spatial dimensions typically contain all the relevant information to address the challenges. Nevertheless, they also contain artefacts from the three actors that play a fundamental role in the measurement: the light source, the sample, and the camera (detector). When photons of light penetrate a tissue, they behave differently depending on the composition and texture of the sample (Amigo & Santos, 2020). Therefore, radiation that exits the sample and reaches the camera contains chemical information about internal quality and composition. Nevertheless, the spectral information can be negatively influenced by the shape of the sample, which may alter the reflection or penetration of the light into the sample. Moreover, the sensors are subject to instrumental noise, having different sensitivity to the photons. All these artefacts must be minimised to obtain accurate chemical signatures and models with sufficient reliability and stability.

The methodologies employed are mainly based on spectral preprocessing, although spatial preprocessing can also be performed (Amigo & Santos, 2020). Both spatial and spectral preprocessing methods are interconnected, so whatever action is performed in the spatial dimension impact the spectra and vice-versa. Therefore, special care is needed to not over-preprocess a sample. Nevertheless, the methodologies to correct spectral signatures are the most popular ones and can be classified depending on the type of artefact they are aimed at correcting:

- *Smoothing/De-noising*: Smoothing methods are based on selecting a small window around a specific point and calculating its projection onto a polynomial fitting of the points from the window (Rinnan et al., 2009). It is one of the most straightforward ways of removing/minimising noise. Nevertheless, the selection of the spectral window must be optimised since large window sizes might smooth informative peaks, while small ones increase the noise.
- *Scatter correction*: Scattering is one of the major issues in reflectance. It is typically visualised as a drift in the baseline that can be additive or multiplicative, and it is strongly correlated with the shape and the roughness of a surface. See, for

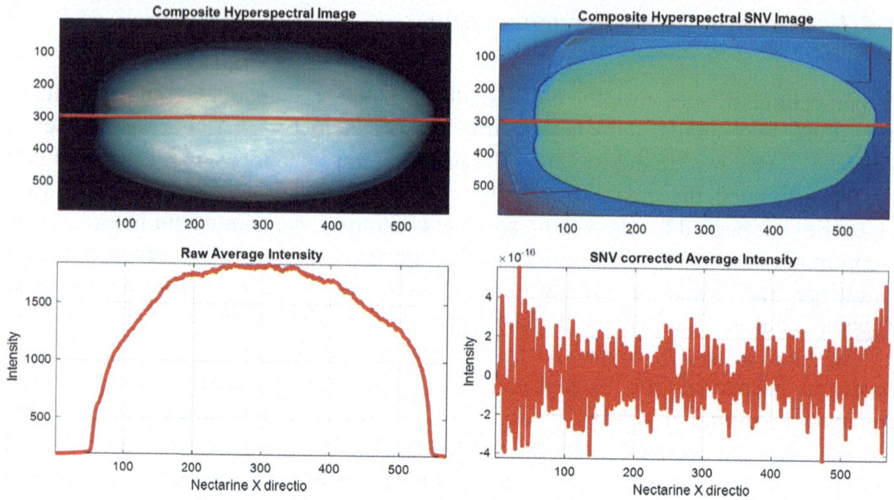

Fig. 4.3 Depiction of the mean spectral raw intensity of a line of a nectarine (top left, composite Vis-NIR image of a nectarine; bottom left, mean spectral raw intensity) and the mean spectral intensity after applying standard normal variate (top right, composite Vis-NIR image of a nectarine; bottom right, mean spectral intensity after using standard normal variate)

instance, Fig. 4.3(left), where the mean spectral raw intensity of one line of a nectarine shows how the shape of the nectarine influences the signal. Two main methods for removing/minimising the scattering effects are standard normal variate (SNV) and multiplicative scatter correction (MSC). SNV subtracts the mean of the spectrum and divides it by the standard deviation, whereas MSC projects the spectrum of one pixel against a reference spectrum. Then the corrected spectrum is obtained by subtracting the offset from the original spectrum divided by the slope. Figure 4.3(right) shows how the influence of the shape of the nectarine is minimised after applying SNV.

– *Derivatives*: The first (1D) and second (2D) derivatives are methods that help to (1) minimise the spectral noise and (2) highlight spectral areas with minor but essential variance (Rinnan et al., 2009).

4.4.2 Analysis Methods

Hyperspectral images are inherently linked to machine learning methodologies that help extract the relevant analytical information from the bulk information. Many different actions can be performed in combination with preprocessing. These actions can be summarised in the four major building blocks described below.

4.4.2.1 Pattern Recognition/Exploration

Spectral pattern recognition applied to spectral data is an important technique for understanding spectral information and its close relationship with the surface of the sample (Marini & Amigo, 2020). The exploratory analysis offers a first overview of the primary sources of variance (variability) in the images. Principal component analysis is, arguably, the most common one, being able to elucidate the complex nature of HSI and MSI multicomponent systems. Even though they are based on variance, they can be extremely useful. For example, to a certain extent, they can be used coupled with other analytical methods to detect hidden damage, composition, or extract features such as the estimated shelf-life of fresh commodities (Cruz-Tirado et al., 2021).

4.4.2.2 Curve Resolution Methods/Spectral Unmixing

Curve resolution or spectral unmixing methods resolve mixtures that each pixel might contain, given the correct number of constituents (Olmos et al., 2017). The main difference with explorative methods is that curve resolution methods do not aim to study the primary sources of variation in the data but to give more accurate information about the hidden physicochemical behaviour of each constituent in each pixel. Their usage in horticultural products is limited since fruits and vegetables are usually a complex mixture of chemical compounds that generally give a broad spectral signature that is not unique.

4.4.2.3 Regression and Classification

Calculating the concentration of several compounds in an image (regression) (Torres & Amigo, 2020) and, especially, the classification of elements in different well-defined categories (classification) is one of the major goals in HSI and MSI analysis when fruits and vegetables are measured. Regression and classification are supervised methods since a robust and reliable set of samples of well-known attributes (concentration or predefined category) is mandatory to validate and train the model (calibration process). Once the training has been performed perfectly, the properties or the classes are predicted in new images pixel-by-pixel manner (Torres & Amigo, 2020). Many algorithms for linear or non-linear training models like partial least squares (PLS), multilinear regression (MLR), support vectors machine (SVM), or artificial neural networks (ANN) can be found in regression (Chaudhry et al., 2020a, b; Munera et al., 2017) and classification applications (e.g. PLS-discriminant analysis). Moreover, many algorithms are specifically designed for a purpose, like the single-class approach offered by SIMCA (soft independent modelling of class analogy) (Munera et al., 2018, 2019b). Being supervised methods, the core of their reliability depends on the mandatory validation step. Validation (internal cross-validation or external validation) is the only tool able to give a real account of the model's ability to predict.

4.5 Applications for Fruit Quality Assessment

Over the past decade, HSI has been extensively studied at the laboratory level as a non-destructive tool for evaluating the quality and safety of horticultural produce (Lorente et al., 2012; Lu et al., 2020). The identification of damage is a fundamental challenge for postharvest quality control systems, as the type of damage can influence its commercialisation. For instance, cosmetic damage that only affects the appearance of the fruit may still be sold in certain markets at a lower price, while rot that destroys the fruit would render it unsellable. Yadav et al. (2022) used a Vis-NIR hyperspectral system to identify seven different types of defects in citrus. They employed principal component analysis (PCA) to select bands and achieved results close to 100% accuracy by using five bands to train a convolutional neural network (CNN). Their findings are consistent with earlier studies, such as Blasco et al. (2007), who used a multispectral system based on three sensors (NIR, RGB, and fluorescence) to distinguish among 11 different types of damage, including invisible rot. One of their conclusions was that invisible rot was only detected in 65% of cases using colour, but the success rate increased to 95% by incorporating multispectral information.

However, the most critical damage in postharvest is caused by fungal infections, which are challenging to detect in their early stages. When not treated or properly managed, they quickly spread to other fruits, causing considerable economic losses and quality and safety issues, as some fungal species can produce mycotoxins. Strawberry fruits are extremely perishable due to their high metabolic rate, making them susceptible to mechanical damage and fungal infections. Traditional screening methods include analytical techniques such as gas and liquid chromatography, which are expensive, time-consuming, and require highly qualified personnel. Siedliska et al. (2018) evaluated using Vis/NIR/SWIR (400–1000/1000–2500 nm) hyperspectral imaging in reflectance mode to detect *Botrytis cinerea* and *Colletotrichum acutatum* infections in strawberry fruit. They reduced the dimensionality of the hyperspectral images (whole fruits) based on the second derivative, and further supervised classification models were developed using the backpropagation neural network. This strategy allowed a successful (>97% accuracy) early detection of fungal infections after 24 h of inoculation when they were not yet visible. Previously, Gómez-Sanchis et al. (2013) and Folch-Fortuny et al. (2016) also successfully classified sound and diseased citrus fruit infected by *Penicillium digitatum* and *Penicillium italicum*—two of the most prevalent pathogens in citrus. Images of whole mandarins were acquired within the 460–1020 nm range before inoculation and after 3 days (Gómez-Sanchis et al., 2013). Using decision trees, the infected fruit was classified between the two types of fungi, with an accuracy of 93%.

4.5.1 Applications for Assessment of Mechanical Damage

Among the main problems in post-harvest fruit management are mechanical damage. Mechanical damage can result from poor harvesting practices or the operation of post-harvest machinery, leading to bumps, cuts, and bruises on the fruit. Fleshy fruits are particularly prone to such mechanical damage. Typically, this damage is invisible at first, causing an imperceptible texture change. Detecting them requires either a tactile examination or a detailed inspection, which proves to be challenging in packinghouses dealing with large productions, so they easily reach the market where they evolve into decay or product losses. Therefore, the use of techniques that get internal information is required (Du et al., 2020). One of the greatest advantages of hyperspectral systems is their ability to detect internal damage invisible to the human eye. The most frequent form of mechanical damage to fruits is bruising during harvesting, handling, and transportation. Bruising is a type of subcutaneous tissue failure caused by an excessive external force on the fruit surface during impact, compression, or vibration, leading to cell breakage but without rupture of the fruit skin (Opara & Pathare, 2014). This includes all impacts that might occur during harvesting and handling operations, such as fruit dropping into picking buckets or sorting operations by electronic sorters that transport and handle the fruit at high speeds (Hussein et al., 2018). In some cases, impacts are visible shortly after they occur, while in others, they evolve internally, deteriorating the tissue and becoming visible only when the consumer peels the fruit, leading to rejection.

HSI systems can detect such damage by analysing the spectral signature of the fruit, which contains information about the chemical composition, structure, and texture. By analysing this spectral data, HSI can identify changes in the internal structure of the fruit that reveal the presence of mechanical damage. Numerous studies have been conducted on different fruits to detect such damage. Typically, experimental designs involve three techniques to reproduce mechanical damage artificially in a comparable way. The first includes hitting the fruit with a weight attached to a pendulum, the second consists of dropping a ball of known weight onto the fruit from a specified height, and the third is to drop the fruit from a certain height. The idea is to cause the same damage to all fruits in the study and then attempt to detect it using a hyperspectral system, employing various analysis techniques.

In general, most of the work has been carried out by applying different forces to the fruit to obtain damage variability, simulating different impact situations. Subsequently, the reflectance spectra of the artificially damaged and healthy areas or control fruit are measured. These spectra are measured before the impact, in the first hour after the impact, and at regular intervals ranging from a few hours to several days. Overall, the results obtained are excellent, with detection levels exceeding 90% in most cases, as well as a selection of the most influential bands for damage detection.

Zhang et al. (2023a) dropped fruit from different heights to induce damage to detect bruises on Red Fuji apples. They used three heights from 30 to 90 cm analysed in the 900–1700 nm spectral range. Spectra were measured at 20 min, 3 h, and 2 days after the damage, with the best classification obtained for drop-induced bruising from a height of 90 cm. Pourdarbani et al. (2023) used 70 lemons that were dropped from a height of 30 cm. They used an HSI system in the 400–1100 nm range to obtain images at the moment of impact and after 8 and 16 h, obtaining high detection results using CNN-based classifiers, but only on 7 fruits. Hu et al. (2016) used the same technique on blueberries, dropping them from a height of 120 cm to induce bruises. Detection was performed using a Vis-NIR HSI system capturing images at the moment of impact and after 12, 24, and 48 h, achieving a 97% detection rate using machine learning techniques. The most important bands were selected through competitive adaptive reweighted sampling (CARS), but no smaller set than 27 bands could be found.

A more precise alternative is to let a ball with a given weight hit the fruit. In this way, the damage caused is not influenced by the weight or shape of the fruit, while the damaged area can be better controlled when obtaining the spectra. Munera et al. (2021) performed this experiment on persimmon fruits. They dropped the ball from 30 cm, impacting the fruit with a force of 0.2 J. Spectral measurements in the Vis-NIR range were performed at 0, 24, 48, and 72 h after causing the damage. PCA was used to detect the damages in the entire images achieving 77.3 of correct detection immediately after the damage and 99.2 in the third day. Then partial least square discriminant analysis (PLS-DA) was applied to predict the moment of the impact, achieving 100% correct classification depending on the day after bruising. Nturambirwe et al. (2021) performed a similar experiment on apples, dropping the ball at 6 different heights to cause variability of impacts. The ball impacted the fruit with forces from 0.013 to 0.2 J. Spectra were measured at 0, 1, 6, 18, 48, and 72 h after bruising, detecting the damage with 85% success using PCA and different classifiers like linear discriminant analysis (LDA), *k*-nearest neighbours (*k*-NN), SVM, or ensemble subspace discriminant (ESD). Pang et al. (2022) also did it on apples employing four balls with weights from 10–40 g instead of dropping the ball from different heights. The success, in this case, was 100% using PCA and YOLOv3 (you only look once) algorithm in the 1000–2500 nm range. Finally, Zhang et al. (2023b) also dropped the ball from different heights (20–70 cm) on Red Fuji apples. The aim was to predict the damaged area after 1, 6, 24, and 72 h from causing the damage. Using CARS to select the bands and ANN to predict the area they achieved an R^2 of 0.87.

The third method commonly used to capture the ability of HSI systems to detect early mechanical damage is causing the damage using a pendulum. This is similar to dropping a fruit, but in this case, instead of simulating a drop, the method simulates hitting the fruit during transportation or handling operations. Vélez-Rivera et al. (2014) used this method with a force of 0.12 J to cause lesions in mangoes, capturing hyperspectral images in the 400–1000 nm range from 1 h after the damage and the next 7 days. Results showed that 94.9% of the damaged fruit could be detected after the first day, while on day 7, the detection was 98.1% using five classifiers. Similarly, Su et al. (2022) used a pendulum to cause three levels of damage in pears. They correctly detected the damaged fruit in 98.98% of cases using PLS-DA and SVM in the Vis-NIR range, performing a band selection using CARS.

Despite the success in detecting mechanical damage, and the possibility of selecting the relevant bands, which demonstrate the ability of the HSI systems to detect internal and invisible damages, there are still no known systems implemented at the industry level. This could be explained by the price or complexity of the HSI equipment compared to the benefit that the early detection of these damages is for the industry. An example is the detection of mechanical damage in lemon described by Pourdarbani et al. (2023). Since damage caused by impacts from 30 cm can be worthy of detection for other fruits, such as apple, it has little importance in lemon and almost null effect on its commercialisation.

4.6 In-Line Applications

In the scientific literature, most successful cases of hyperspectral imaging applied to the quality inspection of fruits and vegetables are achieved in laboratory experiments under controlled conditions. These experiments demonstrate the capacity of this technology to obtain information on properties and internal damage but are not reproducible in industry. Unfortunately, attempts to emulate inspection systems often use experimental equipment or designs that cannot be employed in an actual industrial implementation, which requires systems capable of performing in-line inspections at high speed. To meet commercial requirements, the industry needs machines capable of quickly processing large amounts of produce, which implies inspecting all fruits individually on electronic sorters at a very high speed. This real-time in-line inspection has long been achieved with systems for estimating the external properties or appearance of fruit (Cubero et al., 2011). Yet, the challenge of achieving systems capable of monitoring internal quality remains. To maintain competitiveness in the complex agri-food market, producers and machinery makers demand more sophisticated techniques to inspect fresh products beyond the traditional appearance attributes such as size, colour, or external defects. In this regard, spectroscopy or hyperspectral imaging appear as promising alternatives.

The imaging spectrograph is the most advantageous for performing an in-line inspection among the different technologies available for acquiring hyperspectral images. Techniques such as LCTF require the sample to be still while the image is obtained, which can take several seconds. In contrast, imaging spectrographs, when capturing line-by-line, need the sample to travel beneath the camera at a rate synchronised with the acquisition of the images. If the speed is excessive, the complete sample is not captured in the image. In contrast, redundant information is obtained if the speed is too low. Another problem, in the case of some fruits or vegetables, is the need to inspect the whole fruit and not just the side facing the camera, especially when the property to be examined is not homogeneously distributed in the product.

Despite the efforts made by the scientific community, the adoption of this technology by industry is still relatively scarce. Several causes exist, but, in general, most studies carried out to date have been laboratory tests that are difficult to apply in an industrial environment. Pushbroom hyperspectral systems require the movement of the samples synchronised with the line-scan acquisition to complete the hypercubes. However, the fact that the samples are moved by a conveyor belt does not mean a true in-line inspection. Thien Pham and Liou (2022) described an in-line hyperspectral system to detect bruises in jujube, but the best times they reported (16.6 s) are very far from any industrial implementation. Additional difficulties are related to the large amount of data these systems generate, which must be processed quickly. Sometimes hyperspectral systems are used to select crucial wavelengths related to the problem to reduce both the final acquisition time and the amount of data to be processed. For example, Mo et al. (2017) created a system for the in-line inspection of lettuce contaminated with worms and slugs. Using an analysis of variance (ANOVA) on the hyperspectral images, they selected one or two wavelengths to detect contaminated lettuces and distinguish them from healthy ones in-line. Mishra et al. (2022) used a hyperspectral system to develop a multispectral model for the detection of contaminated almonds. They used CARS to select a reduced set of important bands for a faster detection of damaged almonds. Keresztes et al. (2016) developed a system to detect bruises in apples in-line in real-time. However, the apples advanced at a speed of only 0.1 m/s. Although the study is on the right track, it is still far from an acceptable speed for the industry, where rates of 10–15 fruits per second can be reached. This study also emphasises the need to reduce the effect of undesired glares caused by bad illumination or the sphericity of the fruit that can hide bruises.

As shown above, monitoring fresh fruit using hyperspectral imaging is highly complex. Considering the fast-sorting speed required and the high computing time involved, the focus is on using dedicated hardware, such as graphics processing units (GPU). Due to their ability to accelerate parallel applications, GPUs are especially interesting for running deep learning-based applications (Yusuf & Alawneh, 2018), one of the most recent research techniques studied for this problem. Further information on the needs and concerns of the industry to implement commercial fruit and vegetable inspection systems using real-time hyperspectral imaging are described in depth by Kim (2015).

4.7 Future Trends

The awareness of the industry and consumers regarding the quality and safety of food is increasing, and, therefore, interest in the search for non-invasive and non-destructive control techniques is growing. Hyperspectral imaging systems are a promising technology for automatic food inspection since it allows the early detection of mechanical damage or invisible internal bruise and the determination of internal quality and properties. It also makes it possible to predict properties associated with the stage of maturity, which facilitates making optimal decisions on storage and commercialisation, or the final destination of the fruit based on the predicted shelf-life or the presence or absence of damage.

Instrumentation for hyperspectral imaging is becoming increasingly available at a lower cost. This availability has supported a surge in the adoption of postharvest research literature by the industry. These measurement technologies were initially applied to perform fast and non-destructive laboratory analyses. Yet, the reduction in prices, more spatial resolution, and the new advances in hardware and data processing techniques are closer to creating systems capable of performing in-line inspection at high speed. However, hyperspectral image acquisition and processing remain slow, preventing widespread implementation in an industry that needs real-time inspection. One solution can be the selection of key wavelengths to address each problem individually, but sometimes relevant information is lost or limits the potential of the final application. Most of the investigations carried out to date have the objective of detecting these sets of wavelengths to satisfy the specific problem addressed in the tests. Therefore, in most cases, the results depend on the samples used, the laboratory conditions, or the statistical techniques applied and are not strongly related to the intrinsic physicochemical properties of the product that could generalise the results obtained. For example, different studies select different wavelengths as being essential to solve the same problem, thereby causing confusion. For this reason, it is essential to advance in creating robust models tested with truly independent samples that allow us to obtain useful applications for the industry.

The properties of fruits and vegetables depend on many factors, such as the growing place, the handling practices or the soil, the local climatic conditions, the time of harvest, etc. In addition, they evolve after harvest. Fruits of the same variety can have very different characteristics if they come from different lots. Furthermore, detecting a specific property can be related to others that evolve differently. This complicates the creation of robust models to satisfy an industry that handles fruit from different plots and with different stages of ripeness. For this reason, automatic inspection systems sometimes require frequent recalibrations, which are costly and make it challenging to implement these systems. In the case of hyperspectral imaging, the problem is exacerbated, so it is essential to continue developing effective calibration systems and creating robust models.

The application of deep learning techniques to hyperspectral image analysis is currently state of the art. This technique allows very high levels of precision to be achieved on large volumes of information. Deep learning is typically implemented using a neural network architecture. The term "deep" refers to the number of layers in the network: the more layers, the deeper the network. Traditional neural networks typically contain around 2–3 layers, while deep networks can consist of hundreds. Algorithms based on deep convolutional neural networks (DCNNs) are currently some of the most popular for analysing hyperspectral information using deep learning. They are starting to be applied to analyse hyperspectral images in fruit and vegetable inspection systems (Zhang et al., 2021; Fazari et al., 2021). Moreover, creating high-performance GPU-based hardware makes it possible to accelerate the training of the enormous amounts of data necessary for deep learning by reducing training time from weeks to hours.

4.8 Conclusions

Hyperspectral imaging is a promising technique for monitoring the mechanical damage, internal bruise, and quality of fresh produce. Due to the relationships between some wavelengths and some compounds, the presence, absence, or even concentration of these compounds can be determined using this technology. Moreover, comparing the spectral signature of good products with those with internal disorders or hidden damage allows early detection of defective products that nevertheless seem to be of excellent quality. It is, therefore, possible to detect internal damage, unnoticed by the human eye, which can only be discovered when the product is peeled, which prevents it from reaching the consumer or spreads it over the rest of the batch of fruit.

However, most of the studies have been carried out under laboratory conditions. Although they promise to lay the foundations for in-line systems, the truth is that there is still a long way to go before industrial systems can be obtained. Industry requires equipment capable of capturing images in-line at high speed and processing data in real-time. Once the images have been obtained, to detect any internal damage or classify a product, it is necessary to create statistical models that relate the spectral information to the characteristics of the problem to detect. Currently, most studies use chemometric methods to create these models. However, new techniques, such as deep learning, are emerging. To process the large amount of data generated by these systems, important wavelengths can be selected and thus reduce the equipment to a multispectral system. If it is necessary to process the full spectrum, specific hardware can be used, such as GPUs, which can quickly process large amounts of data.

For more information on hyperspectral inspection systems for fresh fruits and vegetables, the reviews performed by Lorente et al. (2012) and by Lu et al. (2020) on hyperspectral imaging, and by Walsh et al. (2020) and Cortés et al. (2019b) on spectroscopy can be consulted.

Acknowledgements This work is co-funded by the projects, GVA-PROMETEO CIPROM/2021/014, IVIA-GVA 52204 and the European Union through the European Regional Development Fund (ERDF) of the Generalitat Valenciana 2021–2027.

References

Amigo, J. M., & Grassi, S. (2020). Configuration of hyperspectral and multispectral imaging systems. In J. M. Amigo (Ed.), *Chapter 1.2 in data handling in science and technology* (Vol. 32, pp. 17–34). Elsevier.

Amigo, J. M., & Santos, C. (2020). Preprocessing of hyperspectral and multispectral images. In J. M. Amigo (Ed.), *Chapter 2.1 in data handling in science and technology, 32* (pp. 37–53). Elsevier.

Amigo, J. M. (2019). Hyperspectral and multispectral imaging: setting the scene. In José Manuel Amigo (ed) Data Handling in Science and Technology. *Elsevier, 32*, 3–16. https://doi.org/10.1016/B978-0-444-63977-6.00001-8

Ariana, D. P., & Lu, R. (2008a). Quality evaluation of pickling cucumbers using hyperspectral reflectance and transmittance imaging: Part I. Development of a prototype. *Sensing and Instrumentation for Food Quality and Safety, 2*, 144–151.

Ariana, D. P., & Lu, R. (2008b). Quality evaluation of pickling cucumbers using hyperspectral reflectance and transmittance imaging—Part II. Performance of a prototype. *Sensing and Instrumentation for Food Quality and Safety, 2*, 152–160.

Blasco, J., Aleixos, N., Gómez, J., & Moltó, E. (2007). Citrus sorting by identification of the most common defects using multispectral computer vision. *Journal of Food Engineering, 83*(3), 384–393. N° orden en su área de interés: 15/114. IF: 1,848. N° citas: 181. https://doi.org/10.1016/j.jfoodeng.2007.03.027

Chaudhry, M. M. A., Amodio, M. L., Amigo, J. M., de Chiara, M. L. V., Babellahi, F., & Colelli, G. (2020a). Feasibility study for the surface prediction and mapping of phytonutrients in minimally processed rocket leaves (*Diplotaxis tenuifolia*) during storage by hyperspectral imaging. *Computers and Electronics in Agriculture, 175*, 105575.

Chaudhry, M. M. A., Amodio, M. L., Babellahi, F., de Chiara, M. L. V., Amigo, J. M., & Colelli, G. (2020b). Hyperspectral imaging and multivariate accelerated shelf life testing (MASLT) approach for determining shelf life of rocket leaves. *Journal of Food Engineering, 238*, 122–133.

Cortés, V., Cubero, S., Blasco, J., Aleixos, N., & Talens, P. (2019a). In-line application of visible and near-infrared diffuse reflectance spectroscopy to identify apple varieties. *Food and Bioprocess Technology, 12*, 1021–1030.

Cortés, V., Blasco, J., Aleixos, N., Cubero, S., & Talens, P. (2019b). Monitoring strategies for quality control of agricultural products using visible and near-infrared spectroscopy: A review. *Trends in Food Science & Technology, 85*, 138–148.

Cruz-Tirado, J. P., Oliveira, M., Filho, M. J., Teixeira, G. H., Barbin, A. J. M., & D.F. (2021). Shelf life estimation and kinetic degradation modeling of chia seeds (*Salvia hispanica*) using principal component analysis based on NIR-hyperspectral imaging. *Food Control, 123*, 107777.

Cubero, S., Aleixos, N., Moltó, E., Gómez-Sanchis, J., & Blasco, J. (2011). Advances in machine vision applications for automatic inspection and quality evaluation of fruits and vegetables. *Food and Bioprocess Technology, 4*(4), 487–504.

Driver, R. D. (2009). Quantification and threshold detection in real-time hyperspectral imaging. *Proceedings of SPIE, 7315*, 73150L.

Du, Z., Zeng, X., Li, X., Ding, X., Cao, J., & Jiang, W. (2020). Recent advances in imaging techniques for bruise detection in fruits and vegetables. *Trends in Food Science and Technology, 99*, 133–141. https://doi.org/10.1016/j.tifs.2020.02.024

Fazari, A., Pellicer-Valero, O. J., Gómez-Sanchıs, J., Bernardi, B., Cubero, S., Benalia, S., Zimbalatti, G., & Blasco, J. (2021). Application of deep convolutional neural networks for the detection of anthracnose in olives using VIS/NIR hyperspectral images. *Computers and Electronics in Agriculture, 187*, 106252.

Goetz, A. F. H. (2000). Short course in hyperspectral imaging and data analysis. In J. W. Boardman (Ed.), *Center for the Study of earth from space*. University of Colorado.

Gómez-Sanchis, J., Blasco, J., Soria- Olivas, E., Lorente, D., Escandell-Montero, P., Martínez-Martínez, J. M., Martínez-Sober, M., & Aleixos, N. (2013). Hyperspectral LCTF-based system for classification of decay in mandarins caused by *Penicillium digitatum* and *Penicillium italicum* using the most relevant bands and non-linear classifiers. *Postharvest Biology and Technology, 82*, 76–86. https://doi.org/10.1016/j.postharvbio.2013.02.011

Gómez-Sanchis, J., Lorente, D., Soria-Olivas, E., Aleixos, N., Cubero, S., & Blasco, J. (2014). Development of a hyperspectral computer vision system based on two liquid crystal tuneable filters for fruit inspection. Application to detect citrus fruits decay. *Food and Bioprocess Technology, 7*, 1047–1056.

Hu, M. H., Dong, Q. L., & Liu, B. L. (2016). Classification and characterisation of blueberry mechanical damage with time evolution using reflectance, transmittance and interactance imaging spectroscopy. Computers and Electronics in Agriculture, 122, 19–28. https://doi.org/10.1016/j.compag.2016.01.015

Hussein, Z., Fawole, O. A., & Opara, U. L. (2018). Preharvest factors influencing bruise damage of fresh fruits – A review. *Scientia Horticulturae, 229*, 45–58. https://doi.org/10.1016/j.scienta.2017.10.028

Jerram, P., Fryer, A., Pratlong, M., Pike, J., Walker, A., Dierickx, A., Dupont, B., & B. Defernez A. (2017). Hyperspectral CMOS imager. Proceedings of the SPIE, 10565. *International Conference on Space Optics—ICSO, 2010*, 1056501.

Kasampalis, D. S., Tsouvaltzis, P., Ntouros, K., Gertsis, A., Gitas, I., & Siomos, A. S. (2021). The use of digital imaging, chlorophyll fluorescence and Vis/NIR spectroscopy in assessing the ripening stage and freshness status of bell pepper fruit. *Computers and Electronics in Agriculture, 187*, 106265.

Keresztes, J. C., Goodarzi, M., & Saeys, W. (2016). Real-time pixel based early apple bruise detection using short wave infrared hyperspectral imaging in combination with calibration and glare correction technique. *Postharvest Biology and Technology, 66*, 215–226.

Kim, M. S. (2015). Online screening of fruits and vegetables using hyperspectral line-scan imaging technique. In A. K. Bhunia, M. S. Kim, & C. R. Taitt (Eds.), *High throughput screening for food safety assessment* (pp. 467–490). Woodhead Publishing.

Kim, M. S., Chen, Y. R., & Mehl, P. M. (2001). Hyperspectral reflectance and fluorescence imaging system for food quality and safety. *Transactions of ASAE, 44*, 721–729.

Kyriacou, M. C., & Rouphael, Y. (2018). Towards a new definition of quality for fresh fruits and vegetables. *Scientia Horticulturae, 234*, 463–469.

Lawrence, K. C., Windham, W. R., Park, B., & Buhr, R. J. (2003). Hyperspectral imaging system for identification of faecal and ingesta contamination on poultry carcasses. *Journal of Near Infrared Spectroscopy, 11*, 261–281.

Liu, Y., Wang, T., Su, R., Hu, C., Chen, F., & Cheng, J. (2021). Quantitative evaluation of color, firmness, and soluble solid content of Korla fragrant pears via iriv and ls-svm. *Agriculture (Switzerland), 11*(8), 731.

Lorente, D., Aleixos, N., Gómez-Sanchis, J., Cubero, S., García-Navarrete, O. L., & Blasco, J. (2012). Recent advances and applications of hyperspectral imaging for fruit and vegetable quality assessment. *Food and Bioprocess Technology, 5*(4), 1121–1142.

Lu, Y., Saeys, W., Kim, M., Peng, Y., & Lu, R. (2020). Hyperspectral imaging technology for quality and safety evaluation of horticultural products: A review and celebration of the past 20-year progress. *Postharvest Biology and Technology, 170*, 111318.

Ma, T., Xia, Y., Inagaki, T., & Tsuchikawa, S. (2021). Non-destructive and fast method of mapping the distribution of the soluble solids content and pH in kiwifruit using object rotation near-infrared hyperspectral imaging approach. *Postharvest Biology and Technology, 174*, 111440.

Marini, F., & Amigo, J. M. (2020). In J. M. Amigo (Ed.), *Unsupervised exploration of hyperspectral and multispectral images. Chapter 2.4 in data handling in science and technology, 32* (pp. 93–114). Elsevier.

Mishra, G., Panda, B. K., Ramirez, W. A., Jung, H., Singh, C. B., Lee, S. H., & Lee, I. (2022). Application of SWIR hyperspectral imaging coupled with chemometrics for rapid and non-destructive prediction of aflatoxin B1 in single kernel almonds. *LWT, 155*, 112954. https://doi.org/10.1016/j.lwt.2021.112954

Mo, C., Kim, G., Kim, M. S., Lim, J., Lee, K., Lee, W. H., & Cho, B. K. (2017). On-line fresh-cut lettuce quality measurement system using hyperspectral imaging. *Biosystems Engineering, 156*, 38–50.

Munera, S., Amigo, J. M., Blasco, J., Cubero, S., Talens, P., & Aleixos, N. (2017). Ripeness monitoring of two cultivars of nectarine using VIS-NIR hyperspectral reflectance imaging. *Journal of Food Engineering, 214*, 29–39.

Munera, S., Amigo, J. M., Aleixos, N., Talens, P., Cubero, S., & Blasco, J. (2018). Potential of VIS-NIR hyperspectral imaging and chemometric methods to identify similar cultivars of nectarine. *Food Control, 86*, 1–10.

Munera, S., Blasco, J., Amigo, J. M., Cubero, S., Talens, P., & Aleixos, N. (2019a). Use of hyperspectral transmittance imaging to evaluate the internal quality of nectarines. *Biosystems Engineering, 182*, 54–64.

Munera, S., Hernández, F., Aleixos, N., Cubero, S., & Blasco, J. (2019b). Maturity monitoring of intact fruit and arils of pomegranate cv. 'Mollar de Elche' using machine vision and chemometrics. *Postharvest Biology and Technology, 156*, 110936.

Munera, S., Rodríguez-Ortega, A., Aleixos, A., Cubero, S., Gomez-Sanchis, J., & Blasco, J. (2021). Detection of invisible damages in 'Rojo Brillante' persimmon fruit at different stages using hyperspectral imaging and chemometrics. *Food, 10*, 2170. https://doi.org/10.3390/foods10092170

Nturambirwe, J. F. I., Perold, W. J., & Opara, U. L. (2021). Classification learning of latent bruise damage to apples using shortwave infrared hyperspectral imaging. *Sensors (Basel), 21*(15), 4990. https://doi.org/10.3390/s21154990

Olmos, V., Benítez, L., Marro, M., Loza-Alvarez, P., Piña, B., Tauler, R., & de Juan, A. (2017). Relevant aspects of unmixing/resolution analysis for the interpretation of biological vibrational hyperspectral images. *TrAC. Trends in Analytical Chemistry, 94*, 130–140.

Opara, U. L., & Pathare, P. B. (2014). Bruise damage measurement and analysis of fresh horticultural produce—A review. *Postharvest Biology and Technology, 91*, 9–24. https://doi.org/10.1016/j.postharvbio.2013.12.009

Pang, Q., Huang, W., Fan, S., Zhou, Q., Wang, Z., & Tian, X. (2022). Detection of early bruises on apples using hyperspectral imaging combining with YOLOv3 deep learning algorithm. *Journal of Food Process Engineering, 45*(2), e13952. https://doi.org/10.1111/jfpe.13952

Pasquini, C. (2003). Near infrared spectroscopy: Fundamentals, practical aspects and analytical applications. *Journal of the Brazilian Chemical Society, 14*(2), 198. https://doi.org/10.1590/S0103-50532003000200006

Pathare, P. B., Opara, U. L., & Al-Said, F. A. J. (2013). Colour measurement and analysis in fresh and processed foods: A review. *Food and Bioprocess Technology, 6*, 36–60.

Polder, G., & Gowen, A. (2021). The hype in spectral imaging. *Journal of Spectral Imaging, 33*(3), 12–14.

Pourdarbani, R., Sabzi, S., Dehghankar, M., Rohban, M. H., & Arribas, J. I. (2023). Examination of lemon bruising using different CNN-based classifiers and local spectral-spatial hyperspectral imaging. *Algorithms, 16*, 113.

Qin, C., Li, Y., Niu, W., Ding, Y., Zhang, R., & Shang, X. (2010). Analysis and characterisation of anthocyanins in mulberry fruit. *Czech Journal of Food Sciences, 28*, 117–126.

Riccioli, C., Pérez-Marín, D., & Garrido-Varo, A. (2021). Optimising spatial data reduction in hyperspectral imaging for the prediction of quality parameters in intact oranges. *Postharvest Biology and Technology, 176*, 111504.

Rinnan, A., van den Berg, F., & Engelsen, S. (2009). Review of the most common preprocessing techniques for near-infrared spectra. *TrAC. Trends in Analytical Chemistry, 28*, 1201–1222.

Seymour, G. B., Ostergaard, L., Chapman, N. H., Knapp, S., & Martin, C. (2013). Fruit development and ripening. *Annual Review of Plant Biology, 64*, 219–241.

Shewfelt, R. (1999). What is quality? *Postharvest Biology and Technology, 15*, 197–200.

Siedliska, A., Baranowski, P., Zubik, M., Mazurek, W., & Sosnowska, B. (2018). Detection of fungal infections in strawberry fruit by VNIR/SWIR hyperspectral imaging. *Postharvest Biology and Technology, 139*, 115–126. https://doi.org/10.1016/j.postharvbio.2018.01.018

Su, C. T., Li, B., Yin, H., Zou, J. P., Zhang, F., & Liu, Y. D. (2022). Identification of damage in pear using hyperspectral imaging technology. *Journal of Spectroscopy, 2022*, 9094249. https://doi.org/10.1155/2022/9094249

Thien Pham, Q., & Liou, N. S. (2022). The development of on-line surface defect detection system for jujubes based on hyperspectral images. *Computers and Electronics in Agriculture, 194*, 106743.

Torres, I., & Amigo, J. M. (2020). An overview of regression methods in hyperspectral and multispectral imaging. In J. M. Amigo (Ed.), *Chapter 2.8 in data handling in science and technology* (Vol. 32, pp. 205–230). Elsevier.

Vélez-Rivera, N., Gómez-Sanchis, J., Chanona-Pérez, J. J., Carrasco, J. J., Millán-Giraldo, M., Lorente, D., Cubero, S., & Blasco, J. (2014). Early detection of mechanical damage in mango using NIR hyperspectral images and machine learning. *Biosystems Engineering, 122*, 91–98. https://doi.org/10.1016/j.biosystemseng.2014.03.009

Walsh, K., Blasco, J., Zude-Sasse, M., & Sun, X. (2020). Visible-NIR' point' spectroscopy in postharvest fruit and vegetable assessment. *Postharvest Biology and Technology, 168*, 111246.

Wang, Y., Reder, N., Kang, S., Glaser, A., & Liu, J. (2017). Multiplexed optical imaging of tumor-directed nanoparticles: A review of imaging systems and approaches. *Nanotheranostics, 1*, 369–388.

Yadav, P. K., Burks, T., Frederick, Q., Qin, J., Kim, M., & Ritenour, M. A. (2022). Citrus disease detection using convolution neural network generated features and Softmax classifier on hyperspectral image data. *Frontiers in Plant Science, 13*, 1043712. https://doi.org/10.3389/fpls.2022.1043712

Yusuf, A., & Alawneh, S. (2018). A survey of GPU implementations for hyperspectral image classification in remote sensing. *Canadian Journal of Remote Sensing, 44*, 532–550.

Zhang, J., Dai, L., & Cheng, F. (2021). Identification of corn seeds with different freezing damage degree based on hyperspectral reflectance imaging and deep learning method. *Food Analytical Methods, 14*, 389–400.

Zhang, Y., Li, Y., Han, X., Gao, A., Jing, S., & Song, Y. (2023a). A study on hyperspectral apple bruise area prediction based on spectral imaging. *Agriculture, 13*, 819. https://doi.org/10.3390/agriculture13040819

Zhang, P., Wang, H., Ji, H., Li, Y., Zhang, X., & Wang, Y. (2023b). Hyperspectral imaging-based early damage degree representation of apple: A method of correlation coefficient. *Postharvest Biology and Technology, 199*, 112309. https://doi.org/10.1016/j.postharvbio.2023.112309

Chapter 5
Hyperspectral Imaging and Related Machine Learning for Postharvest Bruise Damage Detection and Analysis of Fresh Food Produce

Umezuruike Linus Opara, Ekene Emmanuel Okere, and Alemayahu Ambaw

Abstract During the past couple of decades, the technique of hyperspectral imaging (HSI) has developed rapidly and successfully applied as a novel non-destructive tool for fresh produce quality measurement and analysis, including plant and animal food materials. Bruise damage contributes to the incidence of postharvest losses and food waste as well as downgrading of products along the value chain. In this chapter, we discuss hyperspectral imaging technology and related machine learning for the detection and assessment of bruise damage. Applications of HSI and related machine learning techniques to detect, measure, and assess bruising on different types of fresh food products are presented. Future prospects for fast in-line sorting of bruised produce and the need to overcome the challenges of physical and biological variability of produce and high equipment cost are highlighted.

Keywords Bruising · Hyperspectral imaging · Machine learning · Damage detection · Bruise severity

5.1 Introduction

Fruits and vegetables are critical to the daily nutritional requirements of humans (Mahanti et al., 2022; Opara et al., 2021a, b). Fruit and vegetable in-take have been shown to supply multiple nutrients to consumers (Opara et al., 2021b), even better

---

U. L. Opara (✉)
SARChI Postharvest Technology Research Laboratory, Africa Institute for Postharvest Technology, Faculty of AgriSciences, Stellenbosch University, Stellenbosch, South Africa

UNESCO International Centre for Biotechnology, Nsukka, Enugu State, Nigeria
e-mail: opara@sun.ac.za

E. E. Okere · A. Ambaw
SARChI Postharvest Technology Research Laboratory, Africa Institute for Postharvest Technology, Faculty of AgriSciences, Stellenbosch University, Stellenbosch, South Africa

P. B. Pathare, U. L. Opara (eds.), *Mechanical Damage in Fresh Horticultural Produce*, https://doi.org/10.1007/978-981-99-7096-4_5

than manufactured biofortified foods and supplements (Keatinge et al., 2010). It is, therefore, vital that such food produce meet standard quality measures in terms of physical appearance, taste, and texture (Özdoğan et al., 2021). One of the factors militating the effective utilization of fruit and vegetables is its susceptibility to high postharvest losses and waste (Opara et al., 2021a, b), and this can be attributed to their high moisture content, high respiration rate, and handling operations. This phenomenon of high incidence of fruit and vegetable losses remains a major challenge to food industry and the actualization of the SDG on food security (Keatinge et al., 2011; Opara et al., 2021a, b).

Bruise damage especially to fruits and vegetables is becoming cause of great concern to the consumers, growers, and to those in agribusiness in general (Nagata et al., 2006). The major reason for this is that they are not easily detected and controlled immediately after they occur.

Different factors have been attributed to influence bruise damage in literature (Hussein et al., 2019a; Opara & Pathare, 2014). Overall, bruise severity is majorly linked to the amount of mechanical energy that a fruit produce is exposed to during impact, compression, or vibration in the postharvest chain (Opara & Pathare, 2014).

Bruise damage often results in quick quality deterioration of the bruised fruit, which causes degrading of the fruit as consumers lose confidence in utilizing them for concerns of food safety, thereby cumulating to economic losses (Liu et al., 2018; Nagata et al., 2006). There is therefore urgent need to quickly identify and sort bruised fruit from healthy samples before they are transported or during the postharvest handling stages (Liu et al., 2018; Opara & Pathare, 2014).

Technologies can aid quick and automated quality assessment such as sort fruit for external features such as appearance, texture, blemish and spots, and internal features such as internal decays, bruises, taste, flavor, and/or nutritive value, etc. Such technologies would provide rapid detection, screening and sorting operations in farms and packhouses, thereby assuring fruit quality and consistency (Hu et al., 2015), enhance consumer confidence in consumption, and promote profitability to growers and industry operators (ElMasry et al., 2007).

At the moment, human and manual approach has still been employed for bruise identification and detection in the fruit industry (ElMasry et al., 2007). Such activities necessitate human judgments that are entirely dependent on the external quality characteristics of the fruit specimen. This manual procedure seems time-consuming, cumbersome, and not reliable as bruise damage often does not manifest on the surface of the fruit immediately after occurrence. Studies have shown that bruise evaluation can be effectively carried out both manually and non-destructive using several non-destructive approaches (Opara & Pathare, 2014).

5.2 HSI for Fresh Produce Quality Analysis: An Overview of Principles

Hyperspectral imaging (HSI) was initially introduced for remote sensing applications in geographic field for landscape monitoring and vegetation control. HSI combines conventional imaging and spectroscopy (Opara & Pathare, 2014; Su & Sun, 2018). One advantage HSI has over other non-destructive techniques is that it has the capability to provide spatial and spectral data on the sample analyzed (Okere et al., 2021; Su & Sun, 2018). It works by simultaneously acquiring huge amounts of monochromatic images with wavebands, where a full spectrum can be extracted for each image pixel (Okere et al., 2021). The large size of acquired data usually requires to be filtered via machine learning to find the most representative wavelengths (Dian et al., 2019; Li et al., 2018b). The difference between multispectral imaging (MSI) and HSI is that while HSI has the capability to provide data on an object in the range of 400–1000 nm or wider, MSI only obtains data of several fixed wavelengths (Dian et al., 2019).

For most bruise studies, different devices have been developed for bruise simulation. This is because bruises occur in different ways and at different times during postharvest handling of fruits and vegetables. Figure 5.1 illustrates the different setups that have been developed by several researchers for bruise simulation.

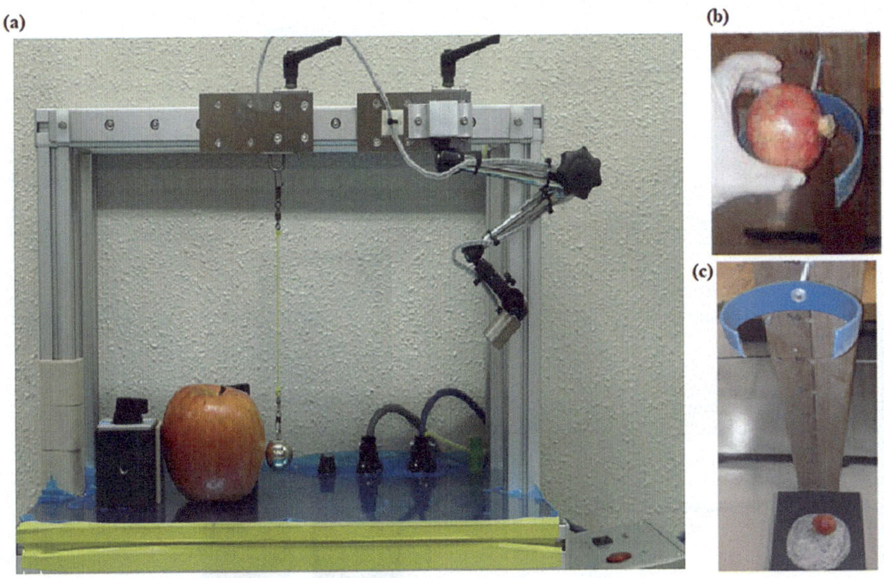

Fig. 5.1 Different bruise simulation device for impact and compression bruise damage on (**a**) apple fruit, (**b**, **c**) pomegranate fruit. (Adapted from Luo et al. (2012), (**b**) Okere et al. (2023))

5.3 HSI for Pome Fruit (Apple and Pear) Bruise Analysis

Apple (*Malus communis* L.) is undeniably the most famous fruit commodity globally, and it is cultivated across the different climatic regions. Apple fruit is grown for its texture, flavor, visual appeal, and nutritive content (Fan et al., 2020). It is estimated that annual global production stands at around 133 million tons (Çetin et al., 2022). Apple fruit suffer considerable postharvest losses due to its susceptibility to surface defects, such as bruises, blemishes, and diseases. These factors usually reduce fruit quality and affect consumer purchasing decisions (Fan et al., 2020).

Bruise detection in apple is one of the most reported investigations for non-destructive application. Though apple bruise detection is relatively automated, the system in place for automatic sorting and grading for bruises detection still lacks precision. It is reported that manual sorting method is still used (Baranowski et al., 2012). Typical steps in hyperspectral imaging technique application for apple as well as most fruit are summarized in Fig. 5.2.

Researchers are finding several non-destructive ways to improve bruise detection of apply and hyperspectral imaging technique is one of the deployed techniques for this investigation (Baranowski et al., 2012). Apple early bruise detection and intensity were investigated under three different wavelength ranges; visible and near-infrared (400–1000 nm), short wavelength infra- red (1000–2500 nm) and in

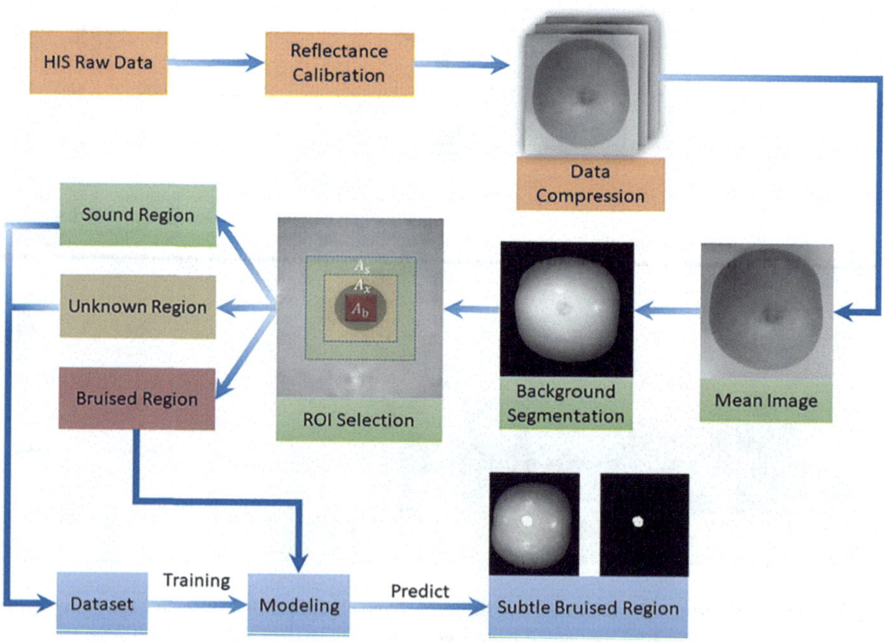

Fig. 5.2 Bruise analysis workflow for apple bruise detection. (Adapted from Che et al., 2018)

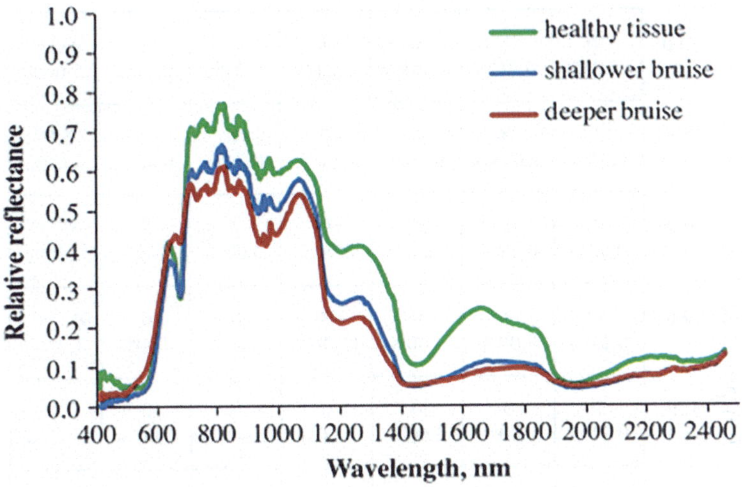

Fig. 5.3 VNIR and SWIR spectral characteristic curves extracted from the pixels in the region of interest (ROI) of the hyperspectral images representing healthy (not bruised), shallowly bruised, and deeply bruised tissue of "Topaz" apples. (Adapted from Baranowski et al., 2012)

mid-wavelength infrared (3500–5000 nm) to incorporate thermal imaging camera (Baranowski et al., 2012). The authors compared the performance of three different supervised classifiers: soft independent modeling of class analogy (SIMCA), linear discriminant analysis (LDA), and support vector machines (SVM). Spectral plot for their study showed that bruise profile for healthy, shallower, and deep bruise samples exhibited different profiles (Fig. 5.3). All three classifiers showed varying classification success with test results from LDA and SVM models showing better performance compared to SIMCA when all three spectral ranges were combined. The highest classification accuracy achieved in their study was 95%, and this result shows that bruise can be detected for apple using hyperspectral imaging combined with the appropriate machine learning techniques.

In a different study, bruise presence was investigated with respect to time of bruise (Baranowski et al., 2013). In this study, bruise was simulated on five different apple cultivars and bruise development was monitored using HSI under five supervised machine learning approaches and within the vis-NIR spectral and SWIR spectral range. Their result showed bruise can be detected to an overall average accuracy of 97.71%. It can be observed that bruise accuracy improved from (95%) on the second day to (100%) on the 14th day after bruise impact. A confusion matrix usually is used to show this summary in classification performance for a classifier (Fig. 5.4). The authors found that the most efficient time after bruise classification was based on the second derivative transformation pre-processing method of the spectral data and together with backpropagation neural networks and logistic regression classifiers.

Siedliska et al. (2014) applied HSI to investigate the possibility for bruise and cultivar discrimination of apple fruit samples. Using the CFS algorithm, the authors

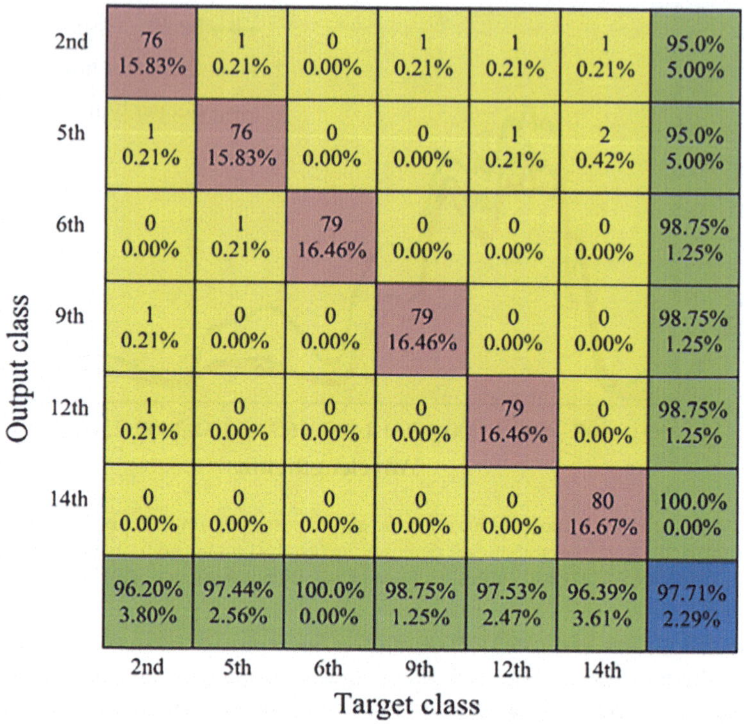

Fig. 5.4 The confusion matrix obtained shows classification performance for the 6 different days interval of bruise development with respect to time for apple fruit. Results were developed using the linear logistic regression model. (Adapted from Baranowski et al., 2013)

found six effective wavelengths (506.3 nm, 581.3 nm, 1131.0 nm, 1376.5 nm, 1382.8 nm, 2202.3 nm) to be optimum wavelengths for bruise detection in apple. The classification accuracy for the ten (10) different classifiers investigated ranged from 73.3 to 90% with three models (SVM, SLOG, SMO) yielding 90% validation accuracy. Similar to studies by Baranowski et al. (2013), the authors found that the hyperspectral system's CSF algorithm and second derivative pre-treatment were most useful for creating apple bruise detection models.

Che et al. (2018) employed the pixel extracted from hypercubes after imaging bruised and healthy "Fiji" apple samples to develop classification model to detect bruise and bruise progression with respect to time after bruise. In this study, three periods after bruise regime (0 h, 12 h, 18 h after bruising) were selected to simulate periods usually when apple fruits are exposed to bruise damage (harvest, transport, and storage). The authors applied random forest (RF) classifier after extracting pixels from 60 samples of imaged fruit obtained in the Vis-NIR range (400–1000 nm) for the classification into bruised and healthy samples. They found that two characteristic wavebands, around 675 nm and 960 nm, were effective to built an RF model, which achieved a CA of 99.90%, and thereby demonstrated the effectiveness of the

HSI technique as a nondestructive approach for implementation on apple quality sorting and grading system.

A one-dimensional signal hyperspectral image reduction approach (hyperspectrogram) was explored for model development to discriminate bruised from unbruised apple samples. In this study, two apple cultivars: "Golden Delicious" and "Pink Lady" datasets were investigated for bruise presence during 6 days interval to detect early bruise presence. The authors applied iPLS-DA and PLS-DA discriminant algorithms to distinguish bruised from unbruised samples. Their results showed the prediction accuracy of 92.42% and 94.04% for the "Golden Delicious" and on the "Pink Lady" datasets, respectively. The different interval size, spectra pre-treatment methods, and cultivar dataset yielded relatively satisfactory results, though the auto-scale pre-processing method with an interval side of 10 showed the best overall performance of 91.27% for recent bruise detection and 98.44% for old bruises (6 days after bruise).

Reseachers have monitored the presence of glare, stem, and bruise damage on three apple cultivars (Keresztes et al., 2017) in a step-wise pixel-based detection system using an online scan hyperspectral imaging in the shortwave infrared (SWIR) wavelength range. The "Jonagold," "Kanzi," and "Joly Red" cultivars of apple fruit were impacted with a swinging pendulum device to simulate bruises and bruised samples were imaged and monitored for 36 h. The authors utilized the PLS-DA classifier to develop classification models and found accuracy to range between 73.2 and 96.5%. The authors reported model classified the "Kanzi" cultivar more accurately (96.5%) compared to the other two cultivars (73.2%) for "Jonagold" (J), and (80.1%) "Joly Red" (JR), respectively. They found that model showed higher recognition rate for stem (94.37%), than the glare (90.03%), and also identified bruise samples (91.90%) over sound apples (87.83%).

In a similar study, Keresztes et al. (2016) after pre-processing hyperspectral data applied three different spectral correction methods: the pixel-based reflectance (PBR), pixel based with absolute reflectance correction (PBARC), and the reference pixel method (RPM). The authors reported the mean centered together with the first derivative pre-processing method to achieve the highest classification accuracy of 96.25%, thus, confirming that the pixel-based real-time SWIR hyperspectral imaging is feasible and more reliable and flexible for HSI-based sorting solutions for bruise detection. Similar studies in the SWIR spectral range also reported high accuracy for early bruise detection (Nturambirwe et al., 2021). The authors reported 100% accuracy for the first hour after bruise for "Golden Delicious" apple (Table 5.1). They found that the spectral range of 952–1377 nm and 1550–1850 nm showed increasing gap between spectra of bruised and sound tissues and attributed it to the changes in chemical composition in bruised areas, further validating that the SWIR spectral arrange can efficiently detect bruise damage in apple fruit.

Luo et al. (2012) in the vis-NIR (380–1000 nm) spectral range investigated effective wavelength for optimum bruise detection study, the authors applied PLS-DA as classifier to distinguish bruised from control samples. They found the optimum wavelength ranges for the four cultivars studied to be, respectively, 760 nm, 772 nm, 762 nm and 742 nm for "Fuji," "Jonagold," "Orin," and "Sinano

Table 5.1 Overview of application of HSI for bruise detection of different fruits

Produce	Bruise type	Data analysis	Model accuracy (%)	References
Apple	Bruise severity/ presence	SIMCA, LDA, SVM	CA range between 67 and 95 Overall CA was 97.71	Baranowski et al. (2012) Baranowski et al. (2013)
	Bruise time	LLR, BNN, LMT, LINE, SL, SMO, FT	CA ranged from 73.3 to 90	Siedliska et al. (2014)
	Bruise presence	LLR, BNN, LMT, LINE, SL, SMO, FT, LOG	Overall CA was 99.90	Che et al. (2018)
	Bruise time	PCA, RF	CA ranged from 83.87 to 96.23	Ferrari et al. (2015)
	Bruise presence/time	iPLS-DA, PLS-DA	CA ranged from 73.2 to 96.5	Keresztes et al. (2017)
	Bruise presence/time	iPLS-DA, PLS-DA	Overall CA was 96.25	Keresztes et al. (2016)
	Bruise presence	PLS-DA, PBARC, PBR, RPM	CA ranged from 81 to 99	Luo et al. (2012)
	Bruise presence/time	PLS-DA	Overall CA over 85	Nturambirwe et al. (2021)
	Bruise development	ESD, k-NN, SVM, LDA	Overall CA was 99.1	Tan et al. (2018a)
	Bruise degree	CARS-CCM-GS-SVM, GS-SVM	Overall CA was 92.5	Tan et al. (2018b)
	Bruise grading	CARS-CCM-GS-SVM, GS-SVM	Overall CA was 97.33	Tang et al. (2020)
	Bruise presence	PA, SPA, GA, KPCA, PCA	CA ranged from 80.75 to 92.86	Pan et al. (2018)
	Bruise degree/time	RF, LDA, GBDT, ERT, SGD, SVM, DT	CA ranged from 86.7–100%	Chiu et al. (2015)
	Bruise presence	ELM, PLS-DA, CART, MNF, SNV, mRMR	Overall CA over 92.9	Zhu and Li (2019)
	Bruise presence/time	MSC, CFS, ICA	CA ranged from 80.56 to 100	Zhang and Li (2018)
Blueberry	Bruise presence	FCN, SVM	CA ranged from 75.4 to 81.1	Zhang et al. (2020)
	Bruise presence	CARS, LS-SVM, CARS-LS- SVM	CA ranged from 77.5 to 98	Fan et al. (2018)
	Bruise presence	LR, CARS, SNV, MPBP	CA ranged from 74.4 to 95.2	Hu et al. (2016)

(continued)

Table 5.1 (continued)

Produce	Bruise type	Data analysis	Model accuracy (%)	References
	Bruise presence	ResNet, ResNeXt, SMO, LR, RF, Bagging, MLP	CA ranged from 71.13 to 88.44	Wang et al. (2018)
Peaches	Bruise	PCA, PLSR, LS-SVM, PLS-DA, ANN, SVM	CA ranged from 85.3 to 100	Li et al. (2018a) and Pan et al. (2016)
Pear	Bruise presence	ELM, RELIEF, SPA, CARS	CA ranged from 85 to 97.5	Feng et al. (2019)
	Bruise presence	k-NN, SVM	–	Dang et al. (2012)
	Bruise presence	PLS-DA	CA ranged from 97.8 to 100	Jiang et al. (2016)
	Bruise presence	*F*-values, PLS, PCA	CA achieved was 92	Lee et al. (2014)
Pomegranate	Bruise presence/time	ANN	CA ranged from 80 to 90	Okere et al. (2022b)
	Bruise severity	ANN	CA ranged from 80 to 100	Okere et al. (2023)
Strawberry	Bruise presence	LDA, ND, ANN	CA ranged from 84.6 to 99.9	Nagata et al. (2006)
	Bruise presence	PLS-DA and BBNN, SPA	CA ranged from 84.6 to 99.9	Liu et al. (2018)
	Bruise presence	CNN	CA was over 99%	Liu et al. (2018)

Gold" cultivar. These wavelengths are similar to those reported in previous studies on apple (Che et al., 2018; Keresztes et al., 2017). Tan et al. (2018b) identified seven effective wavelengths (797, 813, 818, 871, 876, 983, and 989 nm), in the vis-NIR spectral range as the characteristic wavelengths for further bruise detection on apple and reported an average early detection accuracy of 99.1%. The authors found that by combining the CARS-CCM, accuracy was improved from 92.5% of the original data to 97.5% when the CARS-CCM was applied. In a different study to detect bruise degree, Tan et al. (2018a) investigated the effect of bruise degree on classification results on bruise damage. The authors bruised and grouped samples into five classes, six classes, seven classes, and eight classes of bruising degrees and developed classification model to recognize different classes of bruise degree, they reported classification accuracy of 80%, 78.3%, 67.1%, 62.5%, respectively, for the different classes. The average CA achieved was 92.5% using the SNV pre-processing method.

Several authors have reported studies in the vis-NIR range and reported satisfactory results (Chiu et al., 2015; Huang et al., 2015b; Pan et al., 2018; Tang et al.,

2020; Zhu et al., 2016; Zhu & Li, 2019). These studies summarized in Table 5.1 further demonstrate the robustness and elaborate application of different machine learning algorithms, pre-processing methods, wavelength selection approaches for different bruise detection studies, and the results prove hyperspectral imaging to be an effective detection method.

Pear fruit is another economically important fruit that is consumed globally (Fang et al., 2019). Bruise damage has also been found to be the cause of major quality losses of pears and often occurs at all stage in the postharvest chain (Dang et al., 2012; Fang et al., 2019). Some of the resultant consequences of bruise damage to pears is that the damaged tissue easily becomes susceptible to decay due to the growth of rot fungi, the presence of fungi rot on damaged samples usually infects healthy samples as it easily spreads across to unbruised samples in the surrounding.

Fang et al. (2019) applied HSI to detect the presence of bruise damage on "Korla" pears. In this study, effect of wavelength selection as well as the integrated approach was compared for classification model development using the extreme learning machine (ELM) algorithm method. Three wavelengths selection methods showed varying recognition success for bruise detection accuracy with the RELIEF method showing better performance 92.5% accuracy. They found the integrated approach to yield higher wave numbers (37) compared to the three methods considered and even better classification accuracy of 97.5%, thus showing the efficacy of HSI as a bruise detection of pear fruit. In a similar study on early bruise detection on pear, two classification algorithms, viz. k-NN and SVM were compared for bruise detection on pears. The authors found SVM to show inconsistent assignment of bruised pixels to the accurate point in the training set as compared to the k-NN classifier. They reported the SVM algorithm to have assigned more pixels as bruise points, on the first day after bruising though the decision can vary according to the training data set. Their study concluded that both methods are appropriate for application for bruise early detection study for pears.

Lee et al. (2014) investigated bruise damage detection below pear skin using NIR in the range of 950–1650 nm. In this study, hyperspectral images of pears were acquired before and after bruise simulation on samples, and the authors developed calibration model for bruise detection within 1 h after the initial bruising, bruise development was monitored for 2, 3, 4, 7, 9, and 11 days after the first day of bruise damage. They found the model to yield a highest classification accuracy of 92% when the selected optimal NIR reflectance image ratio (1074 nm and 1016 nm: R1074/R1016) was applied, thereby demonstrating the ability of hyperspectral imaging to identify bruise damage affected on pear fruit surface.

5.4 HSI for Pomegranate Bruise Analysis

Almost 500 distinct varieties of the pomegranate (Punica granatum L.), a very old Middle Eastern perishable fruit, are produced throughout the world. Due to the rising demand that has been connected to the nutritional and health advantages of

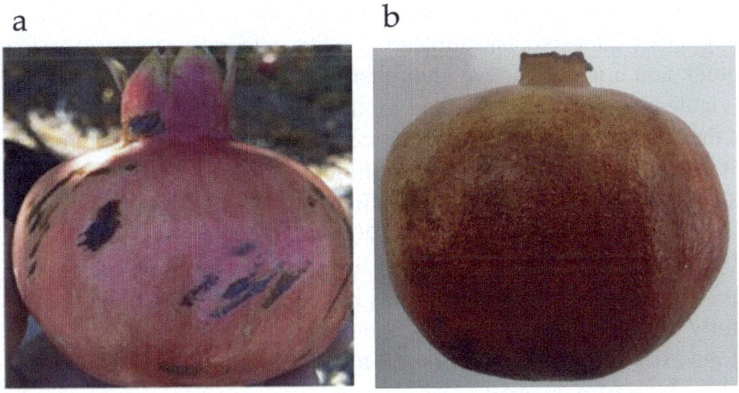

Fig. 5.5 The manifestation of bruise damaged pomegranate fruit includes: (**a**) physical injury resulting in peel hyperpigmentation; (**b**) peel browning as a side effect of bruised or damaged peel

pomegranates, the sector has seen an increase in sales and production (Al-Said et al., 2009; Fawole & Opara, 2013; Lansky & Newman, 2007; Opara et al., 2009; POMASA, 2021). The avoidance of inflammatory disorders, the induction of anti-proliferative effects, and the enhancement of anti-metastatic side effects in humans are only a few of the health benefits linked to its use (Pareek et al., 2015).

Bruise is the most common type of postharvest mechanical injury affecting pomegranate fruit. Opara et al. (2021a, b) reported a strong link between the presence of fruit damage and rotting/spoilage. This study also showed that damaged fruit were more vulnerable to insect attack because the affected area of the fruit skin becomes fragile as a result of the bruise. The presence of bruising degrades fruit quality, results in postharvest losses and lower profits (Hussein et al., 2019a, b; Mohammad Shafie et al., 2017; Opara & Pathare, 2014) because damaged fruit does not fulfill both local and export market quality standards. Some of the symptoms of bruise damage on pomegranate fruit include discoloration, scars, and peel browning (Fig. 5.5).

Pomegranate fruit responds physiologically to bruise damage, according to scientific findings (Hussein et al., 2019b). Increased peel electrolyte leakage (PEL), a greater browning score, and enhanced polyphenol oxidase (PPO) enzyme activity are a few of the modifications connected to bruising damage. There have also been reports of a high respiratory rate following bruising injury (Hussein et al., 2019b). Once fruit is bruised, physico-chemical changes such as those in total soluble solids (TSS), titratable acidity (TA), Brix-to-acid ratio (TSS: TA), and vitamin A have also been noticed (Hussein et al., 2019b).

Early identification of bruising on pomegranate fruit is challenging because of the strong and leathery skin of this fruit, in contrast to other fruit having delicate tissues as well as thin rind/peel such as apples and pears (Hussein et al., 2019a). Pomegranate fruit bruise is not instantly obvious after the hit (Hussein et al., 2019a, b). Bruises are often found by visual examination by qualified panels or line operators and physically removed in the industry. This method of diagnosing bruises is time-consuming, labor-intensive, and subjective. Thus, alternative technology is required for the quick and non-destructive identification of early bruising damage.

As a viable non-destructive method for the assessment of produce quality and for grading, the use of spectroscopic technique has gained considerable momentum in recent times (Arendse et al., 2018; Jamshidi et al., 2019; Khodabakhshian et al., 2017). X-ray (Hussein, 2019a, b), thermal imaging (TI) (Zeng et al., 2020), magnetic resonance imaging (MRI) (Razavi et al., 2018), fluorescence imaging (FI) (Everard et al., 2016; Chiu et al., 2015) as well as hyperspectral imaging (Dian et al., 2019; Zhu et al., 2016) are among the numerous image-based technologies that have been successfully applied in bruise sensing and analysis. Recently, Okere et al. (2022a) reported an extensive review on non-destructive approaches for quality measurement and analysis of pomegranates.

In the last several years, hyperspectral imaging (HSI) has become a potent non-destructive examination technology in the agricultural, biosecurity diagnostic, and food industries. Spectroscopy and imaging are combined into one system by the non-destructive/non-invasive HSI approach (Wu & Sun, 2013).

It is generated by producing images from many electromagnetic wavelength spectrum components from the same area of an object and at the same scale (Su & Sun, 2018). Three-dimensional (3D) structures made up of two spatial and one spectral dimensions make up the data retrieved (hypercube) from HSI systems (Gowen et al., 2007, 2015). For the purpose of identifying various fruit flaws, this non-destructive method has been presented (Arendse et al., 2020; Okere et al., 2021).

The efficacy of HSI as a potential non-destructive technique for bruise detection for pomegranate fruit was explored in literature (Okere et al., 2022b, 2023). To investigate the potential of early bruise detection on pomegranate fruit, Okere et al. (2022a) developed a classification model to distinguish bruised from unbruised fruit immediately after bruise damage.

The researchers collected samples from 60 pomegranate fruits (cv. Wonderful), divided them into 30 each of bruised and unbruised sample groups, and then subjected the specimens to three distinct after-bruise time regimes (immediate, 7 days, and 14 days). The spectral and spatial data of samples from the various groups were collected using two hyperspectral imaging setups: visible and near-infrared (400–1000 nm) and short wavelength infrared (1000–2500 nm). The typical spectrum of the whole fruit surface (WFS) as well as the region of interest (ROI) of every specimen were used as contribution toward a two feed artificial neural network (ANN) machine learning algorithm for the development of a classification model after spectral pre-processing. On the very same day of bruising, the authors reported classification model accuracies ranging between 80 and 90%. For bruised specimens, precision was 86.7%, while for healthy samples it was 90%.

In a different study to monitor bruise severity, Okere et al. (2023) applied the artificial neural network (ANN) algorithm to develop classification model to distinguish different bruise severity level of impact bruise damage of pomegranate fruit sample. According to the authors, the accuracy of bruise intensity categorization varied from 80 to 96.7%. Their findings show model accuracy to increase with increase in bruise drop height. Both studies have shown the capability of hyperspectral imaging technology in identifying and monitoring early bruises and to monitor bruise development as well as the severity of impact of the bruise damage on pomegranate fruit.

5.5 HSI for Strawberry Bruise Analysis

Strawberries (*Fragaria ananassa Duch.*) are some of the most consumed fruits in the world due to their nutritional compositions, such as vitamins, minerals, and phytochemicals (Afrin et al., 2016; Liu et al., 2018). It is arguably the most economically important fruit considering the different ways it can be consumed (ElMasry et al., 2007; Liu et al., 2018).

Fresh mechanical bruised strawberries fruits are often downgraded and are usually made into processed products like jams or juices in fruit industry (Liu et al., 2018). However, it is difficult to identify the mechanical bruised strawberry fruit by naked eyes because the damage occurs under the pericarp of the fruit making it difficult to be noticed (Liu et al., 2018; Nagata et al., 2006). It is therefore of importance that non-destructive detection methods to evaluate bruise presence are explored during the postharvest processing to enable distinguishing the bruised ones from the sound (unbruised) ones.

In a study to investigate bruise presence on strawberries, Nagata et al. (2006) deployed hyperspectral imaging technique to develop classification model to distinguish bruised from unbruised samples. The authors identified 825 nm and 980 nm wavebands to be the most effective wavebands for classifying between bruised and non-bruised on strawberries. Their results showed accuracies ranging from 84.6 to 99.9% using three different classifiers. The linear discriminant analysis (LDA) was reported to perform best for classification purposes when compared to the artificial neural network (ANN) and the normalized difference (ND) method. These findings show that by choosing the best classifier and most efficient waveband during image analysis, model results can be greatly improved as well as time of analysis.

Lui et al. (2018) identified eight optimal wavelengths (560, 604, 650, 675, 720, 771, 890, and 995 nm) by the successive projection algorithms (SPA) from the whole range of wavelength when they investigated the presence of bruise damage on strawberry (*var. Hongyan*) using hyperspectral image data. The authors reported classification accuracies ranging from 77.78 to 92.59%. The highest overall identification accuracy achieved was 92.59% using the SVM classifier which performed better than the PLS-DA and BBNN methods. Similar range of effective wavelengths has been found for different quality attributes of strawberry (ElMasry et al., 2007).

5.6 HSI for Bruise Analysis of Vegetables (Cucumber, Mushroom)

The quality of mushroom mostly depends on the manufacturing processes, mode of transportation, and its storage conditions (Gormley, 1987; Gowen et al., 2010). Mushrooms are rated for their white appearance, making any browning on the vegetable an index of poor quality (Fig. 5.6). Bruise damage manifests on mushrooms in the form of browning on the surface. This is often after undergoing excessive physical impact during the harvest, transportation, and storage period.

To discriminate surface tissues condition of mushroom (*Agaricus bisporus*), hyperspectral imaging technique was compared to the performance of RGB machine vision system (Taghizadeh et al., 2011). The authors developed partial least squares discriminant analysis (PLS-DA) models to classify mushroom tissue into casing soil damaged, vibrational damaged and undamaged class. The authors found hyperspectral imaging to yield satisfactory discriminate result with an accuracy of 81.08%, while the RGB model yielded an accuracy of 81.79% for the combined discriminant of all three classes. HSI model achieved an accuracy of 89.83% which

Fig. 5.6 Images illustrating different quality grades of mushroom. (**a**) Mushroom damaged using casing soil, (**b**) samples damaged by vibration force, (**c**) undamaged samples, (**d**) samples developing enzymatic browning during after storage after bruise. (Adapted from Taghizadeh et al., 2011)

proves its capability of identifying undamaged regions, casing soil and enzymatic damaged areas on mushrooms.

In another study, Gowen et al. (2008) developed PCA model using hyperspectral imaging data for damage assessment of the white mushroom cap. The authors utilized a HSI push-broom line-scanning device, operating in wavelength range of 400–1000 nm to image-controlled vibration damage to imitate bruise damage caused by transportation of mushroom samples. Results showed that PCA model classified unbruised samples to an accuracy of 96% and bruised samples to an accuracy of 83%. To also assess bruise severity, samples bruised by shaking at 400 rpm for 20 min showed higher classification accuracy (83%) as compared to those bruised at 200 rpm for 200 min (79%). These studies indicate the possibility of HSI combined with PCA to non-destructively identify bruise damage on mushrooms. Bruise development over time has also been investigated for simulated bruise conditions for white mushroom for different times (0, 60, 120, 300, and 600 s) after bruise damage (Gowen & O'Donnell, 2009). The authors reported model to show satisfactory result as bruise progressed with time.

Cucumber, however, is another vegetable that is also susceptible to bruise damage leading to high economic losses. Ariana et al. (2006) developed bruise detection models to discriminate bruised from unbruised samples of cucumber using capture hyperspectral images from pickling cucumbers in the spectral region of 900–1700 nm. The authors simulated mechanical bruise damage on samples by dropping or rolling samples under load and imaged bruised samples in 3 days interval, immediately (0 day), 3 days and 6 days after bruise. PCA, band ratio, and band difference were applied to acquired hypercubes during image analysis to distinguish between bruised and healthy cucumbers and author reported classification accuracies ranging from 75 to 95%. Their result shows that both PCA and band ratio method yielded classification model that accurately classified bruised samples from healthy ones immediately after bruise damage to an accuracy of 94.6%. As days after bruise damage progressed, model showed lower accuracies and the authors attributed it to the self-healing characteristic of cucumber tissue.

Cen et al. (2014) demonstrated that LED illuminated samples with a reduced moving speed (85 mm/s) during image acquisition of cucumber samples would yield better classification results over samples illuminated with the tungsten halogen lights running at the conveyor speed of 110 mm/s. The authors reported improved result 82.8% from 79.9% when model was developed using the optimum two band ratio of 87/837 nm after lowering the speed of the conveyor belt from 165 mm/s to 85 mm/s. They found the minimum redundancy–maximum relevance (MRMR) wavelength selection method to be more effective for band selection compared to the PCA method and reported the highest classification accuracy of 94.2%. This study thereby provides evidence that different configurations during image acquisition can impact the outcome of model performance.

By far, the majority of studies on HSI applications on vegetables have focused on mushrooms. A summary of selected such applications for bruise detection on mushrooms and cucumber is presented in Table 5.2. Given their high perishability, high incidence of postharvest losses and waste, and importance on human diet and

Table 5.2 Summary of application of hyperspectral imaging for bruise detection for vegetables

Sample	Spectral range (nm)	Data analysis	Accuracy (%)	References
Mushroom	400–1000	PCA	79–100	Gowen et al. (2008)
Mushroom	400–1000	PCA	79–100	Gowen and O'Donnell (2009)
Mushroom	445–945	PCA, PLS-DA	Up to 89.83	Taghizadeh et al. (2011)
Mushroom	900–1700	ANN	91	Rojas-Moraleda et al. (2017)
Cucumber	900–1700 400–1000	BR, PCA PCA, MRMR	75–95 66.4–94.2	Ariana et al. (2006) Cen et al. (2014)

Table 5.3 Summary of application of hyperspectral imaging for bruise detection pork meat and potatoes, showing the spectral range and classifier models used

Food material	Spectral range (nm)	Classifier	Accuracy (%)	References
Meat (pork)	1280, 1440, and 1660	BP-ANN		Huang et al. (2015a)
Potatoes	400–2500	SIMCA, PLS-DA	Above 93	López-Maestresalas et al. (2016)

nutrition, more studies are necessary to further explore the cost-effectiveness and commercial applicability of HSI and related chemometrics to detect the presence of mechanical injuries and onset of spoilage of fresh vegetables.

5.7 HSI for Potato Bruise Analysis

Mechanical damage of potatoes due to impacts often referred to as "blackspot" occurs during harvesting and postharvest operations and is a major source of considerable economic losses in the industry annually. In the USA, it was reported that about 70 percent of such damage occurred during harvest (Peters, 1996). In one particular bad season, financial losses due to potato damage ranging from US$20 to US$60 million were reported in Washington State alone, the second major producer in the USA (Mathew and Hyde 1997).

Potato tubers do not show any external symptoms immediately after impact, which makes bruise detection in potatoes not straightforward. Therefore, a non-destructive and accurate method capable of identifying bruised tubers is needed. Hyperspectral imaging (HSI) has been shown to be able to detect other subsurface defects such as bruises in apples. This method is non-destructive, fast, and can be fully automated.

López-Maestresalas et al. (2016) applied three different classifiers (PCA, SIMCA, and PLS-DA) to build classification model to recognize bruised from unbruised samples of potatoes (Table 5.3). Two HSI setups, visible near-infrared

(Vis-NIR) (400–1000 nm) and short wave infrared (SWIR) (1000–2500 nm) were applied to image 188 samples of three different varieties of potatoes. The authors induced and monitored bruise development for 4 h regime after bruise (1, 5, 9, and 24 h) and reported an average of 94% accuracy. SWIR was reported to show better overall accuracy. The authors reported poor accuracy for the bruises after 1 h, with a misclassification rate of 65.22% which goes to show potatoes can only be detected for bruises after 5 h of impact.

5.8 HSI for Bruise Analysis of Meat

Food produce such as meat constitute important and fundamental source of human nutrition and dietary requirement for survival (Kozan et al., 2016). The quality of meat is often defined around three attributes, viz. physical attributes, chemical and biological attributes. Quality aspects such as texture, color, marbling, tenderness make up the physical meat attributes. Chemical components affecting meat quality and nutritional value include fat, moisture, protein, pH, and drip loss. Bacterial count and freedom from infection make up the biological attributes (Huang et al., 2015a; Kozan et al., 2016).

Meat quality assessment is still heavily manual, involving countless hours in laboratory for destructive and eco-unfriendly analysis to ascertain quality of good quality from inconsumable ones (Kozan et al., 2016). Growing global population and subsequent consumers demand for high quality meat produce require more effective meat quality assessment that is rapid, non-invasive, and environmental friendly testing technique so as to meet this quality need and assure food safety concerns (Table 5.3) (Huang et al., 2015a).

5.9 Conclusions and Future Prospects

The development and use of hyperspectral imaging techniques in non-destructive fruit and vegetable quality evaluation have accelerated in past decades. The ideas, developments, and implementations of hyperspectral imaging technology for inspecting the quality of fruits and vegetables have been addressed in this chapter with particular emphasis on the bruise damage measurement and analysis. This chapter also includes information on the key elements, fundamental theories, and pertinent processing and analytical techniques. Hyperspectral imaging faces both challenges and opportunities for application in quick in-line sorting. These difficulties include the impact of biological and physical variability, full surface detection, differentiating flaws from stems and calyxes, detecting hidden defects, resilience of the components and algorithms, and quick development of multispectral imaging systems. The difficulties listed above will remain for some time into the future, even though numerous solutions have been put forth to address the complicated challenges in fruit and vegetable quality management by using hyperspectral imaging method. Specific attention will be required to optimize HSI for the detection, analysis, and control of mechanical damage during postharvest handling of specific types of fresh foods, including fruit and vegetables.

Acknowledgement This work is based on the research supported in part by the National Research Foundation of South Africa (Grant Numbers: 64813). The opinions, findings, and conclusions or recommendations expressed are those of the author(s) alone, and the NRF accepts no liability whatsoever in this regard.

References

Afrin, S., Gasparrini, M., Forbes-Hernandez, T. Y., Reboredo-Rodriguez, P., Mezzetti, B., Varela-López, A., Giampieri, F., & Battino, M. (2016). Promising health benefits of the strawberry: A focus on clinical studies. *Journal of Agricultural and Food Chemistry, 64*(22), 4435–4449. https://doi.org/10.1021/acs.jafc.6b00857

Al-Said, F. A., Opara, L. U., & Al-Yahyai, R. A. (2009). Physico-chemical and textural quality attributes of pomegranate cultivars (Punica granatum L.) grown in the Sultanate of Oman. *Journal of Food Engineering, 90*(1), 129–134. https://doi.org/10.1016/j.jfoodeng.2008.06.012

Arendse, E., Fawole, O. A., Magwaza, L. S., Nieuwoudt, H., & Opara, U. L. (2018). Fourier transform near infrared diffuse reflectance spectroscopy and two spectral acquisition modes for evaluation of external and internal quality of intact pomegranate fruit. *Postharvest Biology and Technology, 138*(October 2017), 91–98. https://doi.org/10.1016/j.postharvbio.2018.01.001

Arendse, E., Nieuwoudt, H., Magwaza, L. S., Fredric, J., Nturambirwe, I., Fawole, O. A., & Opara, U. L. (2020). Recent advancements on vibrational spectroscopic techniques for the detection of authenticity and adulteration in horticultural products with a specific focus on oils, juices and powders. *Food and Bioprocess Technology, 14*, 1. https://doi.org/10.1007/s11947-020-02505-x

Ariana, D. P., Lu, R., & Guyer, D. E. (2006). Near-infrared hyperspectral reflectance imaging for detection of bruises on pickling cucumbers. *Computers and Electronics in Agriculture, 53*(1), 60–70. https://doi.org/10.1016/j.compag.2006.04.001

Baranowski, P., Mazurek, W., Wozniak, J., & Majewska, U. (2012). Detection of early bruises in apples using hyperspectral data and thermal imaging. *Journal of Food Engineering, 110*(3), 345–355. https://doi.org/10.1016/j.jfoodeng.2011.12.038

Baranowski, P., Mazurek, W., & Pastuszka Woźniak, J. (2013). Supervised classification of bruised apples with respect to the time after bruising on the basis of hyperspectral imaging data. *Postharvest Biology and Technology, 86*, 249–258. https://doi.org/10.1016/j.postharvbio.2013.07.005

Cen, H., Lu, R., Ariana, D. P., & Mendoza, F. (2014). Hyperspectral imaging-based classification and wavebands selection for internal defect detection of pickling cucumbers. *Food and Bioprocess Technology, 7*(6), 1689–1700. https://doi.org/10.1007/s11947-013-1177-6

Çetin, N., Karaman, K., Kavuncuoğlu, E., Yıldırım, B., & Jahanbakhshi, A. (2022). Using hyperspectral imaging technology and machine learning algorithms for assessing internal quality parameters of apple fruits. *Chemometrics and Intelligent Laboratory Systems, 230*(August), 104650. https://doi.org/10.1016/j.chemolab.2022.104650

Che, W., Sun, L., Zhang, Q., Tan, W., Ye, D., Zhang, D., & Liu, Y. (2018). Pixel based bruise region extraction of apple using Vis-NIR hyperspectral imaging. *Computers and Electronics in Agriculture, 146*(February), 12–21. https://doi.org/10.1016/j.compag.2018.01.013

Chiu, Y. C., Chou, X. L., Grift, T. E., & Chen, M. T. (2015). Automated detection of mechanically induced bruise areas in golden delicious apples using fluorescence imagery. *Transactions of the ASABE, 58*(2), 215–225. https://doi.org/10.13031/trans.58.10578

Dang, H. Q., Kim, I., Cho, B. K., & Kim, M. S. (2012). Detection of bruise damage of pear using hyperspectral imagery. In *International conference on control, automation and systems* (pp. 1258–1260).

Dian, R., Li, S., Fang, L., & Wei, Q. (2019). Multispectral and hyperspectral image fusion with spatial-spectral sparse representation. *Information Fusion, 49*(December 2017), 262–270. https://doi.org/10.1016/j.inffus.2018.11.012

ElMasry, G., Wang, N., ElSayed, A., & Ngadi, M. (2007). Hyperspectral imaging for nondestructive determination of some quality attributes for strawberry. *Journal of Food Engineering, 81*(1), 98–107. https://doi.org/10.1016/j.jfoodeng.2006.10.016

Everard, C. D., Kim, M. S., & Lee, H. (2016). Assessment of a handheld fluorescence imaging device as an aid for detection of food residues on processing surfaces. *Food Control, 59*, 243–249. https://doi.org/10.1016/j.foodcont.2015.05.030

Fan, S., Li, C., Huang, W., & Chen, L. (2018). Data fusion of two hyperspectral imaging systems with complementary spectral sensing ranges for blueberry bruising detection. *Sensors (Switzerland), 18*(12), 4463. https://doi.org/10.3390/s18124463

Fan, S., Li, J., Zhang, Y., Tian, X., Wang, Q., He, X., Zhang, C., & Huang, W. (2020). On line detection of defective apples using computer vision system combined with deep learning methods. *Journal of Food Engineering, 286*(April), 110102. https://doi.org/10.1016/j.jfoodeng.2020.110102

Fang, Y., Yang, F., Zhou, Z., Lin, L., & Li, X. (2019). Hyperspectral wavelength selection and integration for bruise detection of Korla pears. *Journal of Spectroscopy, 2019*, 6715247. https://doi.org/10.1155/2019/6715247

Fawole, O. A., & Opara, U. L. (2013). Fruit growth dynamics, respiration rate and physico-textural properties during pomegranate development and ripening. *Scientia Horticulturae, 157*, 90–98. https://doi.org/10.1016/j.scienta.2013.04.004

Feng, L., Zhu, S., Zhou, L., Zhao, Y., Bao, Y., Zhang, C., & He, Y. (2019). Detection of subtle bruises on winter jujube using hyperspectral imaging with pixel-wise deep learning method. *IEEE Access, 7*, 64494–64505. https://doi.org/10.1109/ACCESS.2019.2917267

Ferrari, C., Foca, G., Calvini, R., & Ulrici, A. (2015). Fast exploration and classification of large hyperspectral image datasets for early bruise detection on apples. *Chemometrics and Intelligent Laboratory Systems, 146*, 108–119. https://doi.org/10.1016/j.chemolab.2015.05.016

Gormley, T. R. (1987). Handling, packaging and transportation of fresh mushrooms. In *Proceedings of the 5th National Mushroom Conference*. Malahide.

Gowen, A. A., & O'Donnell, C. P. (2009). Development of algorithms for detection of mechanical injury on white mushrooms (Agaricus bisporus) using hyperspectral imaging. In *Proc. SPIE 7315, Sensing for Agriculture and Food Quality and Safety, 73150G (27 April 2009)*. https://doi.org/10.1117/12.818597

Gowen, A. A., O'Donnell, C. P., Cullen, P. J., Downey, G., & Frias, J. M. (2007). Hyperspectral imaging—an emerging process analytical tool for food quality and safety control. *Trends in Food Science and Technology, 18*(12), 590–598. https://doi.org/10.1016/j.tifs.2007.06.001

Gowen, A. A., O'Donnell, C. P., Taghizadeh, M., Cullen, P. J., Frias, J. M., & Downey, G. (2008). Hyperspectral imaging combined with principal component analysis for bruise damage detection on white mushrooms (Agaricus bisporus). *Journal of Chemometrics, 22*(3–4), 259–267. https://doi.org/10.1002/cem.1127

Gowen, A. A., Taghizadeh, M., & O'Donnell, C. P. (2010). Using hyperspectral imaging for quality evaluation of mushrooms. In *Hyperspectral imaging for food quality analysis and control* (1st ed.). Elsevier. https://doi.org/10.1016/B978-0-12-374753-2.10013-9

Gowen, A. A., Feng, Y., Gaston, E., & Valdramidis, V. (2015). Talanta recent applications of hyperspectral imaging in microbiology. *Talanta, 137*, 43–54. https://doi.org/10.1016/j.talanta.2015.01.012

Hu, M. H., Dong, Q. L., Liu, B. L., Opara, U. L., & Chen, L. (2015). Estimating blueberry mechanical properties based on random frog selected hyperspectral data. *Postharvest Biology and Technology, 106*, 1–10. https://doi.org/10.1016/j.postharvbio.2015.03.014

Hu, M., Dong, Q., & Liu, B. (2016). Classification and characterization of blueberry mechanical damage with time evolution using reflectance, transmittance and interactance imaging spectroscopy. *Computers and Electronics in Agriculture, 122*, 19–28. https://doi.org/10.1016/j.compag.2016.01.015

Huang, Q., Chen, Q., Li, H., Huang, G., Ouyang, Q., & Zhao, J. (2015a). Non-destructively sensing pork's freshness indicator using near infrared multispectral imaging technique. *Journal of Food Engineering, 154*, 69–75. https://doi.org/10.1016/j.jfoodeng.2015.01.006

Huang, W., Li, J., Wang, Q., & Chen, L. (2015b). Development of a multispectral imaging system for online detection of bruises on apples. *Journal of Food Engineering, 146*, 62–71. https://doi.org/10.1016/j.jfoodeng.2014.09.002

Hussein, Z., Fawole, O. A., & Opara, U. L. (2019a). Bruise damage susceptibility of pomegranates (Punica granatum, L.) and impact on fruit physiological response during short term storage. *Scientia Horticulturae, 246*(November 2018), 664–674. https://doi.org/10.1016/j.scienta.2018.11.026

Hussein, Z., Fawole, O. A., & Opara, U. L. (2019b). Determination of physical, biochemical and microstructural changes in impact-bruise damaged pomegranate fruit. *Journal of Food Measurement and Characterization, 13*(3), 2177–2189. https://doi.org/10.1007/s11694-019-00138-z

Jamshidi, B., Mohajerani, E., Farazmand, H., Mahmoudi, A., & Hemmati, A. (2019). Pattern recognition-based optical technique for non-destructive detection of Ectomyelois ceratoniae infestation in pomegranates during hidden activity of the larvae. *Spectrochimica Acta - Part A: Molecular and Biomolecular Spectroscopy, 206*, 552–557. https://doi.org/10.1016/j.saa.2018.08.059

Jiang, H., Zhang, C., He, Y., Chen, X., Liu, F., & Liu, Y. (2016). Wavelength selection for detection of slight bruises on pears based on hyperspectral imaging. *Applied Sciences (Switzerland), 6*(12), 450. https://doi.org/10.3390/app6120450

Keatinge, J. D. H., Waliyar, F., Jamnadas, R. H., Moustafa, A., Andrade, M., Drechsel, P., Hughes, J. D., Kadirvel, P., & Luther, K. (2010). Relearning old lessons for the future of food-by bread alone no longer: Diversifying diets with fruit and vegetables. *Crop Science, 50*(April), S-51–S-62. https://doi.org/10.2135/cropsci2009.09.0528

Keatinge, J. D. H., Yang, R. Y., Hughes, J., Easdown, W. J., & Holmer, R. (2011). The importance of vegetables in ensuring both food and nutritional security in attainment of the millennium development goals. *Food Security, 3*(4), 491–501. https://doi.org/10.1007/s12571-011-0150-3

Keresztes, J. C., Goodarzi, M., & Saeys, W. (2016). Real-time pixel based early apple bruise detection using short wave infrared hyperspectral imaging in combination with calibration and glare correction techniques. *Food Control, 66*, 215–226. https://doi.org/10.1016/j.foodcont.2016.02.007

Keresztes, J. C., Diels, E., Goodarzi, M., Nguyen-Do-Trong, N., Goos, P., Nicolai, B., & Saeys, W. (2017). Glare based apple sorting and iterative algorithm for bruise region detection using shortwave infrared hyperspectral imaging. *Postharvest Biology and Technology, 130*(April), 103–115. https://doi.org/10.1016/j.postharvbio.2017.04.005

Khodabakhshian, R., Emadi, B., Khojastehpour, M., Golzarian, M. R., & Sazgarnia, A. (2017). Non-destructive evaluation of maturity and quality parameters of pomegranate fruit by visible/near infrared spectroscopy. *International Journal of Food Properties, 20*(1), 41–52. https://doi.org/10.1080/10942912.2015.1126725

Kozan, H. I., Sariçoban, C., Akyürek, H. A., & Ünver, A. (2016). Hyperspectral imaging technique as a state of art Technology in meat science. *Green Chemistry & Technology Letters, 2*(3), 127–137. https://doi.org/10.18510/gctl.2016.232

Lansky, E. P., & Newman, R. A. (2007). Punica granatum (pomegranate) and its potential for prevention and treatment of inflammation and cancer. *Journal of Ethnopharmacology, 109*(2), 177–206. https://doi.org/10.1016/j.jep.2006.09.006

Lee, W., Kim, M. S., Lee, H., Delwiche, S. R., Bae, H., Kim, D., & Cho, B. (2014). Hyperspectral near-infrared imaging for the detection of physical damages of pear. *Journal of Food Engineering, 130*, 1–7. https://doi.org/10.1016/j.jfoodeng.2013.12.032

Li, J., Chen, L., & Huang, W. (2018a). Detection of early bruises on peaches (Amygdalus persica L.) using hyperspectral imaging coupled with improved watershed segmentation algorithm. *Postharvest Biology and Technology, 135*(July 2017), 104–113. https://doi.org/10.1016/j.postharvbio.2017.09.007

Li, X., Li, R., Wang, M., Liu, Y., Zhang, B., & Zhou, J. (2018b). Hyperspectral imaging and their applications in the nondestructive quality assessment of fruits and vegetables. In *Hyperspectral Imaging in Agriculture, Food and Environment*. IntechOpen. https://doi.org/10.5772/intechopen.72250

Liu, Q., Sun, K., Peng, J., Xing, M., Pan, L., & Tu, K. (2018). Identification of bruise and fungi contamination in strawberries using hyperspectral imaging technology and multivariate analysis. *Food Analytical Methods, 11*(5), 1518–1527. https://doi.org/10.1007/s12161-017-1136-3

López-Maestresalas, A., Keresztes, J. C., Goodarzi, M., Arazuri, S., Jarén, C., & Saeys, W. (2016). Non-destructive detection of blackspot in potatoes by Vis-NIR and SWIR hyperspectral imaging. *Food Control, 70*, 229–241. https://doi.org/10.1016/j.foodcont.2016.06.001

Luo, X., Takahashi, T., Kyo, K., & Zhang, S. (2012). Wavelength selection in Vis/NIR spectra for detection of bruises on apples by ROC analysis. *Journal of Food Engineering, 109*(3), 457–466. https://doi.org/10.1016/j.jfoodeng.2011.10.035

Mahanti, N. K., Pandiselvam, R., Kothakota, A., Ishwarya, S. P., Chakraborty, S. K., Kumar, M., & Cozzolino, D. (2022). Emerging non-destructive imaging techniques for fruit damage detection: Image processing and analysis. *Trends in Food Science and Technology, 120*(December 2021), 418–438. https://doi.org/10.1016/j.tifs.2021.12.021

Mohammad Shafie, M., Rajabipour, A., & Mobli, H. (2017). Determination of bruise incidence of pomegranate fruit under drop case. *International Journal of Fruit Science, 17*(3), 296–309. https://doi.org/10.1080/15538362.2017.1295416

Nagata, M., Tallada, J. G., & Kobayashi, T. (2006). Bruise detection using NIR hyperspectral imaging for strawberry. *Environmental Control in Biology, 44*(02), 133–142.

Nturambirwe, J. F. I., Perold, W. J., & Opara, U. L. (2021). Classification learning of latent bruise damage to apples using shortwave infrared hyperspectral imaging. *Sensors, 21*, 4990. https://doi.org/10.3390/s21154990

Okere, E. E., Arendse, E., Nieuwoudt, H., Fawole, O. A., Perold, W. J., & Opara, U. L. (2021). Non-invasive methods for predicting the quality of processed horticultural food products , with emphasis on dried powders, juices and oils: a review. *Food, 10*, 3061.

Okere, E. E., Ambaw, A., Perold, W. J., & Opara, U. L. (2022a). Non-destructive detection of bruises in pomegranate fruit by Vis-NIR and SWIR hyperspectral imaging. *Acta Horticulturae, 1349*, 381–388. https://doi.org/10.17660/actahortic.2022.1349.52

Okere, E. E., Arendse, E., Tsige, A. A., Perold, W. J., & Opara, U. L. (2022b). Pomegranate quality evaluation using non-destructive approaches: A review. *Agriculture, 12*, 2034. https://doi.org/10.3390/agriculture12122034

Okere, E. E., Tsige, A. A., Perold, W. J., & Opara, U. L. (2023). Vis-NIR and SWIR hyperspectral imaging method to detect bruises in pomegranate fruit. *Frontiers in Plant Science, 14*, 1151697. https://doi.org/10.3389/fpls.2023.1151697

Opara, U. L., & Pathare, P. B. (2014). Bruise damage measurement and analysis of fresh horticultural produce-a review. *Postharvest Biology and Technology, 91*, 9–24. https://doi.org/10.1016/j.postharvbio.2013.12.009

Opara, L. U., Al-Ani, M. R., & Al-Shuaibi, Y. S. (2009). Physico-chemical properties, vitamin C content, and antimicrobial properties of pomegranate fruit (Punica granatum L.). *Food and Bioprocess Technology, 2*(3), 315–321. https://doi.org/10.1007/s11947-008-0095-5

Opara, I. K., Fawole, O. A., Kelly, C., & Opara, U. L. (2021a). Quantification of on-farm pomegranate fruit postharvest losses and waste, and implications on sustainability indicators: South African case study. *Sustainability, 13*(9), 1–20. https://doi.org/10.3390/su13095168

Opara, I. K., Fawole, O. A., & Opara, U. L. (2021b). Postharvest losses of pomegranate fruit at the packhouse and implications for sustainability indicators. *Sustainability, 13*(9), 5187. https://doi.org/10.3390/su13095187

Özdoğan, G., Lin, X., & Sun, D. W. (2021). Rapid and noninvasive sensory analyses of food products by hyperspectral imaging: Recent application developments. *Trends in Food Science and Technology, 111*, 151–165. https://doi.org/10.1016/j.tifs.2021.02.044

Pan, L., Zhang, Q., Zhang, W., Sun, Y., Hu, P., & Tu, K. (2016). Detection of cold injury in peaches by hyperspectral reflectance imaging and artificial neural network. *Food Chemistry, 192*, 134–141. https://doi.org/10.1016/j.foodchem.2015.06.106

Pan, X., Sun, L., Li, Y., Che, W., Ji, Y., Li, J., Li, J., & Xu, Y. (2018). Non-destructive classification of apple bruising time based on visible and near-infrared hyperspectral imaging. *Journal of the Science of Food and Agriculture, 99*, 1709. https://doi.org/10.1002/jsfa.9360

Pareek, S., Valero, D., & Serrano, M. (2015). Postharvest biology and technology of pomegranate. *Journal of the Science of Food and Agriculture, 95*(12), 2360–2379. https://doi.org/10.1002/jsfa.7069

Peters, R. (1996). Damage of potato tubers, a review. *Potato Research, 39*, 479–484.

POMASA. (2021). *The economic contribution of South Africa's pomegranate industry*. Division for Macro & Resource Economics.

Razavi, M. S., Asghari, A., Azadbakh, M., & Shamsabadi, H. A. (2018). Analyzing the pear bruised volume after static loading by magnetic resonance imaging (MRI). *Scientia Horticulturae, 229*(November 2017), 33–39. https://doi.org/10.1016/j.scienta.2017.10.011

Siedliska, A., Baranowski, P., & Mazurek, W. (2014). Classification models of bruise and cultivar detection on the basis of hyperspectral imaging data. *Computers and Electronics in Agriculture, 106*, 66–74. https://doi.org/10.1016/j.compag.2014.05.012

Su, W. H., & Sun, D. W. (2018). Fourier transform infrared and Raman and hyperspectral imaging techniques for quality determinations of powdery foods: A review. *Comprehensive Reviews in Food Science and Food Safety, 17*(1), 104–122. https://doi.org/10.1111/1541-4337.12314

Taghizadeh, M., Gowen, A. A., & O'Donnell, C. P. (2011). The potential of visible-near infrared hyperspectral imaging to discriminate between casing soil, enzymatic browning and undamaged tissue on mushroom (Agaricus bisporus) surfaces. *Computers and Electronics in Agriculture, 77*(1), 74–80. https://doi.org/10.1016/j.compag.2011.03.010

Tan, W., Sun, L., Yang, F., Che, W., Ye, D., Zhang, D., & Zou, B. (2018a). Study on bruising degree classification of apples using hyperspectral imaging and GS-SVM. *Optik, 154*, 581–592. https://doi.org/10.1016/j.ijleo.2017.10.090

Tan, W., Sun, L., Yang, F., Che, W., Ye, D., Zhang, D., & Zou, B. (2018b). The feasibility of early detection and grading of apple bruises using hyperspectral imaging. *Journal of Chemometrics, 32*(10), 1–14. https://doi.org/10.1002/cem.3067

Tang, Y., Gao, S., Zhuang, J., Hou, C., He, Y., Chu, X., Miao, A., & Luo, S. (2020). Apple bruise grading using piecewise nonlinear curve fitting for hyperspectral imaging data. *IEEE Access, 8*, 147494–147506. https://doi.org/10.1109/ACCESS.2020.3015808

Wang, Z., Hu, M., & Zhai, G. (2018). Application of deep learning architectures for accurate and rapid detection of internal mechanical damage of blueberry using hyperspectral transmittance data. *Sensors, 18*, 1126.

Wu, D., & Sun, D. W. (2013). Advanced applications of hyperspectral imaging technology for food quality and safety analysis and assessment: A review—part I: Fundamentals. *Innovative Food Science and Emerging Technologies, 19*, 1–14. https://doi.org/10.1016/j.ifset.2013.04.014

Zeng, X., Miao, Y., Ubaid, S., Gao, X., & Zhuang., S. (2020). Detection and classification of bruises of pears based on thermal images. *Postharvest Biology and Technology, 161*(August 2019), 111090. https://doi.org/10.1016/j.postharvbio.2019.111090

Zhang, M., & Li, G. (2018). Visual detection of apple bruises using AdaBoost algorithm and hyperspectral imaging. *International Journal of Food Properties, 21*(1), 1598–1607. https://doi.org/10.1080/10942912.2018.1503299

Zhang, M., Jiang, Y., Li, C., & Yang, F. (2020). Fully convolutional networks for blueberry bruising and calyx segmentation using hyperspectral transmittance imaging. *Biosystems Engineering, 192*, 159–175. https://doi.org/10.1016/j.biosystemseng.2020.01.018

Zhu, X., & Li, G. (2019). Rapid detection and visualization of slight bruise on apples using hyperspectral imaging. *International Journal of Food Properties, 22*(1), 1709–1719. https://doi.org/10.1080/10942912.2019.1669638

Zhu, Q., Guan, J., Huang, M., Lu, R., & Mendoza, F. (2016). Predicting bruise susceptibility of "Golden delicious" apples using hyperspectral scattering technique. *Postharvest Biology and Technology, 114*, 86–94. https://doi.org/10.1016/j.postharvbio.2015.12.007

Chapter 6
Bruise Damage Susceptibility of Pome Fruit

Pankaj B. Pathare, Rebogile Mphahlele, and Mai Al-Dairi

Abstract Pome fruit like pears and apples are a good source of nutritional elements. Also, they are highly grown and largely consumed around the world. Pome fruit susceptibility to mechanical damage is aggravated by thin peel as it comes in contact with the packaging units, harvesting containers, etc. Bruise damage is the most common sort of mechanical damage in fresh produce. Grading, sorting, packing, and transporting are the main factors that cause bruising to pome fruit during the postharvest supply chain. Bruising leads to poor fresh produce quality which results in a major financial loss after harvest. Many forces are considered to cause mechanical damage in fresh pome fruit including impact, compression, abrasion, vibration, cut, and puncture forces. Several laboratory tests have been conducted to mimic damages experienced by pome fruit at various stages of postharvest handling like drop impact test, pendulum test, compression test, and vibration test. Bruise damage in fresh pome fruit has been conducted by several researchers. Manual measurements can be applied to measure the bruise dimension, to calculate the bruise area and volume. Image analysis is also utilized to detect the bruise in its early stage in pome fruit. This includes magnetic resonance imaging (NMR), time-resolved reflectance spectroscopy (TRS), visible and near-infrared (VNIR) hyperspectral imaging, thermal imaging, etc. Reduction of bruising in pome fruit has been discussed in several studies by using proper packaging and cushioning materials during transport, packing the fruit in the right and safe location inside the packaging, and using several treatments to delay the subsequent damage after transport.

P. B. Pathare (✉)
Department of Soils, Water & Agricultural Engineering, College of Agricultural and Marine Sciences, Sultan Qaboos University, Al-Khod, Oman
e-mail: pankaj@squ.edu.om

R. Mphahlele
Agricultural Research Council (ARC)-Tropical and Subtropical Crops, Nelspruit, South Africa

M. Al-Dairi
Department of Soils, Water and Agricultural Engineering, College of Agricultural & Marine Sciences, Sultan Qaboos University, Al-Khod, Oman

P. B. Pathare, U. L. Opara (eds.), *Mechanical Damage in Fresh Horticultural Produce*, https://doi.org/10.1007/978-981-99-7096-4_6

Keywords Apple · Pear · Bruising · Postharvest handling · Impact · Non-destructive techniques

6.1 Introduction

Apples and pears belong to the family of Rosaceae and the subfamily Pomoideae (the pome fruit) (Jackson, 2003). Apples belong to the subtribe Pyrinae and are ranked third after citrus and banana as the most important fruit worldwide (Gharaghani et al., 2016). Globally, the total harvestable area and production of apples were four million hectares and 93 million tons, respectively (FAOSTAT, 2022), where China ranked first, followed by the USA, Turkey, and Poland (Celik et al., 2021). Pear (Pyrus spp.) is a very valuable fruit worldwide and is highly harvested in the temperate areas of both the southern and northern hemispheres. Also, China is considered the largest producer of pear fruit in the world (Gharaghani et al., 2016). In 2021, the harvested area and the quantity produced under pear production in the world reached one million hectares and 25 million tons, respectively (FAOSTAT, 2022).

Pome fruit like apples and pears are a rich source of nutritional elements. Apples are rich in phenolics, vitamins, organic acids, fibers, and antioxidants for the human diet (Celik et al., 2021). Also, the pear fruit gained attention in food research and industry as a valuable fresh produce because of the availability of several beneficial compounds like anthocyanins, vitamin C, phenols, antioxidants, etc. Pears are also essential due to the increased utilization of antioxidant compounds, involving the advantages of avoiding diseases like cardiovascular disease and cancer (Azadbakht et al., 2019).

Consumers evaluate the fresh produce quality on the bases of visual appearance. Consumers prefer juicy fruits, a good firm state, tasty, and intact. Thus, fresh fruits provided to the stores should follow the international quality standards for freshness, firmness, intactness, and other quality parameters (Komarnicki et al., 2016). The production, distribution, and transportation of these fruits need special care and attention to minimize the damage. The globalization of markets and trends of several countries to specialize in a particular fruit encourage intrastate and domestic transportation (Unuigbe & Onuoha, 2013).

Pear fruit are also among the extremely perishable horticultural produce. Compared to other pome fruit like apples, pears have a lesser storability. Besides, the complex biological structure of plant material generates mechanical properties of the fruit to be affected by turgor, temperature, shape, size, ripeness, fruit orientation, and microstructure (Stropek & Gołacki, 2022). Pome fruit is highly sensitive to unsuitable conditions during postharvest operations including loading, reloading, transport, handling, storage, etc. (Komarnicki et al., 2016). Postharvest mechanical damage in pome fruit is responsible for fruit quality damage and deterioration. Mechanical damage is the main reason for the damage and loss due to impact that could occur because of vibration or sudden drop of fresh produce from a certain height (Unuigbe & Onuoha, 2013).

For the majority of fruit, involving pome fruit, bruising is the most known form of mechanical injury. Bruise damage is generally described as a collection of tissues with brown color directly under the peel of the fruit. Bruise damages in fruit are sort of subcutaneous tissue failure which shows no rapture of the skin where the discoloration observed in the damaged tissues distinguishes the injured area/spot (Rostampour & Motlagh, 2018). Several studies have been conducted to evaluate the bruise susceptibility and mechanical properties to bruising of fresh produce like apples and pears by many researchers (Azadbakht et al., 2019; Celik et al., 2021; Pathare & Al-Dairi, 2021; Unuigbe & Onuoha, 2013). These studies showed that bruising can alter the quality attributes of pome fruit including color, flavor, textural properties, weight, etc. which highly affects the consumer perception and acceptance of the fruit. Bruising causes fungal and bacterial contamination that leads to reduced fruit shelf-life (Nguyen et al., 2022). Using non-destructive techniques to evaluate bruising in pome fruit has been widely used in recent years. These methods include thermal imaging (Doosti-Irani et al., 2016), inverse adding-doubling method (Zhang et al., 2017), computer vision system (Fu et al., 2023), magnetic resonance imaging (MRI) (Razavi et al., 2018), etc. Therefore, this chapter discusses the importance of bruising in the pome fruit industry, mechanisms of bruise damage, evaluation and analysis method (bruising), and current status and trends related to bruising in pome fruit.

6.2 Mechanisms of Bruise Damage

Mechanical damage is unavoidable along the postharvest supply chain of fresh produce. Mechanical damage incidence in pome fruit or other fresh produce during postharvest handling can be observed because of different sorts of forces including impact, compression, cut, abrasion, and vibration stresses. In most cases, apples are categorized by soft flesh, high water content, complex, and unstable interior structure (Opara & Fadiji, 2018). During postharvest operations, pome fruits undergo mechanical injury generated by the impact, quasi-static, or dynamic loads. Other causes of mechanical damage are unsuitable conditions during reloading, transporting, and storage (Komarnicki et al., 2017a). Nguyen et al. (2022) reported that apples are highly susceptible to exterior impact from distortion or falling in different stages and methods which are related to harvesting and postharvest handling.

Mechanical damage generated by impact is the main problem leading to huge costs in fresh processing (Al-Dairi et al., 2022). Impact stress on fruit can be generated by several reasons involving a sudden drop of fresh fruits from a particular height, vibration during transportation, as well as unsuitable cushioned material of packed produce during handling. The impact can appear as external damage, bruise, or cracks on the fruit's surface that can reduce the quality attribute of the produce

(Fadiji et al., 2016a). Additionally, Van Zeebroeck et al. (2007a) mentioned that the most known damage caused by impact force mostly occurs in free drop of fruit from the tree to the ground during the harvesting time and the dynamic impacts between each fruit, other fruits, and containers/packages. Besides, Lu et al. (2010) found that the percentage of impact damage in apples especially in one or double-wall corrugated containers elevated as the height of the drop increased during the processes of handling and transporting. In addition, they observed that the damage in apple fruits packed in the top layer of both corrugated containers was remarkably greater than in the lower one.

Van Zeebroeck et al. (2008) simulated the impact damage in apples within the path of the vehicle over a speed bump and recorded that the higher load of the vehicle led to the low formation of bruising. Also, they found that apples inside bulk bins and kept behind the rear axle experienced more injury than fruit inside bins in front of the rear axle. According to Gołacki et al. (2009), impact damage may occur at each phase of the technological procedure of handling fresh fruit and is difficult to remove in large-scale distribution and production of horticultural commodities.

Compression is another form of mechanical damage that can occur everywhere during storage, loading, transportation, etc. and can generate mechanical injuries when fresh produce is overfilled and stacked. Besides, the incidence of compression force can occur due to the extreme depth of the bin or even the stack height of cartoons (Opara & Pathare, 2014). Shu et al. (2020) stated that the resulting damage from vibration is a form of accumulative failure that is highly generated by dynamic vibrations that caused an interior injury to fresh produce. Quantitatively, this kind of deterioration is hard to detect and evaluate due to the lack of obvious wounds. Besides, Wang et al. (2022) confirmed that the mechanical injury of fruit caused by shock, vibration, and compression is difficult to prevent in fresh produce during postharvest handling. Nowadays, transport for long distances becomes more required and familiar. Constant random vibration results in mechanical damage to fresh fruit if it lacks suitable packaging. The abrasion force occurs when the fruit moves against other fruits leading to damage the external layer of the produce which is highly observed when fresh fruit is transported at high speed (Li & Thomas, 2014).

The incidence of cut forces/stresses can be observed when the fresh produce is pushed and rotated near the sharp edge of the packaging unit. Like other fresh produce, pome fruit suffered from puncture force which can occur during the process of harvesting. Puncture injuries lead to a decline in the value of the fruit, pathogen invasion, decreased shelf-life, increasing susceptibility of fruit to water loss, and considerable economic losses (Li & Thomas, 2014).

6.3 Importance of Bruising in Pome Fruit Industry

Pome fruit are largely grown and extensively consumed around the world for their nutritional properties. Generally, pome fruit are comprised of 88% water and 12% fiber which make them susceptible to mechanical bruising during handling

(Berardinelli et al., 2005; Eissa et al., 2012). The susceptibility to bruising is further aggravated by thin skin/peel as they come in contact with harvesting containers and packaging materials. Bruises on the fruit manifest as brown patches which extend further into the epidermis of fruit (Opara & Pathare, 2014). Bruising results in poor product quality which leads to major financial losses after harvest (Pang et al., 1992; Menesatti et al., 1999). Approximately 30–40% of fresh fruits and vegetables are lost because of mechanical injury which is aggravated by secondary infections (Barchi et al., 2002). It is important to understand how bruises occur in order to prevent or reduce bruising.

Sorting, packing, grading, and transportation are among factors that need to be carefully controlled to minimize bruising incidence in order to maintain fruit quality. Factors including vibration, impact force, and compression load cause physical bruising of the fruit (Ahmadi, 2012; Babarinsa & Ige, 2012; Lu et al., 2012; Fadiji et al., 2016a; Opara & Fadiji, 2018) which are further aggravated by fruit maturity status, fruit falls distance, impact number, impact energy, and the sort of impact surface materials (Opara, 2007). Package dimension and paperboard combination have an influence on the resistance of the package to compression loading (Fadiji et al., 2016b). Injuries lead to total breakdown of the structure of the affected cells which in turn lead to increased respiration rate, ethylene production, serve as an entry point for pathogens and ultimately faster ripening resulting in shortened shelf-life (Miller, 2003; Sargent et al., 1992).

The severity of injury is directly related to ripeness and harvest date of the fruit, temperature, irrigation, and weather (Van Zeebroeck et al., 2007b). Microscopic factors such as the strength of the cell wall, cell shape, and elasticity can affect bruise sensitivity (Devaux et al., 2005; Alamar et al., 2008). According to Opara et al. (2007), firmness, size, shape (curvature radius, mass), and turgor also contribute to bruise susceptibility. Alternative method of reducing bruising during transportation involves use of soft materials (foam, foil) inside the truck as well as vibration-damping system (Pathare et al., 2012; Fadiji et al., 2016a; Komarnicki et al., 2017b).

Various laboratory tests have been performed to mimic injuries experienced by fruit at various stages of postharvest handling. For instance, drop test or free fall method, which allows the fruit to hit the surface at a specific height (Gancarz, 2016; Hussein et al., 2019), pendulum tests which give a better control of impact course, whereas compression test depicts load experienced by fruit during transportation and storage (Komarnicki et al., 2016; Azadbakht et al., 2018; Rasli et al., 2019). In drop test method, the device poorly controls the impact site on the fruit surface, thus resulting in additional friction (Stropek & Gołacki, 2020).

Assessment of bruising in fruits requires high-tech equipment. Although measurement of bruising has proven to be complicated, several advanced technological methods of measuring bruise in fruit have been used by several researchers in the last 5 years. For instance, near-infrared detection method, spectral imaging detection method, optical detection method, acoustic and ultrasonic detection method, thermal sensitivity, and nuclear magnetic resonance imaging method were reported (Clark & Macfall, 2003; Zhang et al., 2017). Even though they are deemed costly and unpractical to be fitted in the sorting lines, methods including spectroscopy, time-resolved reflectance spectroscopy (TRS), magnetic resonance imaging (MRI), spatial frequency-domain imaging (SFDI), X-ray, and spatially resolved hyperspectral (SRS) were tested on bruised pears (Zeng et al., 2020).

6.4 Evaluation and Analysis Method (Bruising)

Several recent studies published in bruise susceptibility of apple and pear fruit are presented in Table 6.1. Bruising techniques employed by several researchers range from drop test (Fig. 6.1a, b) (Pathare & Al-Dairi, 2021), pendulum principles, application of compression force (Cañete et al., 2015; Stropek & Gołacki, 2015; Zhang et al., 2017), and vibration technique (Fadiji et al., 2016a) (Fig. 6.1c, d). Common bruises of horticultural crops occur as a result of compression and impact force are highlighted in Table 6.1. More recently vibration force has gained moment on its effect on farm produce as trucks are used for transportation between farm and retail outlet (Fadiji et al., 2016a). Analysis method used to detect and quantify the bruises includes conventional measurements of the dimensions of bruised tissue to assess the bruise area and bruise volume (Bollen et al., 2001; Opara, 2007; Opara et al., 2007), imaging of bruised area and subsequent utilization of image process analysis software to estimate the damage size (Menesatti & Paglia, 2001), as well as non-invasive methods for in situ identification, evaluation, and analysis (Xing et al., 2006; Dang et al., 2012).

Bruising detection has shifted from a mere traditional method to high-tech instruments for easy analysis. For instance, packers select bruised fruit by visually assessing the cracks, mechanical damage, or burst which may not necessarily detect the internal damage in real time (Van Linden et al., 2008; Opara & Pathare, 2014). Recently, different advanced techniques are used for analysis of bruised fruit by different researchers. A study by Razavi et al. (2018) used non-destructive method of magnetic resonance imaging and image processing techniques to analyze the bruised level in pear by applying different compression force. LMS SCADAS recorder of Siemens (camera) incorporated with the LMS Test. Also, Xpress software was used by Stropek and Gołacki (2015) to measure the impact oscillation by using pendulum principle and the authors concluded that the two separate utilized cameras were not sensitive to impact oscillation that permitted for additional precise displacement measurements.

Detection of mechanical damage at an early stage of fruit harvest has proven difficult especially at sorting lines using conventional methods. Thus, advanced methods were explored to accurately detect and quantify damage in apples and pears. NIR-hyperspectral images differentiated between the bruised and non-bruised Golden delicious' apples (Ferrari et al., 2015). A study by Pan et al. (2019) investigated the use of a non-destructive technique for apple bruising time classification by applying visible and near-infrared (VNIR) hyperspectral imaging and found that the hyperspectral method can be ideally used for predicting bruising time in apples, which in turn will aid to evaluate the interior quality attributes and

Table 6.1 Evaluation and analysis method of bruising in pome fruit

Cultivar	Bruising techniques	Bruise evaluation	References
Pears var. "Dar Gaz"	Compression force	Magnetic resonance image (MIR)	Razavi et al. (2018)
Apples "Fuji"	Impact damage	Optical properties	Zhang et al. (2017)
Apple cvs. "Rubin," "Florina," and "Freedom"	Impact damage	LMS SCADAS recorder of Siemens integrated with the LMS Test. Xpress software	Stropek and Gołacki (2015)
Pears cv. "Korean Niitaka"	Infrared lock-in thermography	A reverse engineering approach, physical material tests, and finite element method (FEM)	Kim et al. (2014)
Apple var. "Rubin," "Rajka," and "Freedom	Pendulum principle	Diameter, restitution coefficient $e = \frac{v_{reb}}{v_{imp}}$	Stropek and Gołacki (2013)
Apple cvs. "Rubin," "Florina," and "Freedom""	Drop test	LMS SCADAS recorder of Siemens company integrated with the LMS Test.Xpress software	Stropek and Gołacki (2015)
Pears cvs. "Lukasówka" and "Xenia"	Pendulum principles	Diameter, restitution coefficient $e = \frac{v_{reb}}{v_{imp}}$	Stropek and Gołacki (2019)
Apples "Golden Delicious"	Electro-dynamic shaker	Elliptical bruise shape $\mathrm{BA} = \frac{\pi}{4W_1W_2}$ $\mathrm{BV} = \frac{\pi^{d_b}}{24\left(3w_1w_2+4d_b^2\right)}$	Fadiji et al. (2016a)
Apples "red Fuji"	Pendulum principle	Hyperspectral image	Tan et al. (2018)
Apple vars. "Jazz," "Envy," and "Granny Smith"	Pendulum principle—Type impact test device	$\mathrm{BA} = \frac{\pi \cdot w_1 \cdot w_2}{4}$ $\mathrm{BV} = \frac{\pi \cdot d}{24}\left(3w_1 \cdot w_2 + 4d^2\right)$	Fu et al. (2017)
Pears var. "D'Anjou"	Drop test		Pathare and Al-Dairi (2021)
Pears cv. Clapp's Favorite	Pendulum mechanism	Computer image analysis	Komarnicki et al. (2016)
Pears var. "Natanz"	Drop test	Finite element method (FEM)	Yousefi et al. (2016)
Apples cv. "Gala"	Drop test	Computer image analysis	Komarnicki et al. (2017b)
Pears "Dar Gazi"	Impact force	Magnetic resonance imaging (MRI)	Razavi et al. (2018)
Apples "Fuji"	Drop test	Inverse adding-doubling method	Zhang et al. (2017)
Pear var. "Shekari"	Compression test	Finite element method (FEM)	Salarikia et al. (2017)

(continued)

Table 6.1 (continued)

Cultivar	Bruising techniques	Bruise evaluation	References
Pear "Korla Fragrant"	Drop test	Thermographic camera	Zeng et al. (2020)
"Golden Delicious" and "Pink Lady" apples	Pendulum	NIR spectral scanner	Ferrari et al. (2015)
Fuji apples	Pendulum	Visible and near-infrared (VNIR) hyperspectral imaging	Pan et al. (2019)
Apples cv "Idared," "Golden Delicious," and "Jonagold" 1	Drop test	Bruise Resistance Index	Stopa et al. (2018)

BA bruise area, *BV* bruise volume, w_1 widths, d_b full depth of bruise

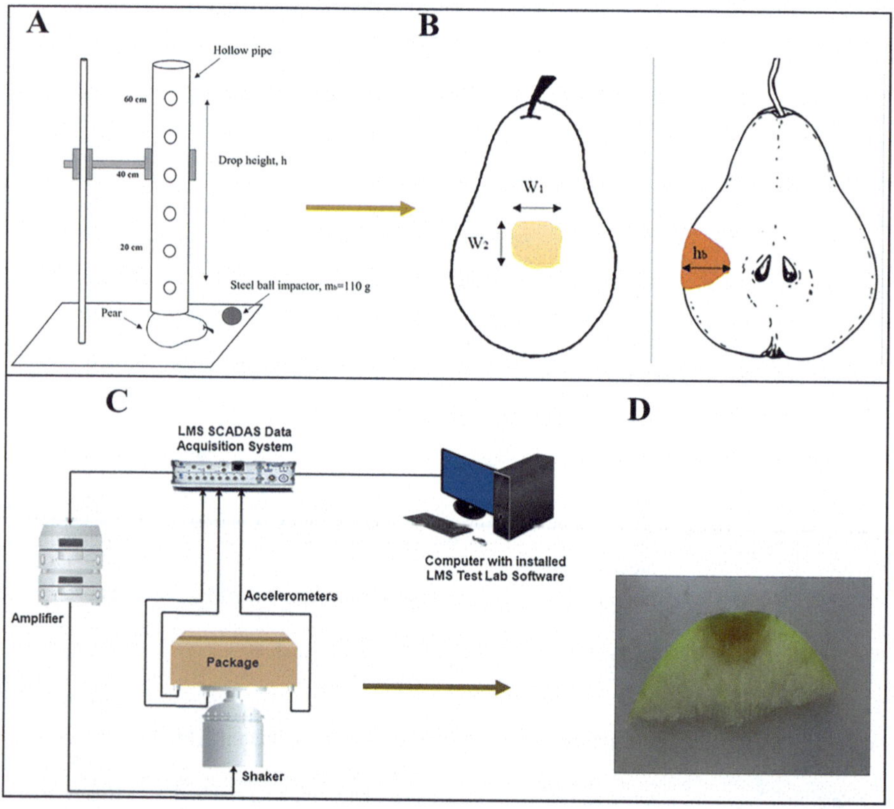

Fig. 6.1 Schematic drop test setup (**a**), bruise measurements of pear (bruise diameter and bruise depth) (Pathare & Al-Dairi, 2021) (**b**), schematic layout of the vibration system (**c**), and section of the apple bruise depth (Fadiji et al., 2016a) (**d**). w_1 *and* w_2 major and minor width, h_b bruise depth

safety of apple fruit. Huang et al. (2020) found that the multichannel Vis/NIR spectroscopic system in semi-transmittance mode included in the Vis/NIR range of 550–1650 nm can help in the detection of localized internal defects in apples. It could be concluded that NIR technique serves as a potential in detecting bruising and can provide solution in detecting bruising in the sorting lines.

The application of optical sensing techniques, including color imaging and hyper/multispectral (Lu et al., 2020), was previously applied for the identification of fresh fruit bruise damage because of their advantages of objective, rapid, and non-invasive features. Also, Sun et al. (2022) used optical property mapping, single snapshot couple with look-up table (SS-LUT), and the three phase couple with curve fitting (TP-CF) to detect bruises in apples at an early stage and they found that TPCF worked better than SS-LUT. However, the author observed that TP-CF was time-consuming in indicating optical properties compared to the SS-LUT, which significantly improved the estimation and the efficiency of processed data.

In another study, Kim et al. (2014) determined the thermal emission signals from pear fruit by utilizing a very sensitive mid-infrared thermal camera and observed that by using lock-in technique to infrared thermography, the measurements sensitivity and signal-to-noise ratio were improved due to the phase-sensitive narrow-band filtering outcome. Komarnicki et al. (2017a) evaluated bruise volume of damaged tissue of pears using computer analysis and bruised resistance index indicators as a correlation between determined bruise volume and surface pressure. The author proposed that the technique can be an efficient device for the evaluation of mechanical injury in apple fruit. However, the literature of Mahanti et al. (2021) concluded that the optical property is unsuitable for inline/online sorting of fruits at the commercial level.

Zhang et al. (2017) investigated the optical properties of apple flesh in the range of 400–1500 nm by using integrating sphere system and the inverse adding-doubling (IAD) algorithm and observed that discriminant model that combines principal component analysis (PCA) with support vector machines (SMV) gave an accuracy as high as 92.5% for bruise degree of apples. They found that a simple thermal image system particularly in the long-wavelength (8–14 μm) was beneficial in capturing cleaner images of bruised pear but gave poorer separation among the invisible or slight differences of bruises with time (Zeng et al., 2020). Doosti-Irani et al. (2016) developed the multiple regression model to estimate the apple's bruise depth using thermal map. The findings of the multiple regression model revealed that the temperature of each apple zone can be utilized to estimate the bruise in different regions on fruit surface. The application of optical images has revealed to have its own challenges but showed to be the promising method for detection of bruises in pome fruit.

6.5 Current Status/Trends

Mechanical damage is the major reason for the losses because of the mechanical impact that is caused by different causes as described earlier. Reducing mechanical damages can elevate the shelf-life of the product and decrease microbial invasion to the barest minimum. An evaluation of different sources of loss in pome fruit is important if the quality of the produce is to be guaranteed. It is necessary to evaluate the bruising level that occurs in fresh pome fruit under a particular condition during a postharvest handling operation (Unuigbe & Onuoha, 2013). Several current studies have been carried out to show the impact of different techniques to reduce mechanical damage in pome fruit.

In logistics, packaging plays a very important role in isolating vibration and absorbing energy to decrease the energy shifted to fruit (Wang et al., 2022). Fadiji et al. (2016b) conducted a simulated transport study to evaluate the susceptibility of apple fruit inside different ventilated corrugated paperboard packages (aka. MK4 and MK6). These packages are commonly utilized for packaging fresh apples in industries. They stated that appropriate packaging and careful transportation and handling practices can minimize fresh pome fruit loss that can be generated because of mechanical damages. Also, they recommended the application of cushioning materials that are beneficial and economical in reducing damage suffered by fresh fruit. The outcomes of their research can be of great significance to packaging handlers and designers of different fresh fruit at various distribution and transportation phases. This can minimize damage and losses caused by mechanical damage particularly because of injury generated by vibration and might also ensure that fresh fruits and vegetables are supplied to the consumers with good quality.

A study by Wang et al. (2022) assessed the effect of mechanical damage of 'Huangguan' pear as subjected to random vibration as affected by vibration level, package unit type, and vibration time. They found that keeping the fruit in the middle of the package can decrease the incidence of bruise damage in pear fruit. Expandable polyethylene (EPE) net + expanded polystyrene (EPS) tray net cover provided the optimal cushioning effect for pear fruit. Nowadays, several techniques are applied to keep the biological materials throughout transport, involving the utilization of a special truck supplied with containers and a particular vibration-damping system that contains specific soft material which links the biological material as stated by Stopa et al. (2018).

Yan et al. (2022) used cross-linked CS that was prepared by cross-linking reaction with diepoxy-poly (ethylene glycol) (PEG) and applied it for the purpose of coating protection against vibration in injured pear fruit. They found that cross-linked CS-PEG films were highly successfully prepared and utilized as coating preservation in pears damaged by vibration, particularly at room temperature. The films maintain the physical appearance and provide a remarkable hardness in damaged pears. Also, Shu et al. (2020) studied the influence of chlorogenic acid (CGA) treatment on the energy metabolism and physiological indicators of apple fruit damaged by vibration stored at 23 °C for more than 28 days. They recorded that the treatment with chlorogenic acid decreased the physiological quality alternations generated by vibration consequences in apples. They also found that CGA treatment declines the quality of apple fruit loss after being damaged by vibration by reducing softening and delaying the reduction of soluble solid contents, electrolyte leakage, titratable acidity, and the accumulation of malondialdehyde. In another study by Shu et al. (2022), they concluded that ensuring adequate energy levels can be considered a promising approach for monitoring the adverse physiological changes generated by vibration and as a potential technique to control the quality of the fruit.

Besides, Du et al. (2020) mentioned that the imaging techniques listed earlier exhibited potential to determine the damaged tissue, bruise area, and volume in real time. Also, imaging techniques are used to overcome tediousness, fallibility, time consumption, and destructiveness that are the most common difficulties faced in conventional approaches. For future research, they stated that combining these methods, developing particular algorithms for the recognition of bruises, and trying to utilize them for commercial value might be the research hotpots. Besides, Li and Thomas (2014) recommended the use of logistic regression modeling of mechanical damages which can be improved by linking it with hyperspectral imaging, nuclear magnetic resonance, ultrasonic techniques, or X-rays. Current sensory assessment techniques are slightly subjective and the usage of objective, new tools can provide a better and clearer understanding of how handling produces bruising.

6.6 Conclusion

The need to supply high-quality fresh produce without cuts, bruises, blemish, and other physiological disorders is essential for consumers. Horticultural products like pome fruit are highly susceptible to mechanical damage during postharvest handling resulting in a subsequent reduction in the fruit quality. Bruising is a common form of mechanical damage in pome fruit. This damage has been simulated in laboratories by using an impact drop test, pendulum test, vibration test, and compression test. Manual measurements and image processing techniques have been implemented to describe the degree of damage in pome fruit. Using modeling that is combined with non-destructive techniques can help to estimate the intensity of the damage. Applying advanced techniques, proper packaging units during transport, and suitable treatments after transport can help to minimize the damage and delay the quality alterations of fresh pome fruit.

Acknowledgements We would like to thank Sultan Qaboos University, Oman for their financial support under the project codes: CL\SQU\QU\AGR\23\02 and IG/AGR/SWAE/19/03.

References

Ahmadi, E. (2012). Bruise susceptibilities of kiwifruit as affected by impact and fruit properties. *Research in Agricultural Engineering, 58*, 107–113.

Alamar, M. C., Vanstreels, E., Oey, M. L., Molto, E., & Nicolai, B. M. (2008). Micromechanical behaviour of apple tissue in tensile and compression test: Storage conditions and cultivar effect. *Journal of Food Engineering, 86*, 324–333.

Al-Dairi, M., Pathare, P. B., Al-Yahyai, R., & Opara, U. L. (2022). Mechanical damage of fresh produce in postharvest transportation: Current status and future prospects. *Trends in Food Science and Technology, 124*(1), 195–207.

Azadbakht, M., Vahedi Torshizi, M., & Mahmoodi, M. J. (2018). The use of CT scan imaging technique to determine pear bruise level due to external loads. *Food Science & Nutrition, 7*(1), 273–280.

Azadbakht, M., Mahmoodi, M. J., & Vahedi, T. M. (2019). Effects of different loading forces and storage periods on the percentage of bruising and its relation with the qualitative properties of pear fruit. *International Journal of Horticultural Science and Technology, 6*(2), 177–188.

Babarinsa, F. A., & Ige, M. T. (2012). Strength parameters of packaged Roma tomatoes at breakpoint under compressive loading. *International Journal of Scientific and Engineering Research, 3*, 240–247.

Barchi, G. L., Berardinelli, A., Guarnieri, A., Ragni, L., & Totaro Fila, C. (2002). Damage to loquats by vibration-simulating intra state transport. *Biosystems Engineering, 82*, 305–312.

Berardinelli, A., Donati, V., Giunchi, A., Guarnieri, A., & Ragni, L. (2005). Damage to pears caused by simulated transport. *Journal of Food Engineering, 66*, 219–226.

Bollen, A. F., Cox, N. R., Dela Rue, B. T., & Painter, D. J. (2001). A descriptor for damage susceptibility of population of produce. *Journal of Agricultural Engineering Research, 78*(4), 391–395.

Cañete, M. L., Hueso, J. J., Pinillos, V., & Cuevas, J. (2015). Ripening degree at harvest affects bruising susceptibility and fruit sensorial traits of loquat (*Eriobotrya japonica* Lindl.). *Scientia Horticulturae, 187*, 102–107.

Celik, H. K., Ustun, H., Erkan, M., Rennie, A. E., & Akinci, I. (2021). Effects of bruising of 'Pink Lady' apple under impact loading in drop test on firmness, colour and gas exchange of fruit during long term storage. *Postharvest Biology and Technology, 179*, 111561.

Clark, C. J., & Macfall, J. S. (2003). Quantitative magnetic resonance imaging of 'Fuyu' persimmon fruit during development and ripening. *Magnetic Resonance Imaging, 21*, 679–685.

Dang, H. Q., Kim, I., Cho, B. K., & Kim, M. S. (2012). Detection of bruise damage of pear using hyperspectral imagery. In *2012 12th International Conference on Control, Automation and Systems (ICCAS)* (pp. 1258–1260). IEEE.

Devaux, M. F., Barakat, A., Robert, P., Bouchet, B., Guillon, F., Navez, B., & Lahaye, M. (2005). Mechanical breakdown and cell wall structure of mealy tomato pericarp tissue. *Postharvest Biology and Technology, 37*, 209–221.

Doosti-Irani, O., Golzarian, M. R., Aghkhani, M. H., Sadrnia, H., & Doosti-Irani, M. (2016). Development of multiple regression model to estimate the apple's bruise depth using thermal maps. *Postharvest Biology and Technology, 116*, 75–79.

Du, Z., Zeng, X., Li, X., Ding, X., Cao, J., & Jiang, W. (2020). Recent advances in imaging techniques for bruise detection in fruits and vegetables. *Trends in Food Science and Technology, 99*, 133–141.

Eissa, A., Alghannam, A., & Azam, M. (2012). Mathematical evaluation changes in rheological and mechanical properties of pears during storage under variable conditions. *Journal of Food Science and Engineering, 2*, 564–575.

Fadiji, T., Coetzee, C., Chen, L., Chukwu, O., & Opara, U. L. (2016a). Susceptibility of apples to 397 bruising inside ventilated corrugated paperboard packages during simulated transport damage. *Postharvest Biology and Technology, 118*, 111–119.

Fadiji, T., Coetzee, C., Pathare, P., & Opara, U. L. (2016b). Susceptibility to impact damage of apples inside ventilated corrugated paperboard packages: Effects of package design. *Postharvest Biology and Technology, 111*, 286–296.

FAOSTAT. (2022). *Food and agriculture organization corporate statistical database*. Rome, Italy: FAO.

Ferrari, C., Foca, G., Calvini, R., & Ulrici, A. (2015). Fast exploration and classification of large hyperspectral image datasets for early bruise detection on apples. *Chemometrics and Intelligent Laboratory Systems, 146*, 108–119.

Fu, L., Liu, S. Y., & Xu, F. X. (2017). Effect of treatment with MeJA on physiology and biochemistry in blueberry fruit. *Journal of Shenyang Normal University (Natural Science Edition), 35*, 473–475.

Fu, H., Du, W., Yang, J., Wang, W., Wu, Z., & Yang, Z. (2023). Bruise measurement of fresh market apples caused by repeated impacts using a pendulum method. *Postharvest Biology and Technology, 195*, 112143.

Gancarz, M. (2016). Correlation between cell size and blackspot of potato tuber parenchyma tissue after storage. *Postharvest Biology and Technology, 117*, 161–167.

Gharaghani, A., Solhjoo, S., & Oraguzie, N. (2016). A review of genetic resources of pome fruits in Iran. *Genetic Resources and Crop Evolution, 63*, 151–172.

Gołacki, K., Bobin, G., & Stropek, Z. (2009). Bruise resistance of apples (melrose variety). *Teka Komisji Motoryzacji i Energetyki Rolnictwa 9*, 40–47

Hussein, Z., Fawole, O. A., & Opara, U. L. (2019). Bruise damage susceptibility of pomegranates (*Punica granatum* L.) and impact on fruit physiological response during short term storage. *Scientia Horticulturae, 246*, 664–674.

Jackson, J. E. (2003). *The biology of apples and pears*. Cambridge University Press.

Kim, G., Kim, G.-H., Park, J., Kim, D.-Y., & Cho, B.-K. (2014). Application of infrared lock-in thermography for the quantitative evaluation of bruises on pears. *Infrared Physics & Technology, 63*, 133–139.

Komarnicki, P., Stopa, R., Szyjewicz, D., & Młotek, M. (2016). Evaluation of bruise resistance of pears to impact load. *Postharvest Biology and Technology, 114*, 36–44.

Komarnicki, P., Stopa, R., Kuta, L., & Szyjewicz, D. (2017a). Determination of apple bruise resistance based on the surface pressure and contact area measurements under impact loads. *Computers and Electronics in Agriculture, 142*, 155–164.

Komarnicki, P., Stopa, R., Szyjewicz, D., Kuta, Ł., & Klimza, T. (2017b). Influence of contact surface type on the mechanical damages of apples under impact loads. *Food and Bioprocess Technology, 10*(8), 1479–1494.

Li, Z., & Thomas, C. (2014). Quantitative evaluation of mechanical damage to fresh fruits. *Trends in Food Science and Technology, 35*(2), 138–150.

Lu, F., Ishikawa, Y., Kitazawa, H., & Satake, T. (2010). Impact damage to apple fruits in commercial corrugated fiberboard box packaging evaluated by the pressure-sensitive film technique. *Journal of Food, Agriculture and Environment, 8*(2), 218–222.

Lu, F., Ishikawa, Y., Kitazawa, H., & Satake, T. (2012). Assessment and prediction of repetitive impact damage to apple fruit using pressure-sensitive film technique. *Journal of Food, Agriculture and Environment, 10*, 156–160.

Lu, Y., Saeys, W., Kim, M., Peng, Y., & Lu, R. (2020). Hyperspectral imaging technology for quality and safety evaluation of horticultural products: A review and celebration of the past 20-year progress. *Postharvest Biology and Technology, 170*, 111318.

Mahanti, N. K., Pandiselvam, R., Kothakota, A., Ishwarya, P., Chakraborty, S. K., Kumar, M., & Cozzolino, D. (2021). Emerging non-destructive imaging techniques for fruit damage detection: Image processing and analysis. *Trends in Food Science and Technology, 120*, 418–438.

Menesatti, P., & Paglia, G. (2001). Development of drop damage index of fruit resistance to damage. *Journal of Agricultural Engineering Research, 80*(1), 53–64.

Menesatti, P., Beni, C., Paglia, G., Marcelli, S., & Dandrea, S. (1999). Predictive statistical model for the analysis of drop impact damage on peach. *Journal of Agricultural Engineering Research, 73*, 275–282.

Miller, R. A. (2003). Harvest and handling injury: Physiology biochemistry and detection. In J. A. Bartz & J. K. Brecht (Eds.), *Postharvest physiology and pathology of vegetables* (pp. 177–208). Marcel Dekker.

Nguyen, C.-N., Van-Linh, L., Le, P.-H., Ho, H.-T., & Nguyen, C.-N. (2022). Early detection of slight bruises in apples by cost-efficient near-infrared imaging. *International Journal of Electrical and Computer Engineering, 12*(1), 349–357.

Opara, U. L. (2007). Bruise susceptibilities of 'Gala' apples as affected by orchard management practices and harvest date. *Postharvest Biology and Technology, 43*, 47–54.

Opara, U. L., & Fadiji, T. (2018). Compression damage susceptibility of apple fruit packed inside ventilated corrugated paperboard package. *Scientia Horticulturae, 227*, 154–161.

Opara, U. L., & Pathare, P. B. (2014). Bruise damage measurement and analysis of fresh horticultural produce-a review. *Postharvest Biology and Technology, 91*, 9–24.

Opara, U. L., Al-Ghafri, A., Agzoun, H., Al-Issai, J., & Al-Jabri, F. (2007). Design and development of a new device for measuring susceptibility to impact damage of fresh produce. *New Zealand Journal of Crop and Horticultural Science, 35*, 245–251.

Pan, X., Sun, L., Li, Y., Che, W., Ji, Y., Li, J., Li, J., Xie, X. U., & Xu, Y. (2019). Non-destructive classification of apple bruising time based on visible and near-infrared hyperspectral imaging. *Journal of the Science of Food and Agriculture, 99*(4), 1709–1718.

Pang, W., Studman, C. J., & Ward, G. T. (1992). Bruising damage in apple-to-apple impact. *Journal of Agricultural Engineering Research, 52*, 229–240.

Pathare, P. B., & Al-Dairi, M. (2021). Bruise susceptibility and impact on quality parameters of pears during storage. *Frontiers in Sustainable Food Systems, 5*, 658132.

Pathare, P. B., Opara, U. L., Vigneault, C., Delele, M. A., & Al-Said, F. A. J. (2012). Design of packaging vents for cooling fresh horticultural produce. *Food and Bioprocess Technology, 5*, 2031–2045.

Rasli, M. A., Nawi, N. M., Ahmad, D., & Yahaya, A. (2019). The effect of crop parameters on mechanical properties of oil palm fruitlets. *Scientia Horticulturae, 250*, 352–358.

Razavi, M. S., Asghari, A., Azadbakh, M., & Shamsabadi, H.-A. (2018). Analyzing the pear bruised volume after static loading by Magnetic Resonance Imaging (MRI). *Scientia Horticulturae, 229*, 33–39.

Rostampour, V., & Motlagh, A. M. (2018). Evaluation of the bruising susceptibility of apple in transport conditions. *Bulgarian Journal of Agricultural Science, 24*(5), 902–908.

Salarikia, A., Miraei Ashtiani, S.-H., Golzarian, M. R., & Mohammadinezhad, H. (2017). Finite element analysis of the dynamic behavior of pear under impact loading. *Information Processing in Agriculture, 4*, 64–77.

Sargent, S. A., Brecht, J. K., & Zoellner, J. J. (1992). Sensitivity of tomatoes at mature-green and breaker ripeness stages to internal bruising. *Journal of American Society for Horticultural Science, 117*, 119–123.

Shu, C., Zhang, W., Zhao, H., Cao, J., & Jiang, W. (2020). Chlorogenic acid treatment alleviates the adverse physiological responses of vibration injury in apple fruit through the regulation of energy metabolism. *Postharvest Biology and Technology, 159*, 110997.

Shu, C., Cao, J., & Jiang, W. (2022). Postharvest vibration-induced apple quality deterioration is associated with the energy dissipation system. *Food Chemistry, 386*, 132767.

Stopa, R., Szyjewicza, D., Komarnickia, P., & Kuta, Ł. (2018). Determining the resistance to mechanical damage of apples under impact loads. *Postharvest Biology and Technology, 146*, 79–89.

Stropek, Z., & Gołacki, K. (2013). The effect of drop height on bruising of selected apple varieties. *Postharvest Biology and Technology, 85*, 167–172.

Stropek, Z., & Gołacki, K. (2015). A new method for measuring impact related bruises in fruits. *Postharvest Biology and Technology, 10*, 131–139.

Stropek, Z., & Gołacki, K. (2019). Impact characteristics of pears. *Postharvest Biology and Technology, 147*, 100–106.

Stropek, Z., & Gołacki, K. (2020). Bruise susceptibility and energy dissipation analysis in pears under impact loading conditions. *Postharvest Biology and Technology, 163*, 111120.

Stropek, Z., & Gołacki, K. (2022). Response of apple flesh to compression under the quasi-static and impact loading conditions. *Materials, 15*(21), 7743.

Sun, Z., Hu, D., Xie, L., & Ying, Y. (2022). Detection of early stage bruise in apples using optical property mapping. *Computers and Electronics in Agriculture, 194*, 106725.

Tan, W., Sun, L., Yang, F., Che, W., Ye, W., Ye, D., & Zhang, D. (2018). The feasibility of early detection and grading of apple bruises using hyperspectral imaging. *Journal of Chemometrics, 32*, e3067.

Unuigbe, O., & Onuoha, S. (2013). Assessment of impact damage to apple fruits. *Nigerian Journal of Technology, 32*(1), 137–140.

Van Linden, V., Sila, D. N., Duvetter, T., De Baerdemaeker, J., & Hendrickx, M. (2008). Effect of mechanical impact-bruising on polygalacturonase and pectin methylesterase activity and pectic cell wall components in tomato fruit. *Postharvest Biology and Technology, 47*, 98–106.

Van Zeebroeck, M., Darius, P., De Ketelaere, B., Ramon, H., & Tijskens, E. (2007a). The effect of fruit properties on the bruise susceptibility of tomatoes. *Postharvest Biology and Technology, 45*(2), 168–175.

Van Zeebroeck, M., Van Linden, V., Darius, P., De Ketelaere, B., Ramon, H., & Tijskens, E. (2007b). The effect of fruit factors on the bruise susceptibility of apples. *Postharvest Biology and Technology, 46*, 10–19.

Van Zeebroeck, M., Lombaert, G., Dintwa, E., Ramon, H., Degrande, G., & Tijskens, E. (2008). The simulation of the impact damage to fruit during the passage of a truck over a speed bump by means of the discrete element method. *Biosystems Engineering, 101*(1), 58–68.

Wang, L.-J., Zhang, Q., Song, H., & Wang, Z.-W. (2022). Mechanical damage of 'Huangguan' pear using different packaging under random vibration. *Postharvest Biology and Technology, 187*, 111847.

Xing, J., Bravo, C., Moshou, D., Ramon, H., & De Baerdemaeker, J. (2006). Bruise detection on 'Golden Delicious' apples by vis/NIR spectroscopy. *Computers and Electronics in Agriculture, 52*, 11–20.

Yan, X., Dong, T., Yun, X., Liu, B., & Meng, J. (2022). Enhancing the functionality of cross-linked chitosan coating on vibration damaged Nanguo pears. *LWT, 162*, 113454.

Yousefi, S., Farsi, H., & Kheiralipour, K. (2016). Drop test of pear fruit: Experimental measurement and finite element modelling. *Biosystems Engineering, 147*, 17–25.

Zeng, X., Miao, Y., Ubaid, S., Gao, X., & Zhuang, S. (2020). Detection and classification of bruises of pears based on thermal images. *Postharvest Biology and Technology, 161*, 111090.

Zhang, S., Wu, X., Zhang, S., Cheng, Q., & Tan, Z. (2017). An effective method to inspect and classify the bruising degree of apples based on the optical properties. *Postharvest Biology and Technology, 127*, 44–52.

Chapter 7
Mechanical Damage in Fresh Stone Fruits: Measurement and Analysis

Ebrahim Ahmadi and Rashid Gholami

Abstract Stone fruits are some of the important agricultural products, which are widely used by humans. These products are highly sensitive during harvest and post-harvest processing. Therefore, like other agricultural products, knowing and having information about the physical, mechanical, and chemical properties of stone fruits are necessary. The physical properties include shape, size, geometrical mean diameter, arithmetic mean diameter, surface area, sphericity, and color indices, mechanical properties comprise penetration, impact, compression and cutting forces, elastic modulus, deformation, and chemical properties like pH, TSS, TA, antioxidant activity, etc. Knowing about these and many other properties will help researchers and farmers improve the quality and quantity of their products. One of the parameters which describe stone fruits' quality is bruise damage. The bruise damage will occur because of the impacts during harvesting, handling, transporting, and sorting. Therefore, having information about the behavior of the fruits under different impact forces will help us to know how we can manage and design the equipment for these steps.

Keywords Stone fruits · Mechanical damage · Bruising · Mechanical properties

7.1 Introduction

Stone fruits get their name from the pit or "stone" in their center that is encased in a fleshy outer area. Stone fruits often referred to as drupes typically have thin, fuzzy, or smooth skins. The most popular stone fruits, which have a significant economic

E. Ahmadi (✉)
Department of Biosystem Engineering, Faculty of Agriculture, Bu-Ali Sina University, Hamedan, Iran
e-mail: eahmadi.eahmadi@basu.ac.ir

R. Gholami
Sonqor Agriculture Faculty, Department of Agricultural Machinery Engineering, Razi University, Kermanshah, Iran

P. B. Pathare, U. L. Opara (eds.), *Mechanical Damage in Fresh Horticultural Produce*, https://doi.org/10.1007/978-981-99-7096-4_7

Fig. 7.1 Some stone fruits

impact on the entire world, are peach, nectarine, plum, and apricot (Fig. 7.1). They are also known as high-perishable and short-market life fruits (Manganaris & Crisosto, 2020).

Stone fruits can be eaten fresh or processed, and some, including plums, apricots, and peaches, can be dried. They contain various organic and inorganic substances that benefit the human diet and health. Water, carbohydrates like sugars, organic acids, fiber, proteins, minerals (Ca, P, Fe, and K), and significant amounts of non-nutritive or bioactive substances, including β-carotene, lutein + zeaxanthin, phenols, flavonoids, anthocyanins, and other vitamins (A, B_1, B_2, K, E (-α-tocopherol), niacin, and pantothenic acid) are available in all stone fruits (Milosevic & Milosevic, 2020).

7.2 Physical Properties of Stone Fruits

Knowing and having information about the physical properties of stone fruits such as other agricultural products is necessary for all agricultural processes. Shape, size, volume, geometrical mean diameter, arithmetic mean diameter, surface area, density, porosity, sphericity, color indices, and appearance are some of the physical properties which are important to solve problems related to the design and optimization of agricultural machines, also analysis of the behavior of the fruits and agricultural products during harvest and post-harvest process. For example, knowing the main dimensions, areas, and even the shape of the desired product play a major role in the design of harvesting and sorting machines, design of packaging, and transportation conditions (Mohsenin, 1986). The physical properties of stone fruits are important for the design of equipment for harvesting and post-harvesting like: storing, transporting, cleaning, separating, sorting, sizing, packaging, and processing it into different food. Some currently used systems have been generally designed without considering these; therefore, the resulting designs lead to inadequate applications. The design of equipment without having information about these criteria results in decreased efficiency and an increase in product loss. Therefore, determination, consideration, and evaluation of these criteria have an important role in designing this equipment for stone fruits and controlling mechanical damages.

7.3 Mechanical Properties of Stone Fruits

In addition to knowing the physical properties, it is very important to know the mechanical properties of agricultural products. Mechanical properties are defined as those which affect the behavior of the agricultural and food material under the applied force, most of the time. The mechanical properties are known by the elastic modulus, penetration force, hardness, compressive strength, impact, and shear resistance, and the rheological properties affect the various operations of agricultural and food processing. Having data of these properties will be useful and practical for application in designing equipment for milling, handling, storage, transportation, food processing, etc.

Stone fruits are known for being tender fruits, so physical and mechanical damage should be considered during the harvest and afterward. One of the main concerns of stone fruit producers on the farm is mechanical damage caused by the mechanization of various production and post-production operations, which must be avoided to maintain product quality during harvest and post-harvest. If fruit mechanical damages decrease, farmers' profits can rise (Ahmadi et al., 2014).

For vegetables and fruits, mechanical damages include bruising, crushing, or rupturing. One of the most significant challenges to producing high-quality fruit and vegetables is bruising. The fruits' direct mechanical damage determines the degree of bruising. In the harvesting, processing, transporting, and sorting of fruits and vegetables, fruit bruising is a significant obstacle preventing mechanization and automation. The process of harvesting and post-harvest products is an important factor in reducing the quality and quantity of fruit and has a negative effect on the profitability of the entire fruit industry (Abedi & Ahmadi, 2014). Because the items had previously made forceful mechanical contact with others, dark spots may emerge close to the product surface. The ratio of the volume of the bruise to the loading energy is known as the susceptibility of bruises (Polat et al., 2012).

Fruit bruising can be caused by an impact, vibration, and compression load. The distance the fruit falls, the impact force, the number of impacts, the type of surface, and the size and maturity of the fruit all affect how much damage occurs. Improving harvesting, handling, transporting, sorting, and equipment require understanding the impact conditions that can cause bruises (Ahmadi & Abedi, 2013). Bruising threshold depends on the specimen tissue failing stress, impact-induced stress, elastic modulus, mass, and radius of curvature of the fruit (Baritelle & Hyde, 2001). Although impact sensitivity (bruise susceptibility) in fruit and vegetables varies widely, there is currently no reliable method for determining, regulating, and monitoring this sensitivity (Ahmadi et al., 2010).

When the tissue is damaged above the bio-yield point, the failure of the outer pericarp due to cell rupture or fracture between cells results in external bruising. However, hits below the bio-yield point can still result in internal bruising if the impact energy is transferred to the underlying tissue with enough force to prevent ripening, as seen by a subsequent aberrant color development and increased gel viscosity (Ahmadi & Abedi, 2013). A measure of how well fruit and vegetables respond to external loading, bruise susceptibility depends on many factors, including maturity, variety, water status, firmness, storage condition (like temperature), size, shape, and some internal fruit aspects, such as cell wall strength and elasticity, cell shape, and internal structure (Ahmadi et al., 2010).

Two groups of independent variables, primary and secondary, could be used in bruise prediction models. Variables relating to impact characteristics termed primary (principal) variables include peak, duration of the impact, contact force, impact energy, and drop height. Secondary variables have to do with fruit qualities, including temperature, firmness, storage time, harvest date, and radius of curvature. Both impact energy and peak contact force (PF), all with advantages and disadvantages, are the primary independent variables in the statistical models. PF models have the advantage of being able to generalize for materials with various properties and radii of curvature. Since the impact energy, in this case, is equivalent to that of fruit-impacting metal, impact energy models created for fruit-impacting metal cannot be applied to a fruit impacting another fruit. PF regression models have the limitation that the PF is quite certainly influenced by the fruit parameters themselves (temperature, ripeness, etc.). However, the impact energy is unaffected by the fruit characteristics, making impact energy models more appropriate to study how fruit factors affect bruise susceptibility (Abedi & Ahmadi, 2014).

Bruise volume is typically used to define bruise extension. The peach ripeness has effect on bruise susceptibility and it is the only factor taken into account by statistical models for peach bruise prediction. The impacting velocity, mass, radius of curvature, elastic modulus, and Poisson ratio of an elastic sphere impacting a rigid surface determine the force response. It has been discovered that the hardness of a fruit directly affects the impact force response and that the impact of an elastic sphere may accurately simulate the impact of a fruit on a rigid surface. It would seem that some factors like the mass of fruits, absorbed energy during impact, and impact velocity the latter produced by the viscoelastic structure of fruit tissue would all affect bruising (Ahmadi et al., 2010).

Some experts describe the quality of fruits and vegetables as the collection of characteristics or features that make them valuable food for humans. External forces can harm plant tissue, which results in changes to the crop's physical and chemical characteristics (color, flavor, and texture).

Fresh produce's appearance, shape, and textural qualities influence consumer perception of its quality, influencing consumer purchasing decisions. Customers choose a higher-quality product that is free of pathogens, physiological problems, and bruising. Fruit can absorb mechanical energy from sources produced during harvest and distribution processes. While the fruit can absorb a certain amount of energy without damage, its tissue often dissipates the acquired energy. Failure of fruit tissue is influenced by the kind, severity, and duration of energy. The most significant variables influencing bruise size are impact energy and susceptibility. Numerous studies have discovered strong connections between impact energy and the degree of the bruise. Absorbed energy is one of the most important parameters which can be used to quantify bruise damage in fruits. Fruit bruise damage can be evaluated using the amount of energy absorbed. It can be said that the energy that is absorbed at the highest rate does the most damage. There is an excellent amount of information and research that have been published about the pre-harvest and post-harvest factors that cause fruits and stone fruits to bruise, including techniques for measuring and analyzing bruise damage, some of which are in the following (Ahmadi et al., 2014; Fadiji et al., 2016).

7.3.1 Apricot

Peaches and nectarines are similar to apricots, but apricots are often smaller in size. They have a tart flavor but a rich, creamy texture. Due to the pectin-rich nature of apricot skin, these fruits are frequently used to make jam or for drying. The scientific name of the apricot is *Prunus armeniaca* from the *Rosaceae* family. Apricots are an excellent source of beta-carotene, which turns into vitamin A in the body and helps the health of the eyes, skin, hair, mouth, etc. and helps the body to fight infections. This fruit is consumed dried, compote, and fresh. Its core contains amygdalin, which is toxic when consumed in large quantities. Although apricot is known as a natural supplement, consuming too much of its kernel is harmful. Do not consume more than 1 or 2 of these kernels during the day because it will cause stomach problems and nausea, and consuming too much amygdalin is fatal. (The bitter pith of the apricot kernel is harmful, while the sweet pith is useful and tonic.) Iran is one of the five main producers of apricots in the world (Turkey, Uzbekistan, Italy, Algeria, and Iran), the amount of Apricot cultivation in Iran was about 60,307 hectares, and the amount of production from this cultivation is estimated at 323,019 tons (FAO, 2021).

Stone fruits like apricots are dense, the little space between their cells and the sensitive texture of this product lead them to bruise. Fruit can absorb mechanical energy from sources produced during harvest and distribution processes. While the fruit can absorb a certain amount of energy without damage, its tissue often dissipates the acquired energy. Failure of fruit tissue is influenced by the variety, intensity, and duration of force applied to the sample (Ahmadi et al., 2014).

Apricot tissue failure like other stone fruits depends on the variety, intensity, and duration of force applied to the sample. Some researchers described tissue and the dissipation of energy of fruits during impact tests as a regular arrangement of spherical cells which filled with liquid limited by viscoelastic membranes with air-filled interstitial spaces. At the first step of the compression test, under stress, internal cells will deform from spheres to ellipsoids, and cell divider break would be caused by continuing and more compression force; meaning at high-shearing push locales cells will tend to burst. This phenomenon explains the dissipative energy of the system.

In general, it can be said that mechanical damage is one of the main concerns of apricot producers and farmers. Evaluation of mechanical properties and damages is necessary to ensure about agricultural production and maintain their quality during post-harvest processes like transportation. If fruits' mechanical damages reduce, farmers' profits could be increased.

Many shocks and impacts are inflicted on the agricultural product during the post-harvest processes. Impact energy and bruise susceptibility are the foremost critical components influencing bruise measure. Numerous analysts found noteworthy connections between bruise seriousness and the impact energy that the product endures during the post-harvest period (Diener et al., 1982; Klein, 1987). Absorbed energy can be used to evaluate bruise damage in stone fruits. The results of recent researches showed that the most damage is caused by the highest absorbed energy (Van Zeebroeck et al., 2007). For example, quality indices for two apricot varieties, through the survey method of the acoustic impulse responses and by measuring color parameters were defined by Petrisori et al., 2010. Generally, food coloring can be measured using a spectrophotometer and based on the L^* a^* b^* color system.

The variety of fruits is one the most important factors, which has a significant effect on the physical and mechanical properties. The apricot of the "Ziaolmolki" variety, which is cultivated in Hamedan-Iran, was used in the research entitled " The Effect of Fruit Properties on the Apricot Bruises Susceptibility" by Ahmadi et al. (2014).

To conduct tests on products such as apricots, which are very sensitive, the method of harvesting is very important, in such a way, the product must be harvested carefully so that no damage is caused to the apricot during harvesting. In order to perform laboratory tests, it is suggested that the samples be harvested manually and carefully and transported to the laboratory under special conditions (placed in a plastic container equipped with a shock absorber) to perform the tests. This is done to minimize any damage and bruising to the sample before the main tests. Care must be taken that the samples selected for experiments are not too large or too small. It is suggested that samples within a certain size range (length, width, and thickness), weight be used for performing mechanical tests because size can be one of the most influential factors in changes in the mechanical properties of agricultural products. To perform tests on agricultural products due to the sensitivity and also respiration of these products, in order to reduce possible errors, the product storage conditions and the environmental conditions during the test (in terms of temperature and humidity of the environment) should be controlled. If the samples have been kept in the refrigerator before the tests, at the first step, it is better that their temperature reaches the equilibrium with the ambient temperature and then the tests are performed.

Another thing that is important and necessary in conducting experiments on agricultural products, including apricots, is the repeatability of experiments.

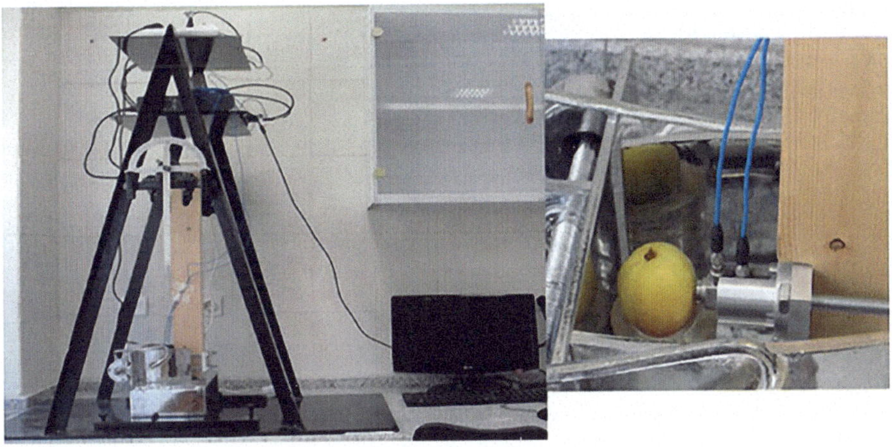

Fig. 7.2 General view of the pendulum device for measuring impact force and impact velocity of the apricot (Ahmadi et al., 2014)

Therefore, it is suggested to conduct experiments with at least three repetitions for each treatment to minimize the error of the experiment.

In performing the impact test, it is important to perform the impact force that is applied to the sample, in other words, the force should be selected in a way that will produce the desired effect on the sample. If the force is low, it will not affect the fruit, and if the force is too high, it will cause rapid damage to the product at the very beginning and will prevent further investigations. For example, Ahmadi et al. (2014) selected three levels of impact for the testing of apricot. At impact level 1 (Impact Energy = 0.029 J and Impact Force = 20 N) the recognition of bruising was little difficult because it was located on the boundary of the bruise detection. But, in two other (2 and 3) levels, bruising was clearly seen and its recognition was not difficult. The value of the energy levels selected for experiments was a little more than the one obtained during mechanical harvest, handling, and transportation so no impact was recorded at level 3.

The use of prediction models can significantly help in reducing the time and costs of conducting tests. In this regard, models can be used to predict mechanical parameters related to apricots. It should be noted that the model must be reliable and first validated and have a high correlation coefficient with experimental data. In the research done by Ahmadi et al. (2014), they introduced models for apricots. In their bruise prediction models, absorbed energy was considered as the dependent variable. The absorbed energy is obtained by the difference between impact energy and rebound energy. According to Ahmadi et al. (2014) kinetic energy before and after the pendulum arm impact (Fig. 7.2) can be used to calculate the impact and elastic energy, respectively. Based on their research the following equations were used:

Table 7.1 Variance analysis of absorption energy as affected by temperature, color, and impact force (Ahmadi et al., 2014)

Source	Mean square
Color	0.027[a]
Temperature	0.587[a]
Impact energy	0.002[a]
Color × temperature	0.000[a]
Color × impact energy	0.146[a]
Temperature × impact energy	0.001[a]
Color × temperature× impact energy	0.001[a]

[a]Significant at 5% level

$$A_{\text{impact}} = E_{\text{kin}}(t - 1)$$

$$A_{\text{elastic}} = E_{\text{kin}}(t + 1)$$

$$A_{\text{absorbed}} = E_{\text{impact}} - E_{\text{elastic}}$$

In the models of bruise estimation, impact energy or peak contact force will be defined as the independent variable. Other independent variables as the inputs to the regression models will be as follows:

Impact energy (E_{i}) (J), peak contact force (PF) (N), apricot temperatures (T) (3 and 20 °C), curvature radius of apricot at the impacted place (R) (mm), apricot acoustical stiffness (S) ($\mathrm{s}^{-2}\ \mathrm{kg}^{2/3}$), apricot color measurements by the colorimetric, parameter b^* (as an indicator of ripeness). The radius of curvature should measure locally at the location of contact by a radius of curvature meter.

Fruit characteristics, including stiffness, temperature, color, and radius of curvature, and impact qualities, including impact energy and peak contact force, were assessed. The apricot's absorbed energy (AE) is defined as a dependent variable in bruise prediction models. Peak contact force (PF), the radius of curvature of the hit apricot (R), apricot color (b^*), and acoustic stiffness were all independent variables in the first model (S). In the second model, energy (E) replaced the role of the impact contact force.

The results of the research on apricot showed that factors such as temperature, color, and impact force had a significant effect on the absorption energy. The research has shown that interaction effects (color × temperature, color × energy, and temperature × energy) have a significant effect on the amount of energy absorbed by apricots in the impact test (as shown in Table 7.1).

Doing an impact test on the apricot was shown that, immediately after the test, the damage caused by this impact is not visible to the eye and only its partial deformation can be seen from a very close distance. But after 24 h, the affected area will gradually change color and will gradually turn brown and start to become softer. The results of the evaluation of this research showed that bruising probability and lesion incidence increased by increasing impact levels. Various models have been proposed to describe this issue.

In the models presented to predict absorbed energy, the main effects of factors like temperature, acoustic stiffness, colorimetric parameter b^*, and radius of curvature were found, as well as some interactions between impact properties (peak impact force or impact energy) and fruit factors (temperature, acoustic stiffness, color parameter b^*, and radius of curvature) at 3 °C. Reported that the absorbed energy increased (when impacted with the same impact energy) with the increase in the ripeness of the apricots, also, with lower stiffness S and an increase colorimetric parameter b^*, the bruise susceptibility to apricot increases. Unlike apricots at 3 °C, those at 20 °C absorb more energy. A smaller radius of curvature enhances energy absorption, increasing susceptibility to bruises.

One of these models which is used to describe bruise estimation models with peak contact force as an independent variable is as follows:

$$\begin{aligned} \mathrm{AE} = 0.0905 + 5.4 \times 10^{-3} \times \mathrm{PF} + 2.217 \times 10^{-5} \times b^* + 0.8659 \times R \\ +8.39 \times 10^{-4} \times S + 0.04 \times \mathrm{PF} \times R – 5.11 \times 10^{-5} \times \mathrm{PF} \times S \\ +8.6 \times 10^{-6} \times \mathrm{PF} \times T \end{aligned}$$

where PF is peak contact force, R is the curvature radius of the apricot at the impacted place, b^* is the color index, S is apricot acoustical stiffness, and T is temperature.

In this model, the absorption energy is predicted by the peak contact as the independent variable. The Ripeness index is important so acoustic stiffness (S) and parameter b^* index are considered as the apricot ripeness index. The results showed that this model with a high correlation coefficient can be used to predict the energy absorbed by apricots. In the figure below, the measured values of absorbed energy are plotted against the values predicted by this model (Fig. 7.3).

One of the other model was presented to evaluate the bruise susceptibility of apricot, such that all main effects exist except temperature.

$$\begin{aligned} \mathrm{AE} = 0.0226 + 1.311 \times E_{\mathrm{i}} + 1.0917 \times 10^{-4} \times b^* + 0.217 \times R + 3.762 \times 10^{-4} \times S \\ - 8.377 \times E_{\mathrm{i}} \times R - 0.0014 \times E \times S + 0.0035 \times E_{\mathrm{i}} \times T \end{aligned}$$

These two models that were presented above are an example of the models related to the absorbed energy in the impact test, in different conditions and considering various other parameters, many suitable models can be made for different conditions and varieties of apricot.

7.3.2 Peach

One of the most popular stone fruits is the peach (*Prunus Persica*), which has a large pit and furry skin. Peaches quickly spread to Europe, where they were once known as "Persian apples". Various types of peaches are grown all over the world, including

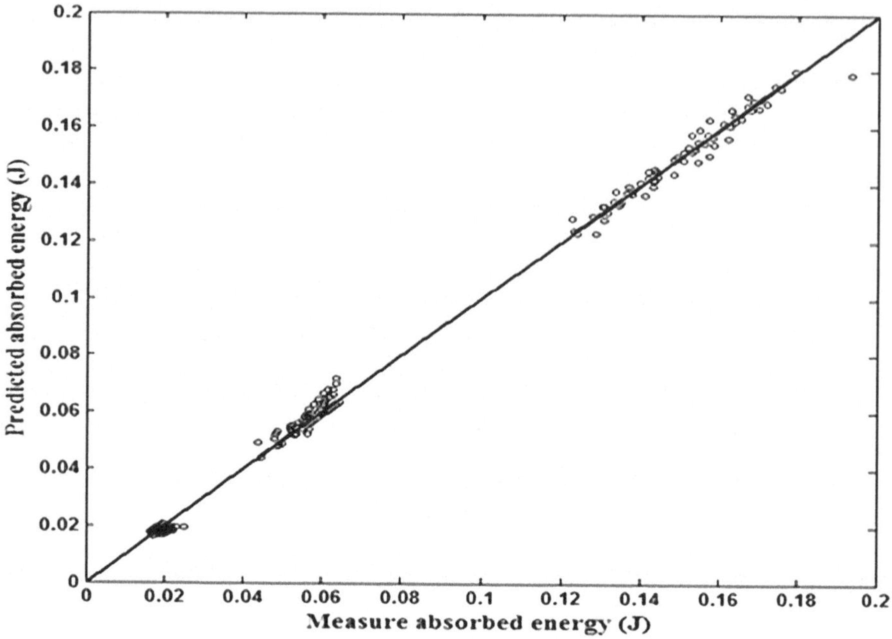

Fig. 7.3 Measured absorbed energy vs. predicted absorbed energy (Ahmadi et al., 2014)

Haj Kazemi, which is widely cultivated in Iran. According to the FAO (2021), the amount of Peaches and Nectarine cultivation in Iran were about 38,700 hectares, and the amount of production from this cultivation is estimated at 68,7213 tons. Peaches are rich in many vitamins and minerals, including vitamin B, potassium, and iron. A small peach can provide 15% of the daily vitamin C your body needs. The properties of peaches are very effective in skin care and bone strength.

Like other stone fruits, there are three factors that can physically cause peach bruising: impact, vibration, and compression load. The severity of the damage to the fruit depends on the distance the fruit falls during transportation, the amount of energy of the impact, the number of impacts, the surface on which the impact is applied, and the size and maturity of the fruit. There is some research done on peaches to evaluate the physical, mechanical, chemical, and other properties of this fruit. Among them, research has been done to predict and model the mechanical behavior of this product. The model for prediction of peach bruise connects the impact characteristics like drop height, and peak contact force, with bruise damage taking into consideration some fruit properties like temperature, and ripeness, determining the bruise sensitivity. It has been determined in peach fruit reviews that the impact of a fruit on a rigid surface can be closely modeled by the impact of an elastic sphere and that the firmness of a fruit has a direct effect on the impact force

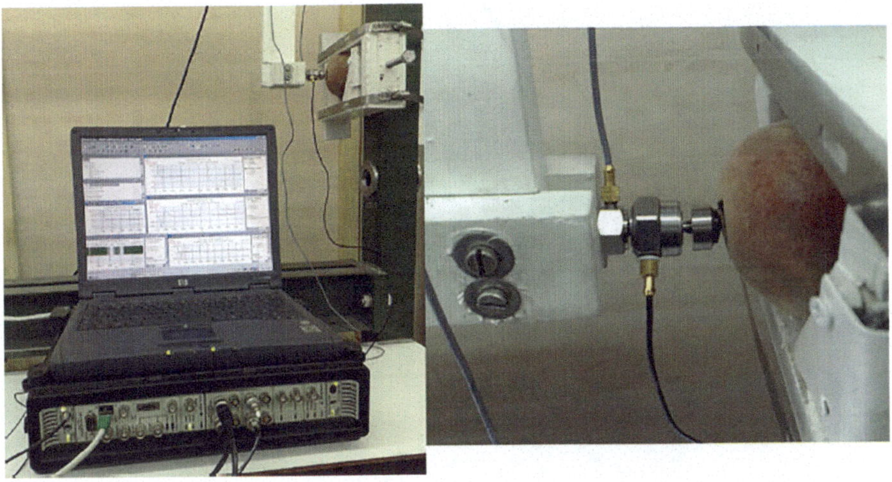

Fig. 7.4 General view of the pendulum device for measuring impact force and impact velocity of the peach fruit (Ahmadi et al., 2010)

response. In the research conducted to model the impact on peaches, we can refer to the research conducted by Ahmadi et al. (2010).

The important point in conducting tests on agricultural and food products is to consider the fact that conducting tests on these products depends on the test conditions and the type of product. In such a way that the temperature changes of the test environment, the laboratory equipment used, the variety of the product, the amount of product maturity, and even the time of day when the test is being conducted can have a significant effect on the results obtained. Therefore, the difference in the results reported by different researchers on the same product will not be far from expected.

Another thing that should be taken into consideration when performing tests on agricultural materials is that the temperature of the sample has reached an equilibrium with the ambient temperature. For this reason, it is suggested that the sample be kept at ambient temperature for a few hours before performing the test so that the temperature balance is established during the test.

Ahmadi et al. (2010) investigated the development of peach bruise prediction models using impact energy or peak contact force along with fruit characteristics like fruit temperature, acoustic stiffness, and radius of curvature of independent variables. The variety used in their research was *Haj Kazemi*.

As shown in Fig. 7.4, their research was done by using a pendulum equipped with a force sensor (type AC20, AP Tech, Netherlands; sensitivity: 1.87 mV N^{-1}) and an incremental optical encoder (RON 275, Heidenhain; resolution: 0.0050) to measure impact energy and impact velocity and then was impacted by a spherical steel indenter ($R = 25$ mm).

Impact energy (E) (J) and peak contact force (PF) (N) as independent variables, peach temperatures (T) (°C), peach radius of curvature (R) at the location of impact

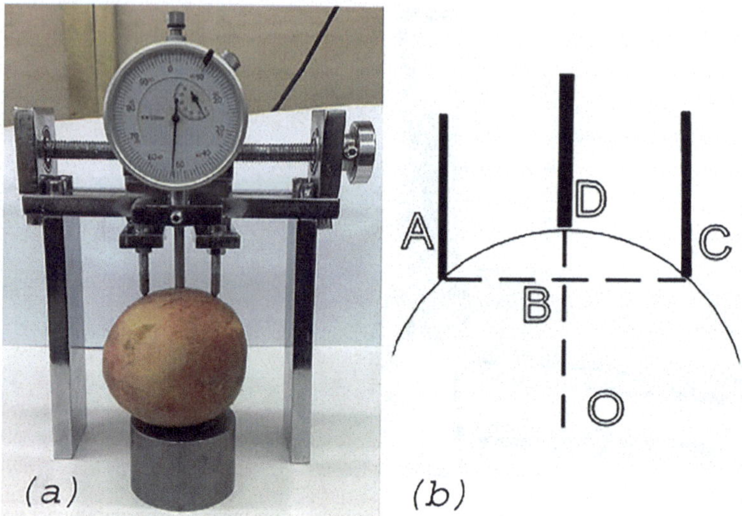

Fig. 7.5 (**a**) General view of the curvature meter and (**b**) schematic representation of geometry to calculate the radius of curvature of the peach fruit (Ahmadi et al., 2010)

(mm), and peach acoustic stiffness (S) (s^{-2} kg $^{2/3}$) as input variables were used for bruise prediction models of peach. It can be used to measure the radius of curvature according to the equation provided by Mohsenin (1986) and the device shown in Fig. 7.5.

$$\text{Radius} = \frac{(\text{AC})^2}{8(\text{BD})} + \frac{(\text{BD})}{2}$$

They presented a bruising model with impact level and fruit properties to describe the process.

The following model was presented by them as a model in which all the factors are included and these factors and their mutual effects have a significant effect at the level of 5% on the amount of bruising. Regression equation of bruise volume (mm^3) of the peach fruit (V) in relation to peak contact force (PF), temperature (T), acoustic stiffness (S), and radius of curvature (R) as independent variables.

$$V = 50.877 + 68.914 \times \text{PF}$$
$$+21.069 \times T – 57.751 \times S – 9.683 \times R – 0.353 \times \text{PF} \times T$$
$$+1218 \times \text{PF} \times S$$

The correlation coefficient of the values obtained from this model vs. the experimental values was 97%, which indicates a good fit between the measured and modeled data (Fig. 7.6).

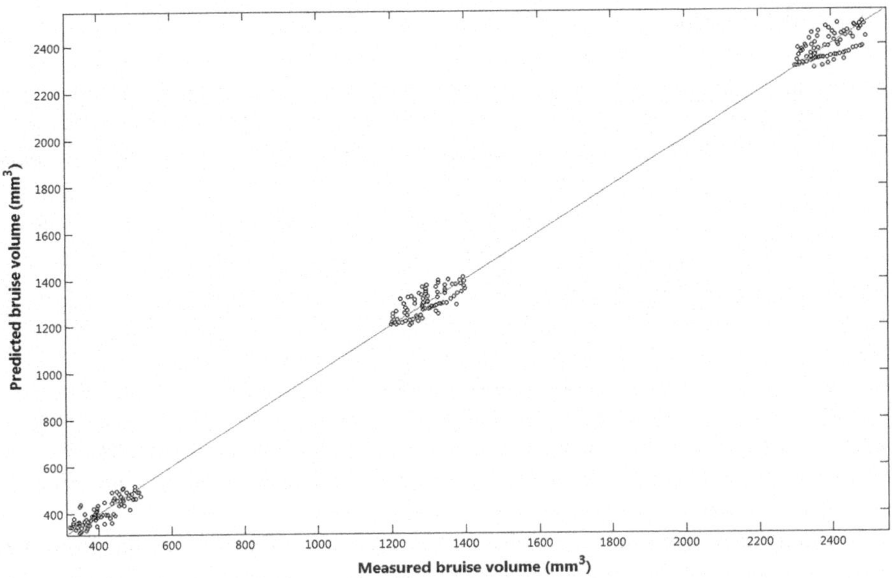

Fig. 7.6 Measured bruise volume (mm^3) vs. bruise volume predicted by model 1 (Ahmadi et al., 2010)

Another model provided by them is as follows: Regression equation of bruise volume (mm^3) of the peach fruit (V) in relation to impact energy (E), temperature (T), acoustic stiffness (S), and radius of curvature (R) as independent variables:

$$V = 530.88 + 24,516 \times E + 9.836 \times T – 20.102 \times S – 1.218 \times R – 97.142 \times E \times T$$

The severity of a peach's bruising damage depends significantly on the impact level and some interactions with fruit properties. According to models, susceptible peaches had smaller radii, were stored at high temperatures, and were at an advanced stage of maturity. A bruise was much more likely to occur when fruit was impacted with a high impact level. High peak contact stresses are a result of the high impact level. The following is a summary of how the peach properties affect bruise volume: (1) Increased bruising was discovered to be a result of peach temperature. (2) The increase in acoustic stiffness resulted in a decrease in bruise volume. (3) Smaller curvature radii caused more bruising damage. (4) There was no discernible difference between the predicted bruise volume and the peak contact force and impact energy of the models.

7.3.3 Nectarine

The scientific name of the nectarine is *Prunus persica var. nucipersica* which is a tree from the Rosaceae family. It is a type of peach that was created by a genetic mutation in one of the peach genes. The nectarine is a type of peach with smoother skin (as a result of mutation, it does not have fuzzy skin, and a special color, aroma, and taste have been created in it) that is found in the warmer regions of the northern and southern hemispheres. The properties of nectarine were discovered more than 4000 years ago in China, and it is interesting to know that peach and nectarine trees are practically similar. The main difference between these two fruits is the absence of plant hairs on the nectarine skin. This fruit has smooth skin and a large hard core. It usually has a sour taste and its flesh may be white, yellow, green, or orange in color. In addition to its pleasant taste, this fruit has special properties. There are different varieties of nectarines around the world, and the most common types in Iran, each of which has different properties, include: Moghan, Red Gold or Shebarang, Shams, and Green nectarine.

Normally, nectarines are manually picked and then quickly removed from the field. From the time the fruit is harvested until it is eaten, physical and mechanical damage might happen at any time. Polat et al. (2012) examined the mechanical characteristics of nectarine fruit and its susceptibility to bruising. As shown in Fig. 7.7, they used a pendulum impactor with a 50-cm arm length for impact tests. The following equation was used to calculate the energy absorption (difference between energy at impact and rebound):

$$E_{\mathrm{a}} = mg(h_1 - h_2)$$

where E_{a} is the energy absorbed in N mm, m is the mass of the fruit in kg, g is a gravitational constant in m s^{-2}, h_1 is drop height in mm, and h_2 is rebound height in mm.

To determine the degree of bruising both in terms of appearance and in terms of measuring the volume of the bruised area, it is better to keep the sample at ambient temperature for 24 h after performing the impact test and then checking the impact area.

Absorbed energy, bruise volume, and bruise susceptibility values were determined by dropping nectarine fruits from 30-, 40-, and 50-cm heights onto a metal surface by Polat et al. (2012).

Their findings showed that nectarines that experienced a hit from the cheek side had larger bruises and were more susceptible to bruises. In contrast, nectarines that sustained an impact from the bottom side were more resistant to impacts. It suggests that a nectarine's various portions have distinct skin thicknesses and tissue structures. A reduction in the percentage of damage, particularly in post-harvest mechanization, was achieved due to the research's identification of nectarines' more sensitive points to mechanical influences.

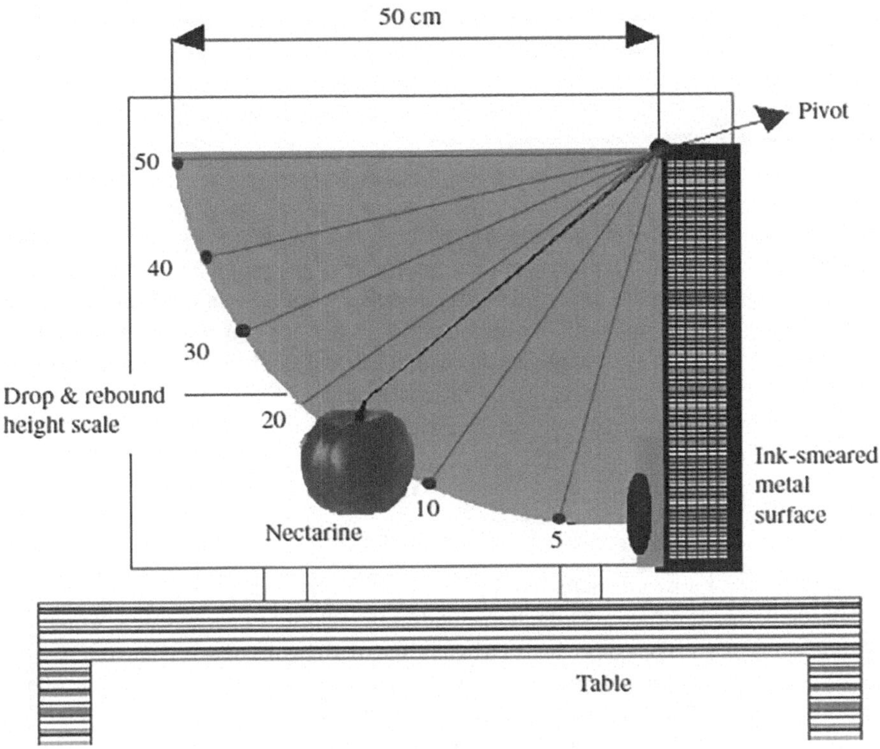

Fig. 7.7 Pendulum impactor for impact tests of the nectarines (Polat et al., 2012)

7.3.4 Plum

The scientific name of the plum is *Prunus domestica* and is from the *Rosaceae* family. Plums are one of the summer fruits that have many nutrients including fiber and antioxidants. You can use the properties of plums to treat constipation and diabetes and even prevent heart disease and cancer. Plums have antioxidant, anti-inflammatory, and memory-improving properties. This summer fruit contains phenols, especially anthocyanin, which are a type of antioxidant. Plum consumption is associated with improved cognitive function, bone health, and heart function. This fruit also has a low glycemic index, so eating it does not increase blood sugar levels.

One of the most significant stone fruit harvests in the world is the plum. More than 2000 different types of plums exist, although only a few are substantial from a commercial perspective. They are grown in the temperate zone, where the major suppliers of plums are China, Romania, and the USA (Birwal et al., 2017).

The sensitive tissue of plums, like that of other members of the stone fruit family, should be taken into account during the harvesting and post-harvesting processes. Plums' quality and commercial worth could be lost due to mechanical damage.

Therefore, it is vital to conduct studies on evaluating and examining the primary factors responsible for plum mechanical damage.

7.3.5 Statistical Analysis

In research and experiments on food and agricultural products, such as the evaluation of physical, mechanical, and chemical properties, the application of statistical software in analyzing the experimental is necessary and essential. There are some software programs in this field, including SPSS, SAS, Design Expert, Minitab, and even MATLAB. It is also suggested to use statistical analysis such as analysis of variance, compare means, and correlation coefficient.

7.4 Conclusion

Stone fruits are known as sensitive and vulnerable products; therefore, bruising is one of the most important factors in evaluating the quality of stone fruits, which can be important and effective in the marketability of these products. For this reason, investigating the factors of bruising of fruits and providing solutions to control and minimize it in post-harvesting processing will ultimately increase the marketability and profitability of the product for farmers and producers. In this section that was studied, an attempt was made to investigate the various aspects and factors on the amount of bruising of stone fruits (apricots, peaches, nectarines, and plums) by evaluating the researches that have been carried out in this field. Different models that have been presented by researchers to predict bruises and the factors affecting bruises were evaluated. The final results show that the models are satisfactory in predicting the amount of bruising under the influence of various factors. Therefore, it is suggested to use statistical models to predict bruise values, which will save time and money.

References

Abedi, G., & Ahmadi, E. (2014). Bruise susceptibilities of Golden Delicious apples as affected by mechanical impact and fruit properties. *Journal of Agricultural Science, 152*, 439–447.

Ahmadi, E., & Abedi, G. (2013). Sensitivity of tomatoes to internal bruising induced by mechanical stress. *European Journal of Horticultural Sciences, 78*(5), 219–224.

Ahmadi, E., Ghasemzadeh, H. R., Sadeghi, M., Moghaddam, M., & Zarifneshat, S. (2010). The effect of impact and fruit properties on the bruising of peach. *Journal of Food Engineering, 97*, 110–117.

Ahmadi, E., Barikloo, H., & Soleimani, B. (2014). The effect of fruit properties on the apricot bruises susceptibility. *Journal of Food Measurement and Characterization, 8*, 46–53.

Anonymous. (2021). *Food and agricultural organization of the United Nations statistics division*. Retrieved from http://FAOstat3.fao.org/download/Q/QC/E.

Baritelle, A., & Hyde, G. M. (2001). Commodity conditioning to reduce impact bruising. *Postharvest Biology and Technology, 21*, 331–339.

Birwal, P., Deshmukh, G., Saurabh, S. P., & Pragati, S. (2017). Plums: A brief introduction. *Journal of Food, Nutrition and Population Health, 1*(1), 1–5.

Diener, R. G., Singha, S., & Petit, j. (1982). The effect of apple maturity and firmness on bruise volume. *Journal of Food Process Engineering, 6*, 85–99.

Fadiji, T., Coetzee, C., Pathare, P., & Opara, U. L. (2016). Susceptibility to impact damage of apples inside ventilated corrugated paperboard packages: Effects of package design. *Postharvest Biology and Technology, 111*, 286–296.

Klein, J. D. (1987). Relationship of harvest date, storage conditions, and fruit characteristics to bruise susceptibility of apple. *Journal of the American Society for Horticultural Science, 112*, 113–118.

Manganaris, G. A., & Crisosto, H. C. (2020). *Stone fruits: Peaches, nectarines, plums, apricots* (pp. 311–322). Elsevier.

Milosevic, T., & Milosevic, N. (2020). Soil fertility: Plant nutrition vis-à-vis fruit yield and quality of stone fruits. In A. K. Srivastava & C. Hu (Eds.), *Fruit crops: Diagnosis and management of nutrient constraints* (pp. 583–606). Elsevier.

Mohsenin, N. N. (1986). *Physical properties of plant and animal materials*. Gordon and Breach Science Publishers.

Petrisori, C., Radu, G. L., Balan, V., & Campeanu, G. (2010). Rapid and nondestructive analytical techniques for measurement of apricot quality. *Romanian Biotechnological Letters, 15*(2), 5213–5216.

Polat, R., Aktas, T., & Ikinci, A. (2012). Selected mechanical properties and bruise susceptibility of nectarine fruit. *International Journal of Food Properties, 15*, 1369–1380.

Van Zeebroeck, M., Van Linden, V., Darius, P., De Ketelaere, B., Ramon, H., & Tijskens, E. (2007). The effect of fruit properties on the bruise susceptibility of tomatoes. *Postharvest Biology and Technology, 45*, 168–175.

Chapter 8
Bruise Damage Susceptibility of Pomegranates

Umezuruike Linus Opara, Zaharan Hussein, and Olaniyi Fawole

Abstract During the past couple of decades, commercial cultivation of pomegranate (*Punica granatum* L.) fruit has become more prevalent throughout semi-temperate and sub-tropical fruit regions of the world. Pomegranate consumption has increased noticeably, in part due to the distinctive sensory and nutritional qualities of the fruit, as well as its potential health and medicinal advantages from its high concentration of antioxidants and health-promoting phytonutrients. Nonetheless, poor storage and handling practices are causing fruit quality loss in the global pomegranate industry. Mechanical damage, specifically bruising that occurs between harvesting and consumption, has become a major contributing factor to the declining market value and quality loss of many fruits, including pomegranates. This chapter will provide an overview of the pomegranate industry, including its global significance, production, trade, consumption, and nutritional value. The coverage has been expanded to include pomegranate fruit bruising—the causes, mechanisms of bruising, and economic significance. This chapter has also shed light on the factors that contribute to pomegranate fruit bruise damage, from preharvest to harvest to postharvest. This chapter also covers the measurement and analysis of pomegranate fruit damage, with a focus on the methods used to investigate into pomegranate fruit damage, with emphasis on methodologies to simulate fruit

U. L. Opara (✉)
SARChI Postharvest Technology Research Laboratory, Africa Institute for Postharvest Technology, Faculty of AgriSciences, Stellenbosch University, Stellenbosch, South Africa

UNESCO International Centre for Biotechnology, Nsukka, Enugu State, Nigeria
e-mail: opara@sun.ac.za

Z. Hussein
Department of Food Science and Technology, College of Agricultural Sciences and Technology, Mbeya University of Science and Technology, Mbeya, Tanzania

Centre for Innovation and Technology Transfer, Mbeya University of Science and Technology, Mbeya, Tanzania

O. Fawole
Postharvest Research Laboratory, Department of Botany and Plant Biotechnology, University of Johannesburg, Johannesburg, South Africa

P. B. Pathare, U. L. Opara (eds.), *Mechanical Damage in Fresh Horticultural Produce*, https://doi.org/10.1007/978-981-99-7096-4_8

bruising, analysis, and measurement procedures, as well as the assumptions made to fit the real-time settings. The review acknowledged recent advancements in operations and handling equipment in the pomegranate industry's modern harvest and postharvest handling system. As a result, future research directions have been outlined in this chapter, including the development of novel techniques to simulate bruising and, later, near-accurate bruise measurement methods to enable real-time simulation and quantification of bruise damage both in situ and ex situ.

Keywords Pomegranate fruit (*Punica granatum L.)* · Bruise damage · Postharvest loss · Bruise measurement · Mechanical damage · Fruit quality

8.1 Introduction

8.1.1 Global Importance of Pomegranates: Production, Trade, Consumption, and Nutritional Value

Pomegranate (*Punica granatum* L.) cultivation has spread throughout much of the world as a commercial fruit. Currently, pomegranate trees are grown in sizable commercial orchards all over the world, from the southern hemisphere's tropics to the northern hemisphere's Mediterranean basin (Hussein et al., 2020b; Opara et al., 2009; Al-Said et al., 2009). Pomegranate production has increased rapidly worldwide; however, commerce has become more localized in the major producing nations (Wang et al., 2020; Opara et al., 2021a, b). Due to the varying seasons of production, pomegranate fruit production is split mostly into two major regions (Northern Hemisphere and Southern Hemisphere) (Rymon, 2011; POMASA, 2019). Countries such as India, China, Iran, and Turkey are the leading producers in the northern hemisphere (accounting for more than 90% of total production) and thus have a larger share of global trade (Rymon, 2011).

There is no comprehensive global data on production of pomegranates, their derivatives and commerce. The reason is that the fruit is regularly produced and gathered from tiny farms around the world, including major producing and exporting countries (India, China, and Iran). The pomegranate production in these countries is characterized by garden setting, where pomegranate fruit is harvested from small, scattered plots with a small number of trees (Kahramanoglu, 2019; Rymon, 2011). Notwithstanding, global output was projected to accelerate from roughly three million tonnes in 2014 to 3.8 million tonnes in 2017 (Kahramanoglu, 2019), and this trend is expected to continue. If production increases by 0.8 to one million tonnes every 3 years, the rough estimate for 2022 production could be 7.8–8.8 million tonnes.

Pomegranate fruit is becoming increasingly popular and in demand around the world due to its multifunctionality and health-promoting effects in the human diet (Magangana et al., 2020). The fruit is consumed fresh or used to make fresh juice, canned beverages, jelly, jam, and paste, as well as flavouring and colouring beverage products (Viuda-Martos et al., 2010; Mousavinejad et al., 2009). Over 500 cultivars are grown worldwide, with some cultivars having different names in different parts of the world (Fawole & Opara, 2013g; Al-Said et al., 2009).

The worldwide pattern of rising demand for pomegranates and its co-products is rising owing to their numerous nutritional and health benefits (Opara et al., 2021a, b; Pareek et al., 2015). Pomegranate juice is rich in vitamins, minerals, and other organic compounds and can be sweet, sour, or acidic depending on the cultivar (Pareek et al., 2015; Fawole & Opara, 2013g). Several studies have connected pomegranate intake to improved health outcomes (de Oliveira et al., 2020; Mphahlele et al., 2016; Faria & Calhau, 2011). The phenolic compounds of pomegranate juice have been found to be excellent anti-inflammatory, anti-oxidative, and anti-carcinogenic chemical compounds, aiding in the reduction of tumour growth and chronic inflammation (Wang et al., 2020; Opara et al., 2017). Furthermore, the hypoglycaemic activity of the pomegranate juice has been shown to aid in the prevention of diabetes mellitus (Katz et al., 2007). Consumption of pomegranate juice has been linked to a lower risk of cardiovascular disease (Wang et al., 2020).

Due to the action of its chemical components, pomegranate fruit are also utilized to heal ulcers, haemorrhage, acidosis, aphthae, diarrhoea, respiratory diseases, dysentery, and microbiological infections (Opara et al., 2009, 2021a, b; Mphahlele et al., 2016; Middha et al., 2013). These active components are also widely used in therapeutic formulations, cosmetics, and food seasonings. Pomegranate aril and peel, which contain such compounds, are used as raw materials in the production of jams, ink, dye, and oil in this regard (Middha et al., 2013).

8.1.2 Pomegranate Fruit Losses, Waste, and Perishability During Postharvest Handling

Fruits and vegetables have a high postharvest waste and loss percentage owing to their high moisture content (usually $> 80\%$), high respiration rate, and susceptibility to mechanical damage during postharvest handling (Opara et al., 2021a, b). In the context of pomegranate fruit industry, postharvest loss refers to fruit with reduced quality due to defects, resulting in downgrading and loss of value, which is typically sold at a low price (hence economic loss) for fresh consumption or used for industrial purposes. Consequently, farmers suffer losses in return on investment due to postharvest losses because fruit is not sold at the best price due to quality defects.

Despite advancements in packaging technology (Ambaw et al., 2021, 2022; Lufu et al., 2021; Mukama et al., 2020a, b, c, 2021) and postharvest handling of pomegranate fruit (Fawole & Opara, 2013a, b, c, d, e, f), postharvest losses and waste have increased (POMASA, 2019). This is due to a variety of pre- and postharvest factors, including disease attack (Munhuweyi et al., 2016), bruise damage (Hussein et al., 2019a, b; 2020a, b), and high rates of respiration and transpiration, resulting in significant weight loss as well as quality loss (Lufu et al., 2019, 2020).

Fig. 8.1 Quality defects in pomegranate fruit that are commonly associated with postharvest losses; superficial injury (**a**), bruises (**b**), sunburn (**c**), blemishes (**d**), crack (**e**), insect damage (**f**), crown rot (**g**), decay (**h**), and Alternaria (**i**). (Modified from Opara et al., 2021a, b)

Pomegranate fruit losses and waste are commonly assessed using three criteria: (1) quality defect issues caused by environmental factors, (2) mechanical damage, and (3) pests and diseases (Opara et al., 2021a, b; Baker et al., 2019). In the study by Opara et al. (2021a), sunburn and cracking were identified as the two major causes of physical loss in pomegranate fruit (Fig. 8.1). Furthermore, the authors revealed that Crown rot, Alternaria, insect damage, blemishes, decay, injury, undersize and misshapen, oversize and bruise are all potential causes of both preharvest and postharvest waste.

Site-level measurements are so difficult and expensive that the reporting of the scale of postharvest losses and their causes often rely on estimates instead. Opara et al. (2021a) assessed the quantity of pomegranate fruit postharvest losses and waste at the packhouse level, reasons for wastage, and the socioeconomic and environmental consequences. Their findings revealed that environmental stress and mechanical damage were the main causes of pomegranate fruit loss. These findings corroborate the findings of numerous other studies on postharvest loss, which established that mechanical damage caused mostly by harvesting technique is a major source of loss at the packhouse (Weerakkody et al., 2010; Bond, 2016; Bantayehu et al., 2017; Bhattarai et al., 2017). According to the authors, improper harvesting may cause superficial injuries, opening wounds for decay, and disease infestation.

Several other causes of postharvest losses have been reported in a number of studies. Reports suggest that most of the losses originate from the farm level but due to oversight from the on-farm fruit sorters, the defects are identified during the packhouse operations (Opara et al., 2021a, b). Details of various causes of postharvest losses and their magnitude are presented in Table 8.1.

8.2 An Overview of Pomegranate Fruit Bruising

8.2.1 Pomegranate Fruit Bruising: Causes and Economic Importance

Harvest and postharvest activities of fresh fruit (such as postharvest handling that includes sorting, grading, packing, and transportation) necessitate extensive mechanical operations (Hussein et al., 2020a, b; Fu et al., 2023). The actions of static and dynamic forces caused by handling operations predispose the fruits to mechanical damage (Opara, 2007; Montero et al., 2009). The terms ‘plastic deformation’, ‘superficial rupture’, and/or ‘destruction’ refer to mechanical harm to plant tissue. According to Polat et al. (2012), the latter includes bruising, crushing, rupture, and abrasion. The majority of mechanical damage can be brought on by rough or incorrect handling, inadequate supervision during fruit handling, bad packaging, or poorly constructed harvesting and handling equipment (Polat et al., 2012: Mohammad Shafie et al., 2017).

Bruising of fresh fruits during harvest and postharvest processing is one of the critical issues in the horticultural industry. According to estimates, 30–40% of fruits and vegetables are mechanically damaged from harvesting to market delivery (Kabas, 2010). Bruise damage alters physiological and metabolic processes, resulting in faster ripening, internal browning, and quality losses (Opara & Pathare, 2014; Hussein et al., 2019a, b; Pathare & Al-Dairi, 2021). Bruise damage causes fruit browning, weight loss, and unsightly shrivelling, all of which result in economic losses (Ergun, 2017; Xia et al., 2020; Pathare & Al-Dairi, 2021). Hussein et al. (2020b) found that bruised pomegranate fruit lost 40% of its weight during long cold storage. Fruit weight, as one of the important quality attributes in fruit production, influences not only consumer preference but also the marketing of fresh fruit (Pathare & Al-Dairi, 2021). Furthermore, impact bruising causes both qualitative internal and minor external changes in fresh produce, such as changes in citric acid content (Hussein et al., 2020b), soluble solid content (Gao et al., 2021), respiration,

Table 8.1 Major causes of pomegranate fruit postharvest losses

Causes of postharvest loss	Specific cause of loss	Remarks
1. Environmental stress	(a) *Sunburn*	Ranked as the highest cause of loss (Opara et al., 2021a)
	'Losses due to sunburn at the pack house originate from the farm where pomegranate fruit are exposed to direct sunlight, which causes discolouration of the rind of the affected fruit, hence downgrading the fruit quality'	
	(b) *Cracks and splits*	
	• Mostly originates from the farm, but spotted in the packhouse especially when fruit sorting at the farm level was not effective	Opara et al. (2021a) reported cracks and splits as the third ranked cause of postharvest loss of pomegranate fruit
	• Increased sunburn would also lead to more cracks and splits because direct sunlight causes the rinds of fruits to harden, which aids cracking and splitting when the moisture content fluctuates	Opara et al. (2021a), Ghanbarpour et al. (2019), Opara et al. (1997)
2. Mechanical and physical damage	(a) *Superficial injuries*	After sunburn, it is the second most common reason for pomegranate fruit loss in the packing plant (Opara et al., 2021a)
	'Minor injuries originating from preharvest and handling technique at the farm level that result into subsequent losses'	
	At the farm level, these types of injuries are typically regarded as unimportant, but when the affected fruit are packed, they do not match market requirements	
	(b) *Bruise damage*	
	• Originated from mechanical damage as a consequence of collision during harvesting, transportation, and handling	Hussein et al. (2020b), Hussein et al. (2018)
	• Bruised fruit normally does not fulfil export or local market criteria, thus it is sold at a reduced price for industrial usage	Ranked the fourth cause of postharvest fruit loss (Opara et al., 2021a)
	(c) *Blemish* • Induced by mechanical damage preceding and following trimming prior to picking pomegranate fruit. It could be caused by a wind-blown scratch against a sharp tree branch, placing blemish markings on the afflicted fruit • Blemish is a significant issue in evaluating the quality of pomegranate fruit for both the export and domestic markets (Czieczor et al., 2018)	Ranked as one of the minimal common reasons of pomegranate fruit loss (Opara et al., 2021a)

3. Biological damage	*Insect damage*	
	• Insect damage occurs when insects consume a tiny amount of the fruit, resulting in a minimal loss of the damaged fruit and a drop in pomegranate fruit quality (Opara et al., 2021a)	Ranked among the least frequent causes of pomegranate fruit loss (Opara et al., 2021a)
	• High incidence of insect damage occurs at farm level. Damaged fruit are discarded at farm, while few that reach at packhouse due to sorters' oversight fail to meet market standards	Insect infestation occurs because bruising softens the damaged area of the fruit skin, making it vulnerable to insect assault (Opara et al., 2021a)
4. Microbial and pathological spoilage	(a) *Decay and rots*	Ranked as one of the least frequent causes of pomegranate fruit loss (Opara et al., 2021a)
	• Microbial pathogens induce decay and rot by causing the rind of the infected fruit to break down, causing minor or total deterioration and rot (Hussein et al., 2019a, b)	
	• Alternaria is a pathogenic disease of pomegranate fruit triggered by Alternaria alternata. The illness enables fruit to decay from the inside out, despite the fact that the diseased fruit's peel seems healthy (Munhuweyi et al., 2016)	
	(b) *Crown rot*	
	• Crown rot is initiated by *Coniella granati*, a fungal infection that particularly affects farm-grown pomegranate fruit. The rind of the infected fruit has pycnidia with rotting crown	Crown rot has a little impact on overall fruit loss (Munhuweyi et al., 2016; Opara et al., 2021a)
	• Farm diseases like crown rot, which do not start in the packhouse, can spread; if affected fruit is found in the packhouse, it could be due to a sorting mistake at the farm	Decay and crown rot pathogens result from wounds created from fruit that had been injured superficially had scratches and wounds on the rind that were different sizes, shapes, and depths (Martinez-Romero et al., 2004)
	(c) *Postharvest fungal diseases* • Pathogens from the genus Penicillium and B. cinerea are the most common pathogens causing postharvest decay of pomegranates (Palou et al., 2010, 2013)	When healthy and contaminated fruit are packaged together, the disease infections can spread through cross contamination. Diseases are frequently linked to postharvest storage losses in pomegranates but are frequently disregarded because of their gradual onset and delayed symptom onset (Munhuweyi et al., 2016)

and ethylene production (Xia et al., 2020), all of which lead to fruit quality loss due to ageing, senescence, and subsequent decay.

8.2.2 Mechanism of Bruise Damage

For most fresh fruits, including pomegranates, bruising is the most frequent sort of mechanical damage that can happen during harvesting, handling, and transportation (Mohammad Shafie et al., 2017; Hussein et al., 2020b). Cell breaking brought on by excessive external force occurring on the fruit surface during collision, compression, or vibration against a rigid body or fruit against fruit results in bruise damage, a kind of fresh produce subcutaneous tissue failure (Kitthawee et al., 2011; Opara & Pathare, 2014; Hussein et al., 2020b). During loading on fruit by impact or compression, bruising will only occur when the stress induced on the fruit surface exceeds the failure stress of the fruit tissue (Opara & Pathare, 2014).

On a biological product, mechanical impact or compression, also known as loading, may create mechanical stress that leads to cell wall and membrane rupture and, as a result, bruising (Ahmadi, 2012). According to Opara & Pathare (2014) and Ahmadi (2012), bruise susceptibility is a measure of a product's reaction to external loading. As a result, the amount of dynamic or static loading that is given to a produce is thought to be the most crucial aspect in bruising, and this loading or absorbed energy is often described in terms of loading (Blahovec, 2006).

All impacts that are likely to happen during harvesting and handling activities, such as fruit crashing into picking buckets or whilst sorting, and also vibrations that predominantly happen during transportation, are considered to be a part of dynamic loading (Komarnicki et al., 2016). Static or compression loading, however, can occur during harvesting, transportation, or storage partly due to poorly designed bins (without cushioned surface) overfilling or stacking, causing the produce in the lower bins to support the weight, potentially causing damage (Lewis et al., 2007; Komarnicki et al., 2016).

Impacts and compression both permanently modify the internal and external components, particularly the fruit peel's mechanical deformation and the flesh cells' spatial deformation. When a cell membrane ruptures following impact or compression, the cytoplasmic enzymes are released into the intercellular spaces, where they react with vacuolar contents (Mitsuhashi-Gonzalez et al., 2010), leading to colour change of damaged tissue/damaged spot of the fruit. The damaged area on a fruit or vegetable is marked by discoloured tissue, which denotes the presence of bruising (Opara & Pathare, 2014).

Change in colour/discolouration of impact-damaged fruit tissue following mechanical loading (impact or compression) is commonly used as an indication for bruising in soft-rind fruit such as apples, litchi, bananas, avocados, and the like. The post-impact/compression colour change on damaged tissue in these fruits takes a few hours to become vivid (Fu et al., 2023). According to reports, it can take up to 48 h for the colour change on impact-damaged pomegranate fruit tissue to become

Fig. 8.2 Bruise damaged spot on pomegranate fruit and arils at different drop impact levels. The higher the impact drop height, the larger the size of bruise damaged spot

vivid (Hussein et al., 2019a). Additionally, the intensity of observable bruise damage spot on fruit depends on the magnitude of loading, such that the higher the impact or compression, the higher the bruise spot (Fig. 8.2) (Hussein et al., 2019a).

Overall, bruising reduces quality and availability of pomegranate fruit in the market while causing significant economic losses. Although we may not be able to see evidence of bruise damage on pomegranate fruit immediately after a severe impact or compression, the arils in the internal part of the fruit are likely to be damaged. Internal bruising on fruit, unlike external bruising, is characterized by hidden damage and thus easily overlooked and difficult to measure. Internal bruising might be regarded as 'latent injury', suggesting that damage is normally incurred at one point of the postharvest process but does not become visible until later in the handling chain (Shewfelt, 1986; Hussein et al., 2018). As a result, despite the fact that the majority of pomegranate fruit postharvest losses occur at the farm or during transportation to the packhouse, significant postharvest losses of pomegranate fruit have always been observed at the packhouse level (Opara et al., 2021a, b).

8.3 Factors Contributing to Bruise Damage of Pomegranate Fruit

8.3.1 Climatic Condition

During the growing season, prevailing climatic conditions such as rainfall, sun exposure (light), temperature, and water availability are all known to impact fresh fruit bruise susceptibility. Long and intense sunlight, as well as the associated high temperature, have a significant impact on the sensitivity of pomegranate fruit to bruising prior to harvest (Hussein et al., 2018). As a result, sun-exposed fruit had a higher chance of developing flesh firmness, which could affect fruit bruise susceptibility. Sun exposure causes the production of thicker and more frequent wax clumps, which contribute to cushioning and hence decreased bruise sensitivity on the sun-exposed side of the fruit compared to the shaded side (Tahir et al., 2009; Hussein et al., 2018)

One of the most critical elements influencing daily impacts on fruit bruise susceptibility is temperature (Mowatt, 1997; Hussein et al., 2020a). Temperature influences the physico-mechanical quality attributes of the fruit as well as its susceptibility to bruising (Hertog et al., 2004; Ahmadi et al., 2010). Temperature's proportionate contribution to bruise susceptibility varies depending on the kind of fruit, cell wall mechanical strength, and physiological condition (Ahmadi et al., 2010). As a result, claims of temperature implications on bruising are widely variable. Nonetheless, the bulk of current investigations have focused on the impact of fruit temperature on bruise susceptibility after harvest and storage (Hussein et al., 2020a).

Before harvest, pomegranate fruit is generally subjected to severe temperature changes ranging from low (cold) nighttime temperatures to hot midday temperatures. This may have a major impact on the fruit physico-mechanical qualities and, as a result, its susceptibility to bruising. Cold temperatures increase the stiffness of the pomegranate fruit, making it more prone to bruising and vice versa (Hussein et al., 2020a). The effect of temperature on bruising could be due in part to higher impact energy absorbed by cold-conditioned fruit, resulting in more bruising.

8.3.2 *Fruit Maturity and Ripening at Harvest*

According to previous research, ripe fruits bruise more easily than immature ones (Canete et al., 2015; Bugaud et al., 2014). Studies on apple and banana fruit, for example, found that as fruit turgor increased, so did bruise damage, while the opposite was true for fruit firmness (Garcia et al., 1995; Bugaud et al., 2014). Through modelling, Garcia et al. (1995) established a good relationship between fruit firmness, turgor, ripening process, and bruise susceptibility. The models proposed that fruit turgidity and firmness changes during ripening affect bruising susceptibility. As a result, the influence of harvest date on fresh fruit bruise damage is reliant on changes in turgor and stiffness that occur during the ripening process. When completely ripe, pomegranate, for instance, has a high turgidity and is less solid. As a result, when this type of fruit is stressed by impact or compression, the cell wall strength is likely to be surpassed, resulting in cell wall disintegration.

8.3.3 Harvest Time

Fresh fruit bruise susceptibility is impacted by both harvest time throughout the season (late or early harvest) as well as harvest time throughout the day. There has been no specific research on the effect of late or early pomegranate fruit harvest on bruising. However, research on other fruits has revealed that late harvested fruits (fruit harvested at the end of the commercial harvest period) are more prone to bruising than those harvested at the start of the season (Garcia et al., 1995; Opara, 2007). The scientific explanation for late harvested fruit's reduced sensitivity to bruising has been attributed to higher pulp firmness as opposed to less mature fruits (Garcia et al., 1995). Furthermore, Studman (1997) stated that the ripening process in fruit is associated with a loss in cell wall strength, such that the cells of riper fruit/ fully mature fruit are less able to withstand external loading, which increases the susceptibility to bruise damage. Although the findings on apple and other fruits cannot be extrapolated to the pomegranate fruit setting, the physico-mechanical properties of these fruits tend to respond similarly during loading. The amount of energy expended during loading may explain any observable difference in response to impact or compression loading between these fruits.

The time of day when fresh fruit is gathered also influences its bruise susceptibility. Fruit shrivelling and wilting are associated with higher temperatures during the hottest part of the day. As a result, relatively shrivelled and wilted fruit have lower turgidity (Garcia et al., 1995; Opara, 2007), which reduces the energy absorbed upon impact or compression, reducing the susceptibility to bruise damage at harvest. As a result of reduced turgor, elevated temperatures, and a combination of these effects, fruit harvested later in the day would be less prone to bruise damage than fruit harvested earlier in the day (the morning hours).

8.3.4 Time After Harvest

To save the fruit from bruise damage during postharvest handling, it is critical to understand the effects of time lapse after harvest on bruising susceptibility (Hussein et al., 2020b). The longer time between harvest and transportation exposes the fruit to field heat for a longer period of time, resulting in faster senescence and loss of turgidity (Bollen, 2005). Additionally, the rise in time passed after harvest is associated with a decrease in tissue fruit turgidity as a result of higher losses of water from fruit after harvest, which increases bruise resistance (Bollen, 2005). In general, newly picked fruits are more prone to bruising than fruit that has been sitting for a few days or weeks.

8.3.5 *Storage Condition and Duration*

Researchers found conflicting results regarding the effect of storage duration on the susceptibility of fresh fruits to bruise damage. According to reports on apple research, stored fruits are less susceptible to bruising than freshly harvested fruit (Pang, 1993; Garcia et al., 1995). This propensity has been attributed to an increase in fruit skin resistance and texture change over storage duration, which presumably reduces the energy absorbed during impact. According to Garcia et al. (1995), turgid fruits (often fresh harvest) deteriorate less than flaccid fruits with the same impact energy, and hence lengthy storage period of fruit might potentially enhance resistance to impact loading.

However, studies on multiple pomegranate fruit cultivars found that extended storage time elevates bruise susceptibility (Hussein et al., 2019a, b). The authors explain an increase in pomegranate fruit bruise size as storage duration rises to a drop in turgor pressure during cold storage. According to Singh et al. (2014), fruit has a high proportion of water, which is responsible for soft tissue turgidity, which tends to reduce during storage owing to constant moisture loss. Reduced fruit turgor implies lower tissue sensitivity to bruise injury (Hussein et al., 2019a, b; Singh et al., 2014).

Generally, the effect of storage duration on fruit bruising could be influenced by storage conditions such as temperature, humidity, and the atmosphere surrounding the stored fruit (Hussein et al., 2020b). Several research have been carried out to explore the effect of different storage conditions and/or treatments on altering the biological and/or physiological makeup of produce and, as a result, variations in its susceptibility to bruising. For example, the efficacy of controlled atmosphere storage (CAS) in conjunction with adequate temperature control might be achieved by regulating humidity in addition to air (oxygen and carbon) concentrations (Hussein et al., 2015). The effects of CAS on changes in fruit attributes during storage could rely on the aforementioned conditions and their relative influence on as physico-mechanical properties of the fruit.

Heat treatment of apples resulted in a cushioning effect that reduced impact pressure by melting skin wax and causing structural changes in the fruit (Roy et al., 1994). Tahir et al. (2009) proposed that combining heat treatment and CAS improves fruit firmness, which influences fruit bruising susceptibility. Furthermore, exogenous polyamines such as putrescine, spermidine, or 1-Methylcyclopropene (1-MCP) have been shown to delay fruit senescence, improve firmness and bruise resistance, and prolong fruit storage (Martínez-Romero et al., 2000; Hussein et al., 2020b).

Fresh produce bruising can be affected by packing, packaging type, as well as produce organization within the box. Packages containing fresh produce are occasionally thrown from certain heights during loading, offloading, or handling, which may result in impact bruising (Jarimopas et al., 2007). Additionally, vibrational movement of broth packages and fruits inside packages during transportation in trucks could also cause impact bruise damage to produce (Fadiji et al., 2016a, b). The energy transformation causes bruising of fruit inside the package, as some of the kinetic energy absorbed by produce causes bruising (Jarimopas et al., 2007; Zarifneshat et al., 2010).

8.4 Measurement and Analysis of Pomegranate Fruit Damage

Understanding bruise damage from fresh fruit in a variety of impact or compression settings, susceptibility or resistance of produce to bruising is critical for developing strategies to reduce or even eliminate this problem (Opara & Pathare, 2014; Fu et al., 2023). Due to its simplicity, the drop test is widely used to collect such information by duplicating bruise damage on a laboratory scale. This emphasizes the significance of conducting a thorough review of the methods used to simulate fruit bruising, analysis, and measurement procedures, as well as the assumptions made to fit the real-time settings, in order to determine the existing gap and find a solution that matches modern harvest and postharvest handling systems.

Current studies on fruit bruising measurement employ both destructive and non-destructive techniques (Opara & Pathare, 2014; Li & Thomas, 2014). A variety of unique and emerging non-destructive methods for measuring bruises in fresh horticulture produce have been described, such as but not limited to near infrared spectroscopy, hyperspectral imaging, thermal imaging, and nuclear magnetic resonance imaging (Opara & Pathare, 2014). The former is more common as it is less expensive but takes more time. Non-destructive approaches, on the other hand, necessitate the use of specialized equipment, which incurs significant costs but yields better test efficiency (Li & Thomas, 2014). There has been an upsurge in interest in non-destructive procedures in recent years, which will most certainly substitute the destructive manual measures now utilized; however, there is controversy regarding their precision.

Three methods are commonly used to categorize the techniques for measuring the severity of bruise damage, including: (a) manual measurement of the dimensions of the bruised tissue to estimate its area and/or volume (Opara, 2007; Opara et al., 2007; Hussein et al., 2019a; Pathare & Al-Dairi, 2021), (b) imaging of the bruised tissue and subsequent use of image analysis software to determine the size of the damage (Menesatti & Paglia, 2001; ElMasry et al., 2008; Pholpho et al., 2011; Baranowski et al., 2012; Fu et al., 2020), and (c) employing non-destructive techniques for in-place detection, measurement, and analysis (Xing et al., 2006; Dang et al., 2012; Mazhar et al., 2015)

Fruit damage is influenced by the drop height at which it falls or the magnitude of loading at a state of compression, the type of impact or compression surface, and the speed at which it hits the surface (Hussein et al., 2020b; Li & Thomas, 2014). To obtain controlled bruise damage on a variety of fresh horticultural produce, several researchers have often employed impact, drop, or compression tests (Ferreira et al., 2008; Kitthawee et al., 2011; Hussein et al., 2019a, b).

A pendulum impactor is frequently used in impact tests and is categorized into (a) fruit-to-fruit impact test, where two pendulums made of thin nylon threads are used to support a pair of fruits that can be released from a certain angle to freely impact each other, or (b) impacts where produce is swung in a pendulum arrangement from various heights onto a hard surface. Overall, both of the aforementioned pendulum impact tests face significant obstacles, such as (1) difficulty in precisely controlling the impact location on the fruit surface and (2) difficulty in estimating the rebound height, which is required to determine the actual impact energy that is absorbed by produce to cause damage.

Another kind of impact test is dropping an even round steel ball of known mass from a variety of standard heights via a vertical hollow PVC pipe onto the fresh produce (Opara, 2007; Pathare & Al-Dairi, 2021). To minimize air resistance which permits a free fall of the steel ball during impact testing in this set up, the opposite sides of the pipe are usually perforated along the pipe length. Similarly, utilizing drop test equipment, an impact test might be performed by dropping a whole produce item of known mass. Samples of entire fruit are dropped from the appropriate drop height onto impact surfaces positioned on the impact plate/or surface in this setup (Hussein et al., 2019a, b).

Compression experiments have also been performed to examine the possibility for bruising injury in fresh horticulture products (Kitthawee et al., 2011; Polat et al., 2012). Fruits are crushed with each fruit sample's stem axis parallel to the horizontal plane. Compression testing might be carried out using various diameter spherical indenters. For each indenter, compression is performed at multiple contact areas on the fruit surface along the equator, followed by retaining the fruit for 24 h or more to allow for bruise formation (Kitthawee et al., 2011). Alternatively, fruits can be compressed between two flat steel plates at a specified loading rate. Different forces can be applied in the direction of diameter and height to cause bruise damage (Polat et al., 2012).

Depending on the kind of fruit, bruise damage is commonly characterized as an area/spot of brown tissue beneath the skin. Bruise severity is often expressed in terms of bruise quantity, diameter, or area, all of which affect produce quality (Opara & Pathare, 2014; Hussein et al., 2019a). Measuring the size of a bruise is difficult and time consuming in order to obtain accurate data. To save time, some assumptions about the bruise shape must be made, and only the dimensions that affect the volume are measured (Holt & Schoorl, 1977; Mohsenin, 1986; Hung & Prussia, 1989). Several models have been proposed to describe the shape of a bruise. Nonetheless, a bruise that is caused by more realistic energies that are encountered in practical handling systems (Pang et al., 1992) rarely extends to the surface of the fruit and has a poorly defined spherical shape. For instance, Bollen et al. (1999) compared the various apple bruise volume estimation methods and concluded that there was no single estimation method that could be used across the range of commercially significant impacts. As a result, the question of whether these volume estimate approaches can properly anticipate bruise volume for bruising increasingly prevalent in current management systems remains unanswered (Bollen et al., 2001). This may highlight the need for more research on non-invasive techniques for detecting and quantifying bruise in all harvest and postharvest settings for a variety of fruits.

Several methods for describing bruise volume have been proposed (Holt & Schoorl, 1977; Mohsenin, 1986; Hung & Prussia, 1989). Since the primary measurements used to characterize bruise size are the diameter and depth of the bruise, the typical measurements usually made to the bruise damaged produce are the bruise depth, d_b and bruise width, w_1 across the major axis of the bruise. Other measurements which could also be required by some other methods are the bruise width across the minor axis, w_2 and the depth to the top of the bruise, d_t. Depending on methods chosen, other important measurements usually taken into accounts include the bruise thickness recorded as the difference between the bruise depth d_b and the depth to the top of the bruise $d_t(d_b - d_t)$, and the radius of curvature of the fruit surface at the bruise site, R. These dimensions could accommodate any arbitrary letters of choice as long as they are well defined. A summary of different methods used to describe and quantify bruise volume is summarized in Table 8.2

8.5 Conclusions and Future Prospects

Fruit quality loss affects the pomegranate sector as a result of improper postharvest handling practices. This chapter has highlighted how mechanical damage, such as bruising, causes major postharvest quality changes in pomegranate fruit at harvest and during storage. Bruise damage leads to a change in fruit colour, weight loss, blemishes, and shrinkage. Furthermore, bruising alters chemical and nutritional/medicinal properties (e.g. changes in TA and TSS, pH, tannins, anthocyanins, antioxidant capacity, and vitamin C). Bruise damage also causes changes in the physiological response of the fruit, such as respiration, weight loss, and decay, all of which lead to reduced sells, loss of premium prices, and physical fruit loss due to decay. These facts could alert the pomegranate fruit industry to the economic loss caused by the problem of bruising. Therefore, developing postharvest handling strategies to preserve fruit quality and reduce postharvest losses is warranted.

The discussion in this chapter has also revealed that a greater degree of fruit bruising occurs during harvest and postharvest handling (particularly at harvest, during packing, and transportation/distribution). Furthermore, during fruit harvesting and postharvest handling, a wide variety of factors known to possibly alter the way fruit responds to mechanical loading have been explored. As a result, controlling all of the aforementioned elements that might impact fruit sensitivity to bruising is one of the tactics for decreasing or eradicating the occurrence of fruit bruising and related economic losses after harvest for orchardists and postharvest handlers/fruit packers.

A set of tactics to lessen impact bruise damage might include modifying handling and postharvest factors, training harvest workers to handle the fruits gently,

Table 8.2 Different methods for calculation of bruise volume

Bruise measurement method	Description/assumptions made	Formula for bruise volume	Shape
1. Bruise thickness method (Mohsenin, 1986) 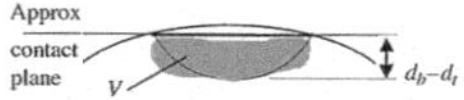	• Bruise volume described as a section of a sphere: (section through the bruise at its mid-point, perpendicular to the skin, appears circular, and the resulting bruise volume is assumed as a section of a sphere)	$V = \frac{\pi(d_b - d_t)}{24}\left(3w_1^2 + 4(d_b - d_t)^2\right)$	Circle
		$V = \frac{\pi(d_b - d_t)}{24}\left(3w_1 w_2 + 4(d_b - d_t)^2\right)$	Ellipse
2. Full depth method (Saltveit, 1984) 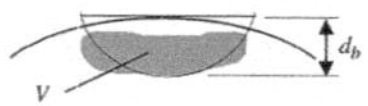	• The assumed shape is a sphere section based on the full depth: (measurement of bruise volume requires measuring the thickness of the bruise at the contact plane, which is difficult because the fruit recovers some of its shape upon rebound. As the result, most of the time, the depth/thickness measurement taken is the full depth from the surface)	$V = \frac{\pi d_b}{24}\left(3w_1^2 + 4d_b^2\right)$	Circle
		$V = \frac{\pi d_b}{24}\left(3w_1 w_2 + 4d_b^2\right)$	Ellipse
3. Enclosed volume method (Holt & Schoorl, 1977) 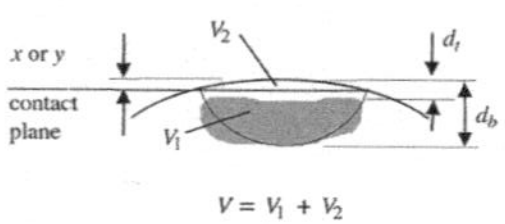	• Modification to the spherical shape model: (suggested that volume be estimated as the sum of two volumes: The volume below the contact plane, bound by the bruise boundary V_1, and the volume bound by the fruit surface V_2. The radius of curvature, R of the fruit surface at the bruise site should be measured in order to calculate the second volume)	$V = \frac{\pi}{24}\left(3w_1^2 d_b + 4(d_b - x)^3 + 4x^3\right)$	Circle
		$V = \frac{\pi}{24}\left(3w_1 w_2 d_b + 4(d_b - y)^3 + 4y^3\right)$	Ellipse
4. Unbruised volume removed method (Bollen & Dela Rue, 1991; Pang et al., 1992) 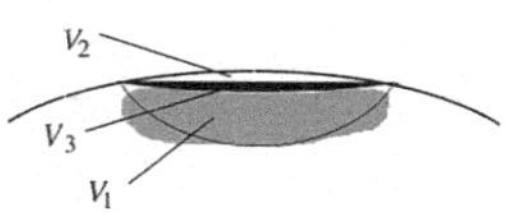	• A modification to the Holt and Schoorl model: (assumes the undamaged section above the bruise has a spherical shape (V_2 + V_3). The bruise volume is calculated by subtracting the unbruised volume from the enclosed volume)	$V = \frac{\pi}{24}\left(3w_1^2(d_b - d_t) + 4(d_b - x)^3 + 4(d_t - x)^3\right)$	Circle
		$V = \frac{\pi}{24}\left(3w_1 w_2(d_b - d_t) + 4(d_b - y)^3 + 4(d_t - y)^3\right)$	Ellipse

5. Ellipsoid method (Hung & Prussia, 1989) 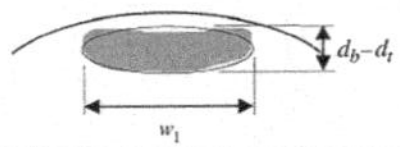	• Bruise assumed as an ellipsoidal shape: (bruise volume is calculated by assuming an ellipsoidal shape and related equation for the volume of an ellipsoid)	$V=\frac{4}{3}\pi\left(\frac{(d_b-d_t)w_1^2}{8}\right)$	Circle
		$V=\frac{4}{3}\pi\left(\frac{(d_b-d_t)w_1w_2}{8}\right)$	Ellipse

V is estimated bruise volume in mm^3; d_b is full depth of bruise in mm; d_t is the depth from fruit surface to top of bruise in mm; R is the fruit radius at bruise in mm; V_1, V_2, and V_3 are volumes used in calculation of V in mm^3; w_1 and w_2 are bruise widths across the major and minor axes in mm; x, height of fruit section above the contact plane (circular bruise) in mm; y, height of fruit section above the contact plane (elliptical bruise) in mm; $x=R-\sqrt{R^2-\frac{w_1^2}{4}}$, $y=R-\sqrt{R^2-\frac{w_1w_2}{4}}$

harvesting at the proper maturity, harvesting dry whenever possible, handling fruits at right storage temperature and humidity, handling each fruit no more than necessary, installing cushion pads inside bulk bins, and avoiding over or under-packing of containers. Additionally, understanding bruise damage in fresh fruit across a variety of handling circumstances might be a key method for decreasing or preventing it. Adoption of innovative postharvest handling methods, such as modifying the design of harvesting machines and plant architecture to avoid bruising, will be critical.

There are a variety of objective indices available to measure the potential for the produce to bruise when subjected to mechanical stress. Although the depth and diameter of a bruise are the most popular methods to define its size, the sensitivity to bruising (measured as the degree of bruise damage per unit of ingested impact/compression energy) is the most suitable indicator for assessing a bruise's severity. Similar to the volume estimation methods reported in this chapter, there are no agreed-upon parameters between research and industry on a single method that can be used for the estimation of bruise volume of fresh horticultural produce across the variety of commercially essential consequences.

Future research should focus on developing new techniques to simulate bruising and, subsequently, near-accurate bruise measurement methods to enable real-time simulation and quantification of bruise damage both in situ and ex situ. This is critically important given the evolving operations and handling equipment in the modern harvest and postharvest handling system. Identifying essential points of control for reducing the frequency and severity of mechanical damage to fresh horticultural produce necessitates research on application of emerging non-inversive technologies for detecting and quantifying bruise damage alone or in combination with destructive methods.

Acknowledgments This work is based on the research supported in part by the National Research Foundation of South Africa (Grant Number: 64813). The opinions, findings, and conclusions or recommendations expressed are those of the author(s) alone, and the NRF accepts no liability whatsoever in this regard.

References

Ahmadi, E. (2012). Bruise susceptibilities of kiwifruit as affected by impact and fruit properties. *Research in Agricultural Engineering, 58*, 107–113.

Ahmadi, E., Ghassemzadeh, H. R., Sadeghi, M., Moghaddam, M., & Neshat, S. Z. (2010). The effect of impact and fruit properties on the bruising of peach. *Journal of Food Engineering, 97*, 110–117.

Al-Said, F. A., Opara, U. L., & Al-Yahyai, R. A. (2009). Physico-chemical and textural quality attributes of pomegranate cultivars (Punica granatum L.) grown in the Sultanate of Oman. *Journal of Food Engineering, 90*(1), 129–134. 169.

Ambaw, A., Fadiji, T., & Opara, U. L. (2021). Thermo-mechanical analysis in the fresh fruit cold chain. A review on recent advances. *Foods, 10*(6), 1357. https://doi.org/10.3390/foods10061357

Ambaw, A., Mukama, M., Fadiji, T., & Opara, U. L. (2022). Fresh fruit packaging design verification through virtual prototyping technique. *Food Packaging and Shelf Life, 32*, 100858. https://doi.org/10.1016/j.fpsl.2022.100858

Baker, G. A., Gray, L. C., Harwood, M. J., Osland, T. J., & Tooley, J. B. C. (2019). On-farm food loss in northern and Central California: Results of field survey measurements. *Resources, Conservation and Recycling, 149*, 541–549.

Bantayehu, M., Alemayehu, M., Abera, M., & Bizuayehu, S. (2017). Postharvest losses assessment of tropical fruits in the market chai of North Western Ethiopia. *Food Science and Quality Management, 66*, 1–12.

Baranowski, P., Mazurek, W., Wozniak, J., & Majewska, U. (2012). Detection of early bruises in apples using hyperspectral data and thermal imaging. *Journal of Food Engineering, 110*(3), 345–355.

Bhattarai, D. R., Subedi, G. D., Gautam, I. P., & Chauhan, S. (2017). Postharvest supply chain study of carrot in Nepal. *International Journal of Horticulture, 7*, 239–245.

Blahovec, J. (2006). Shape of bruise spots in impacted potatoes. *Postharvest Biology and Technology, 39*, 278–284.

Bollen, A. F. (2005). Major factors causing variation in bruise susceptibility of apples (*Malus domestica*) grown in New Zealand. *New Zealand Journal of Crop and Horticultural Science, 33*, 201–210.

Bollen, A. F., Nguyen, H. X., & Dela Rue, B. T. (1999). Comparison of methods for estimating the bruise volume of apples. *Journal of Agricultural Engineering Research, 74*, 325–330.

Bollen, A. F., Cox, N. R., Dela Rue, B. T., & Painter, D. J. (2001). A descriptor for damage susceptibility of a population of produce. *Journal of Agricultural Engineering Research, 78*, 391–395.

Bond, R. (2016). *Carrot loss during primary production: Field waste and pack house waste*. Master's thesis, Hedmark University, Hedmark, Norway.

Bugaud, C., Ocrisse, G., Salmon, F., & Rinaldo, D. (2014). Bruise susceptibility of banana peel in relation to genotype and post-climacteric storage conditions. *Postharvest Biology and Technology, 87*, 113–119.

Canete, M. L., Hueso, J. J., Pinillos, V., & Cuevas, J. (2015). Ripening degree at harvest affects bruising susceptibility and fruit sensorial traits of loquat (*Eriobotrya japonica* Lindl.). *Scientia Horticulturae, 187*, 102–107.

Czieczor, L., Bentkamp, C., Damerow, L., & Blanke, M. (2018). Non-invasive determination of the quality of pomegranate fruit. *Postharvest Biology and Technology, 2018*(136), 74–79.

Dang, H. Q., Kim, I., Cho, B. K., & Kim, M. S. (2012). Detection of bruise damage of pear using hyperspectral imagery. In *2012 12th International Conference on Control, Automation and Systems (ICCAS)* (pp. 1258–1260). IEEE.

de Oliveira, F. L., Arruda, T. Y. P., da Silva Lima, R., Casarotti, S. N., & Morzelle, M. C. (2020). Pomegranate as a natural source of phenolic antioxidants: A review. *Journal of Food Bioactives, 9*, 10–22.

ElMasry, G., Wang, N., Vigneault, C., Qiao, J., & ElSayed, A. (2008). Early detection of apple bruises on different background colors using hyperspectral imaging. *LWT-Food Science and Technology, 41*(2), 337–345.

Ergun, M. (2017). Physical, physiochemical and electrochemical responses of 'Galaxy' apples to mild bruising. *European Journal of Horticultural Science, 82*, 244–250.

Fadiji, T., Coetzee, C., Chen, L., Chukwu, O., & Opara, U. L. (2016a). Susceptibility of apples to bruising inside ventilated corrugated paper-board packages during simulated transport damage. *Postharvest Biology and Technology, 118*, 111–119.

Fadiji, T., Coetzee, C., Pathare, P., & Opara, U. L. (2016b). Susceptibility to impact damage of apples inside ventilated corrugated paperboard packages: Effects of package design. *Postharvest Biology and Technology, 111*, 286–296.

Faria, A., & Calhau, C. (2011). The bioactivity of pomegranate: Impact on health and disease. *Critical Reviews in Food Science and Nutrition, 51*(7), 626-634.9.

Fawole, O. A., & Opara, U. L. (2013a). Changes in physical properties, chemical and elemental composition and antioxidant capacity of pomegranate (cv. Ruby) fruit at five maturity stages. *Scientia Horticulturae, 150*(4), 37–46. https://doi.org/10.1016/j.scienta.2012.10.026

Fawole, O. A., & Opara, U. L. (2013b). Developmental changes in maturity indices of pomegranate fruit: A descriptive review. *Scientia Horticulturae, 159*, 152–161. https://doi.org/10.1016/j.scienta.2013.05.016

Fawole, O. A., & Opara, U. L. (2013c). Effects of maturity status on biochemical content, polyphenol composition and antioxidant capacity of pomegranate fruit arils (cv. 'Bhagwa'). *South African Journal of Botany, 85*, 23–31. https://doi.org/10.1016/j.sajb.2012.11.010

Fawole, O. A., & Opara, U. L. (2013d). Effects of storage temperature and duration on physiological responses of pomegranate fruit. *Industrial Crops and Products, 47*, 300–309. https://doi.org/10.1016/j.indcrop.2013.03.028

Fawole, O. A., & Opara, U. L. (2013e). Fruit growth dynamics, respiration rate and physico-textural properties during pomegranate development and ripening. *Scientia Horticulturae, 157*, 90–98. https://doi.org/10.1016/j.scienta.2013.04.004

Fawole, O. A., & Opara, U. L. (2013f). Harvest discrimination of pomegranate fruit: postharvest quality changes and relationships between instrumental and sensory attributes during shelf life. *Journal of Food Science, 78*(8), S1264–S1272. https://doi.org/10.1111/1750-3841.12176

Fawole, O. A., & Opara, U. L. (2013g). Seasonal variation in chemical composition, aroma volatiles and antioxidant capacity of pomegranate during fruit development. *African Journal of Biotechnology, 12*(25), 4006.

Ferreira, M. D., Sargent, S. A., Brecht, J. K., & Chandler, C. K. (2008). Strawberry fruit resistance to simulated handling. *Science in Agriculture, 65*, 490–495.

Fu, H., Karkee, M., He, L., Duan, J., Li, J., & Zhang, Q. (2020). Bruise patterns of fresh market apples caused by fruit-to-fruit impact. *Agronomy, 10*(1), 59.

Fu, H., Du, W., Yang, J., Wang, W., Wu, Z., & Yang, Z. (2023). Bruise measurement of fresh market apples caused by repeated impacts using a pendulum method. *Postharvest Biology and Technology, 195*, 112143.

Gao, M., Guo, W., Huang, X., Du, R., & Zhu, X. (2021). Effect of pressing and impacting bruises on optical properties of kiwifruit flesh. *Postharvest Biology and Technology, 172*, 111385.

Garcia, J. L., Ruiz-Altisent, M., & Barreiro, P. (1995). Factors influencing mechanical properties and bruise susceptibility of apples and pears. *Journal of Agricultural Engineering Research, 61*, 11–18.

Ghanbarpour, E., Rezaei, M., & Lawson, S. (2019). Reduction of cracking in pomegranate fruit after foliar application of humic acid, calcium-boron and kaolin during water stress. *Erwerbs-obstbau, 61*, 29–37.

Hertog, M. L. A. T. M., Ben-Arie, R., Roth, E., & Nicolai, B. M. (2004). Humidity and temperature effects on invasive and non-invasive firmness measures. *Postharvest Biology and Technology, 33*, 79–91.

Holt, J. E., & Schoorl, D. (1977). Bruising and energy dissipation in apples. *Journal of Texture Studies, 7*, 421–432.

Hung, Y., & Prussia, S. (1989). Effect of maturity and storage time on the bruise susceptibility of peaches (cv. Red Globe). *Transactions of ASAE, 32*, 1377–1382.

Hussein, Z., Caleb, O. J., & Opara, U. L. (2015). Perforation-mediated modified atmosphere packaging of fresh and minimally processed produce–a review. *Food Packaging and Shelf Life, 6*, 7–20.

Hussein, Z., Fawole, O. A., & Opara, U. L. (2018). Preharvest factors influencing bruise damage of fresh fruits–a review. *Scientia Horticulturae, 229*, 45–58.

Hussein, Z., Fawole, O. A., & Opara, U. L. (2019a). Bruise damage susceptibility of pomegranates (Punica granatum, L.) and impact on fruit physiological response during short term storage. *Scientia Horticulturae, 246*, 664–674.

Hussein, Z., Fawole, O. A., & Opara, U. O. (2019b). Bruise damage of pomegranate during long-term cold storage: Susceptibility to bruising and changes in textural properties of fruit. *International Journal of Fruit Science, 20*, 211–230.

Hussein, Z., Fawole, O. A., & Opara, U. O. (2020a). Effects of bruising and storage duration on physiological response and quality attributes of pomegranate fruit. *Scientia Horticulturae, 267*, 109306.

Hussein, Z., Fawole, O. A., & Opara, U. L. (2020b). Harvest and postharvest factors affecting bruise damage of fresh fruits. *Horticultural Plant Journal, 6*(1), 1–13.

Jarimopas, B., Singh, S. P., Sayasoonthorn, S., & Singh, J. (2007). Comparison of package cushioning materials to protect post-harvest impact damage to apples. *Packaging Technology and Science, 20*, 315–324.

Kabas, O. (2010). Methods of measuring bruise volume of pear (*Pyrus communis* L.). *International Journal of Food Properties, 13*, 1178–1186.

Kahramanoglu, I. (2019). Trends in pomegranate sector: Production, postharvest handling and marketing. *International Journal of Agriculture Forestry and Life Science, 3*, 239–246.

Katz, S. R., Newman, R. A., & Lansky, E. P. (2007). Punica granatum: Heuristic treatment for diabetes mellitus. *Journal of Medicinal Food, 10*(2), 213–217.

Kitthawee, U., Pathaveerat, S., Srirungruang, T., & Slaughter, D. (2011). Mechanical bruising of young coconut. *Biosystems Engineering, 109*(3), 211–219.

Komarnicki, P., Stopa, R., Szyjewicz, D., & Mlotek, M. (2016). Evaluation of bruise resistance of pears to impact load. *Postharvest Biology and Technology, 114*, 36–44.

Lewis, R., Yoxall, A., Canty, L. A., & Romo, E. R. (2007). Development of engineering design tools to help reduce apple bruising. *Journal of Food Engineering, 83*, 356–365.

Li, Z., & Thomas, C. (2014). Quantitative evaluation of mechanical damage to fresh fruits. *Trends in Food Science and Technology, 35*, 138–150.

Lufu, R., Ambaw, A., & Opara, U. L. (2019). The contribution of transpiration and respiration processes in the mass loss of pomegranate fruit (cv. Wonderful). *Postharvest Biology and Technology, 157*, 1–10.

Lufu, R., Ambaw, A., & Opara, U. L. (2020). Water loss of fresh fruit: Influencing pre-harvest, harvest and postharvest factors. *Scientia Horticulturae, 272*, 1–16.

Lufu, R., Ambaw, A., & Opara, U. L. (2021). The influence of internal packaging (liners) on moisture dynamics, physical and physiological quality of pomegranate fruit during cold storage. *Foods, 10*, 1388. https://doi.org/10.3390/foods10061388

Magangana, T. P., Makunga, N. P., Fawole, O. A., & Opara, U. L. (2020). Processing factors affecting the phytochemical and nutritional properties of pomegranate (Punica granatum L.) peel waste: A review. *Molecules, 25*(20), 4690.

Martínez-Romero, D., Valero, D., Serrano, M., Burló, F., Carbonell, A., Burgos, I., & Riquelme, F. (2000). Exogenous polyamines and gibberellic acid effects on peach (*Prunus persica* L.) storability improvement. *Journal of Food Science, 65*, 288–294.

Martinez-Romero, D., Serrano, M., Carbonell, A., Castillo, S., Riquelme, F., & Valero, D. (2004). Mechanical damage during fruit postharvest handling: Technical and physiological implications. In *Production practices and quality assessment of food crops* (3rd ed., pp. 233–252). Springer.

Mazhar, M., Joyce, D., Cowin, G., Brereton, I., Hofman, P., Collins, R., & Gupta, M. (2015). Non-destructive 1H-MRI assessment of flesh bruising in avocado (*Persea americana* M.) cv. Hass. *Postharvest Biology and Technology, 100*, 33–40.

Menesatti, P., & Paglia, G. (2001). Development of a drop damage index of fruit resistance to damage. *Journal of Agricultural Engineering Research, 80*, 53–64.

Middha, S. K., Usha, T., & Pande, V. (2013). A review on antihyperglycemic and antihepatoprotective activity of eco-friendly Punica granatum peel waste. *Evidence-Based Complementary and Alternative Medicine, 2013*, 656172.

Mitsuhashi-Gonzalez, K., Pitts, M. J., Fellman, J. K., Curry, E. A., & Clary, C. D. (2010). Bruising profile of fresh apples associated with tissue type and structure. *Applied Engineering in Agriculture, 26*, 509–517.

Mohammad Shafie, M., Rajabipour, A., & Mobli, H. (2017). Determination of bruise incidence of pomegranate fruit under drop case. *International Journal of Fruit Science, 17*(3), 296–309.

Mohsenin, N. N. (1986). Physical properties of plant and animal materials: Structure. In *Physical characteristic and mechanical properties* (1st ed.). Gordon and Breach Science Publishers.

Montero, C. R. S., Schwarz, L. L., Dos Santos, L. C., Andreazza, C. S., Kechinski, C. P., & Bender, R. J. (2009). Postharvest mechanical dam- age affects fruit quality of 'Montenegrina' and 'Rainha' tangerines. *Pesquisa Agropecuaria Brasileira, 44*, 1636–1640.

Mousavinejad, G., Emam-Djomeh, Z., Rezaei, K., & Khodaparast, M. H. H. (2009). Identification and quantification of phenolic compounds and their effects on antioxidant activity in pomegranate juices of eight Iranian cultivars. *Food Chemistry, 115*(4), 1274–1278.

Mowatt, C. M. (1997). *Factors influencing the susceptibility of apples to bruising, horticultural science*. PhD thesis, Massey University, Palmerston North, New Zealand.

Mphahlele, R. R., Fawole, O. A., & Opara, U. L. (2016). Influence of packaging system and long-term storage on physiological attributes, biochemical quality, volatile composition and antioxidant properties of pomegranate fruit. *Scientia Horticulturae, 211*, 140–151.

Mukama, M., Ambaw, A., & Opara, U. L. (2020a). A virtual prototyping approach for redesigning the vent-holes of packaging for handling pomegranate fruit–A short communication. *Journal of Food Engineering, 270*, 1–5.

Mukama, M., Ambaw, A., & Opara, U. L. (2020b). Advances in design and performance evaluation of fresh fruit ventilated distribution packaging: A review. *Food Packaging and Shelf Life, 24*, 100472. https://doi.org/10.1016/j.fpsl.2020.100472

Mukama, M., Ambaw, A., & Opara, U. L. (2020c). Thermophysical properties of fruit—a review with reference to postharvest handling. *Journal of Food Measurement and Characterization, 14*, 2917–2937. https://doi.org/10.1007/s11694-020-00536-8

Mukama, M., Ambaw, A., & Opara, U. L. (2021). Characterisation of ventilated multi-scale packaging designs used for postharvest handling of pomegranate in South Africa. *AMA-Agricultural Mechanization in Asia, Africa and Latin America, 52*(4), 31–38.

Munhuweyi, K., Lennox, C. L., Meitz-Hopkins, J. C., Caleb, O. J., & Opara, U. L. (2016). Major diseases of pomegranate (Punica granatum L.), their causes and management – A review. *Scientia Horticulturae, 211*, 126–139.

Opara, L. U. (2007). Bruise susceptibilities of 'Gala' apples as affected by orchard management practices and harvest date. *Postharvest Biology and Technology, 43*, 47–54.

Opara, U. L., & Pathare, P. B. (2014). Bruise damage measurement and analysis of fresh horticultural produce—A review. *Postharvest Biology and Technology, 91*, 9–24.

Opara, L. U., Studman, C. J., & Banks, N. H. (1997). Fruit skin splitting and cracking. *Horticultural Reviews, 19*, 217–262.

Opara, L. U., Al-Ghafri, A., Agzoun, H., Al-Issai, J., & Al-Jabri, F. (2007). Design and development of a new device for measuring susceptibility to impact damage of fresh produce. *New Zealand Journal of Crop and Horticultural Science, 35*, 245–251.

Opara, U. L., Al-Ani, M. R., & Al-Shuaibi, Y. (2009). Physico-chemical properties, vitamin C content, and antimicrobial properties of pomegranate fruit (Punica granatum L.). *Food and Bioprocess Technology, 2*, 315–321.

Opara, U. L., Hussein, Z., & Caleb, O. J. (2017). Phytochemical properties and antioxidant activities of minimally processed "Acco" pomegranate arils as affected by perforation-mediated modified atmosphere packaging. *Journal of Food Processing and Preservation, 41*(3), e12948.

Opara, I. K., Fawole, O. A., & Opara, U. L. (2021a). Postharvest losses of pomegranate fruit at the packhouse and implications for sustainability indicators. *Sustainability, 13*(9), 5187.

Opara, I. K., Fawole, O. A., Kelly, C., & Opara, U. L. (2021b). Quantification of on-farm pomegranate fruit postharvest losses and waste, and implications on sustainability indicators: South African case study. *Sustainability, 13*(9), 5168.

Palou, L., Guardado, A., & Montesinos-Herrero, C. (2010). First report of Penicillium spp. and Pilidiella granati causing postharvest fruit rot of pomegranate in Spain. *New Disease Reports, 22*(21), 2044–0588.

Palou, L., Taberner, V., Guardado, A., Del Río, M. Á., & Montesinos-Herrero, C. (2013). Incidence and etiology of postharvest fungal diseases of pomegranate (Punica granatum cv. Mollar de Elche) in Spain. *Phytopathologia Mediterranea, 52*, 478–489.

Pang, D. W. (1993). *Prediction and quantification of apple bruising*. Massey University.

Pang, W., Studman, C. J., & Ward, G. T. (1992). Bruising damage in apple-to-apple impact. *Journal of Agricultural Engineering Research, 52*, 229–240.

Pareek, S., Valero, D., & Serrano, M. (2015). Postharvest biology and technology of pomegranate. *Journal of the Science of Food and Agriculture, 95*(12), 2360–2379.

Pathare, P. B., & Al-Dairi, M. (2021). Bruise susceptibility and impact on quality parameters of pears during storage. *Frontiers in Sustainable Food Systems, 5*, 263.

Pholpho, T., Pathaveerat, S., & Sirisomboon, P. (2011). Classification of longan fruit bruising using visible spectroscopy. *Journal of Food Engineering, 104*(1), 169–172.

Polat, R., Aktas, T., & Ikinci, A. (2012). Selected mechanical properties and bruise susceptibility of nectarine fruit. *International Journal of Food Properties, 15*(6), 1369–1380.

Pomegranate Producers Association of South Africa (POMASA). (2019). *Pomegranate industry overview: statistics and information*. Retrieved December 18, 2022, from https://www.sapomegranate.co.za/statistics-and-information/pomegranate-industry-overview/.

Roy, S., Conway, W. S., Watada, A. E., Sams, C. E., Erbe, E. F., & Wergin, W. P. (1994). Heat treatment affects epicuticular wax structure and postharvest calcium uptake in 'Golden Delicious' apples. *HortScience, 29*, 1056–1058.

Rymon, D. (2011). Mapping features of the global pomegranate market. *Acta Horticulturae, 890*, 599–602.

Saltveit, M. E. (1984). Effects of temperature on firmness and bruising of 'Starkrimson Delicious' and 'Golden Delicious' apples. *HortScience, 19*(4), 550–551.

Shewfelt, R. L. (1986). Postharvest treatment for extending the shelf life of fruits and vegetables. *Food Technology Journal, 40*, 70–89.

Singh, F., Katiyar, V. K., & Singh, B. P. (2014). Analytical study of turgor pressure in apple and potato tissues. *Postharvest Biology and Technology, 89*, 44–48.

Studman, C. (1997). Factors affecting the bruise susceptibility of fruit. In G. Jeronimidis & J. F. V. Vincent (Eds.), *Proceedings of the 2nd International Conference of Plant Biomechanics* (pp. 273–281). Centre for Biomimetics, University of Reading.

Tahir, I. I., Johansson, E., & Olsson, M. E. (2009). Improvement of apple quality and storability by a combination of heat treatment and controlled atmosphere storage. *HortScience, 44*, 1648–1654.

Viuda-Martos, M., Fernández-López, J., & Pérez-Álvarez, J. A. (2010). Pomegranate and its many functional components as related to human health: A review. *Comprehensive Reviews in Food Science and Food Safety, 9*(6), 635–654.

Wang, P., Zhang, Q., Hou, H., Liu, Z., Wang, L., Rasekhmagham, R., Kord-Varkaneh, H., Santos, H. O., & Yao, G. (2020). The effects of pomegranate supplementation on biomarkers of inflammation and endothelial dysfunction: A meta-analysis and systematic review. *Complementary Therapies in Medicine, 49*, 1–8.

Weerakkody, P., Jobling, J., Infante, M. M. V., & Rogers, G. (2010). The effect of maturity, sunburn and the application of sunscreens on the internal and external qualities of pomegranate fruit grown in Australia. *Scientia Horticulturae, 124*, 57–61.

Xia, M., Zhao, X., Wei, X., Guan, W., Wei, X., Xu, C., & Mao, L. (2020). Impact of packaging materials on bruise damage in kiwifruit during free drop test. *Acta Physiologiae Plantarum, 42*(7), 1–11.

Xing, J., Bravo, C., Moshou, D., Ramon, H., & De Baerdemaeker, J. (2006). Bruise detection on 'Golden Delicious' apples by vis/NIR spectroscopy. *Computers in Electronics and Agriculture, 52*, 11–20.

Zarifneshat, S., Ghassemzadeh, H. R., Sadeghi, M., Abbaspour-Fard, M. H., Ahmadi, E., Javadi, A., & Shervani-Tabar, M. T. (2010). Effect of impact level and fruit properties on Golden Delicious apple bruising. *American Journal of Agricultural and Biological Sciences, 5*, 114–121.

Chapter 9
Bruise Damage Susceptibility of Tomato

Mai Al-Dairi, Pankaj B. Pathare, and Rashid Al-Yahyai

Abstract Tomato is one of the most highly produced and consumed fresh produce worldwide. Tomato fruit contains a high concentration of beneficial minerals and nutrients which make them favorable for health-conscious people. Tomato fruit is perishable fresh produce that is highly susceptible to external factors during the postharvest supply chain like mechanical damages. Bruising is a mechanical injury that is described as damage to the tissue of the fresh fruit due to different external forces such as impact, cut, abrasion, compression, and vibration. Various studies have been conducted to measure the bruising intensity and susceptibility of tomatoes by using simulated handling practices like drop and pendulum tests. Besides, other studies evaluated the effect of simulated and ground transport on bruising and subsequent quality attributes of tomatoes. Bruising can affect the physico-chemical quality of tomatoes and cause fruit deterioration rendering them unfavorable to consumers. At the market level, consumers prefer high-quality fruit that are usually assessed based on their taste and appearance. Bruise damage reduces the weight and firmness of the fruit. Also, it can accelerate the changes in color, chemical, and nutritional composition of the fruit, thus, reducing the market and economic value. Utilizing techniques for bruise detection may help prevent bruise damage to fruit throughout the supply chain processes. Many studies used non-destructive techniques like visible and near-infrared (Vis-NIR) spectroscopy, hyperspectral imaging, thermal imaging (TI), magnetic resonance imaging (MRI), etc. Bruising susceptibility and its effect on tomatoes and recent technical developments for its detection are discussed.

M. Al-Dairi · P. B. Pathare (✉)
Department of Soils, Water & Agricultural Engineering, College of Agricultural and Marine Sciences, Sultan Qaboos University, Al-Khod, Oman
e-mail: pankaj@squ.edu.om

R. Al-Yahyai
Department of Plant Sciences, College of Agricultural & Marine Sciences, Sultan Qaboos University, Al-Khod, Oman

P. B. Pathare, U. L. Opara (eds.), *Mechanical Damage in Fresh Horticultural Produce*, https://doi.org/10.1007/978-981-99-7096-4_9

Keywords Tomato · Bruising susceptibility · Postharvest handling · Non-destructive techniques · Mechanical damage

9.1 Introduction

Tomato (*Solanum lycopersicum* L.) is one of the most widely grown and important vegetable crops worldwide (Al-Dairi et al., 2021a) and ranks second globally after potato in quantity and area of production (Cherono & Workneh, 2018). Tomato is a vital source of many nutrients like lycopene, folate, potassium, flavonoids, antioxidants, phenolics (Erba et al., 2013), dietary fibers, iron, phosphorus, and other healthy minerals that benefit the human body like vitamins A and C (Sibomana et al., 2015) and contains β-carotene and lycopene which reduce the incidence of cardiovascular diseases and cancer (Al-Dairi et al., 2021b). These numerous nutritional values and health benefits of tomato can link to its high global production which is about 4.8 million ha of harvested areas with 162 million as an estimated production (Sarma, 2018).

Tomato is climacteric, delicate, and one of the highly perishable fresh produce and is susceptible to various issues during storage, transportation, and marketing. Tomatoes as a luscious fruit, are often damaged by excessive vibration during transport and distribution. Tissue damage is the primary cause of tomato decay; even little damage can allow fungal, bacterial, or other pathogens to attack, contaminating and causing rot in the rest of the fruit. Low-grade and low-quality fruit is an unavoidable result of tissue injury (Wu & Wang, 2014). A total of 50% losses were reported during the stages of harvesting, transportation, and consumption of fresh tomatoes (Addo et al., 2015). A percentage of increase in the production of fresh produce is often accompanied by a similar increase associated with postharvest losses (Munhuewyi, 2012).

Unfortunately, fresh produce, such as tomatoes, are highly sensitive to mechanical damage during harvesting, transportation, and packaging, and poor handling which can significantly reduce the fruit quality. In many fruits, bruising is the most common type of mechanical injury (Pathare & Al-Dairi, 2021b). The external and internal quality of tomato fruit is significantly diminished by bruise damage. The bruise damage emerges as a mushy patch on the tomato after some time has passed. In postharvest processing, bruising damage appears to be almost unavoidable (Xing et al., 2006). This chapter provides an overview of bruising, the impact of bruising susceptibility damage on tomato quality, and present case studies on the role of postharvest handling, and bruise detection and analysis.

9.2 Impact of Bruising on Tomato Quality

Consumers prefer high-quality fruit that they usually assess based on their external attributes such as appearance. As a result, fruit are supplied to markets must meet international quality requirements for freshness, firmness, and other quality factors (Lubes & Goodarzi, 2017). Horticultural produce are susceptible to a variety of external factors during the postharvest process. Mechanical damage occurs when these forces exceed the critical value for tissue damage. As a result, fruit can be damaged during handling, harvesting, shipping, and processing, and this damage has been identified as the primary cause of decreasing fruit quality, resulting in lower sales prices and losses for the grower or orchardist (Cui et al., 2018). According to Opara and Pathare (2014), visual quality influences consumer decision (up to 83%) and is heavily influenced by the existence of flaws. Bruising was a more significant barrier to purchasing than price.

In the context of horticultural products, bruising has been characterized as damage to plant tissue caused by external forces, resulting in physical changes in texture and/or subsequent chemical changes in color, flavor, and texture. Bruising is caused by cell membrane breakdown, which results in a loss of cell wall integrity and other subsequent losses in peel resistance owing to mechanical injury (Pathare & Al-Dairi, 2021a). Fresh fruit are susceptible to bruising damage when they impact other rigid surfaces or other fruit during postharvest supply chain operations (Al-Dairi et al., 2022). Idah et al. (2012) mentioned that bruising could also result from either a sudden drop of fresh produce from certain heights or vibration. Additionally, it can occur when the fruit rub against each other, parts of processing equipment, and inside the packaging containers.

The bruise susceptibility of fresh produce is a measure for reactions of other external forces which rely on some other factors like variety, texture, firmness, maturity, water status, temperature, size, and shape; as well as some other internal factors such as cell wall strength, cell shape, internal structure, and elasticity (Cui et al., 2018). Tomatoes' texture is highly reflected by the mechanical features of their pericarp tissue. Cui et al. (2018) stated that the texture variations of tomatoes alter the mechanical properties of the fruit, therefore, affecting the bruise incidence and susceptibility of the fruit. Brusie can hasten the deterioration of the entire crop across the supply chain system, increasing losses by 51%, which is regarded as a serious economic and food safety issue (Li et al., 2017). Even when tomatoes are handled with care, postharvest physical damage is common. Tomatoes have lower storability than other fruits, such as apples, and they are subject to improper harvesting, unloading, and transit conditions (Tran et al., 2017).

A bruise can cause skin cracking and rapture leading to phenolic compounds oxidation and tissue breakdown (Opara & Pathare, 2014). In most fresh fruit, bruising changes the metabolic and physiological processes which cause rapid ripening and quality loss (Costa et al., 2021). Bruising affects not only consumers' preferences and choices but also fresh fruit marketing. It also affects the citric acid

(a) (b) (c) (d) (e)

Fig. 9.1 Tomato damage degrees of (**a**) 1 × g, (**b**) 1.5 × g, (**c**) 2 × g, (**d**) 2.5 × g, and (**e**) 3 × g (Wu & Wang, 2014)

contents, ethylene production, respiration, and total soluble solids of fruit (Pathare & Al-Dairi, 2021b).

Bruising affects the nutritional value of tomatoes. Sablani et al. (2006) reported a reduction in vitamin C in bruised tomato fruit compared to the non-bruised one. Wu and Wang (2014) subjected tomatoes to different vibration intensities (1, 1.5, 2, 2.5, and 3 g). They observed the dark red color of the tomato as the acceleration increased to 2 g and minor bruise damage. A large area of bruising was recorded on the surface of the tomato fruit when subjected to 2.5 g acceleration. This acceleration led to split tomato surface ($n = 4$ out of 56 experimental samples) after 60 min of vibration. When the acceleration reached 3 g, extreme 46.6% damage to tomato samples occurred (Fig. 9.1). They concluded that the sub-surface bruise damage was a result of the exposed vibration. Cui et al. (2018) reported high firmness and weight loss when the impact force level increased. They also found that the bruising and its subsequent quality changes were highly observed at room temperature than at low storage temperature (10 °C). With increasing bruise susceptibility, the metabolic rate is reduced.

Increasing damage can elevate wound respiration which can provide an ideal place for decay organisms, increase overall deterioration, and reduce the visual appearance of tomatoes aspects (Allende et al., 2004). Tomato fruit were dropped from different heights (50, 80, 110, and 140 cm) into different surfaces in a study conducted by Idah et al. (2007). They ascertained that the impact energy on the tomato fruit is greatly affected by the drop height. Tomato fruit dropped from the highest drop height (140 cm) absorbed the greatest energy signifying that the fruit suffered the most impact bruise damage.

Idah et al. (2012) assessed bruising generated due to simulated vibration using different packaging containers (traditional (PM1) and plastic baskets (PM22)). The results showed that 45%, 37.50%, and 40% of the samples of tomatoes from the bottom, middle, and top of the traditional basket, respectively were extremely bruised after 4 h of excitation. The average bruise width of the damaged tomato fruit was 24.36 and 15.18 mm packed in the traditional and plastic containers, respectively. Bruise area increment led to increase the openings of the tomatoes which led to the release of the moisture content from the damaged fruit. Also, bruising exposed tomatoes to invasion by other spoilage organisms, therefore, accelerating the decay rate. Besides, highly bruised tomatoes showed high signs of shrinkage which led to a decrease in the fruit weight (Fig. 9.2).

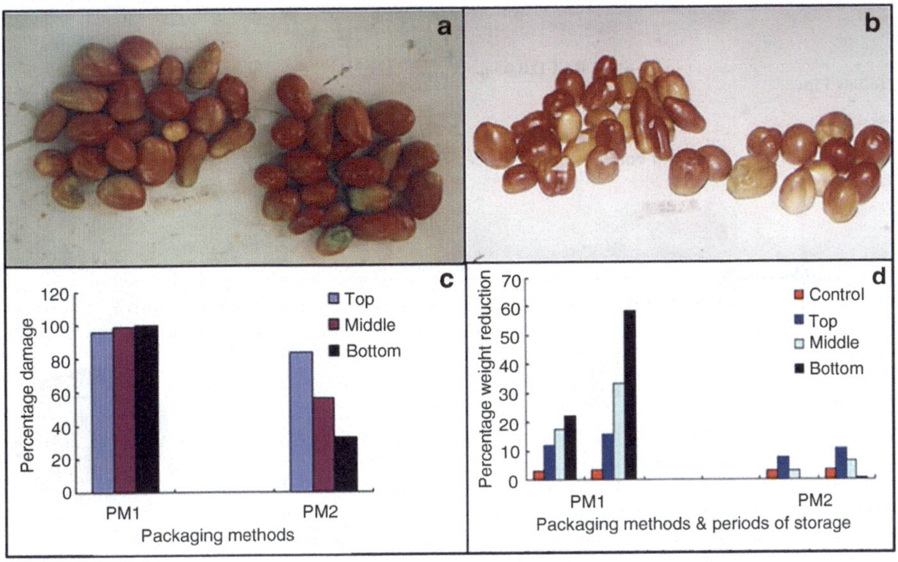

Fig. 9.2 Some samples of the stressed tomato fruits using the traditional (**a**) and plastic (**b**) basket. Percentage damage to fresh tomato fruits 48 h after 4 h of stressing using two packaging methods (**c**). Percentage weight loss in the fresh tomato fruits during the period of storage after vibration (**d**). *MP1* traditional baskets, *MP2* Plastic baskets (Idah et al., 2012)

9.3 Role of Postharvest Handling and Transportation

9.3.1 Handling

Postharvest and economic losses experienced by the horticultural industry annually because of mechanical injuries of fresh produce from harvest to postharvest handling are considerably huge. Bruise susceptibility of fruit is a measure of the reaction to external loading (Hussein et al., 2017). The impact test (Fig. 9.3), which includes dropping a steel ball with a known mass on a tomato has been studied by Pathare and Al-Dairi (2021a). In their study, they examined the bruising extent and other quality changes of tomato fruit as influenced by three different drop impact heights and three different storage conditions during 10 days of storage. They performed multiple linear regression models to ascertain the effect of the studied independent variables on the dependent variables (bruise area, firmness, weight loss, total soluble solids, color redness, and lycopene).

The results of the study showed that pigments content (lycopene and carotenoids) and total soluble solids increased significantly ($p < 0.05$) as drop height (bruising) and storage temperature increased. On the last day of storage (day 10), the highest drop impact level and storage at ambient temperature (22 °C) produced a higher reduction in lightness, firmness, yellowness, and color purity (hue°). Besides, the findings of regression models revealed a significant influence of storage temperature,

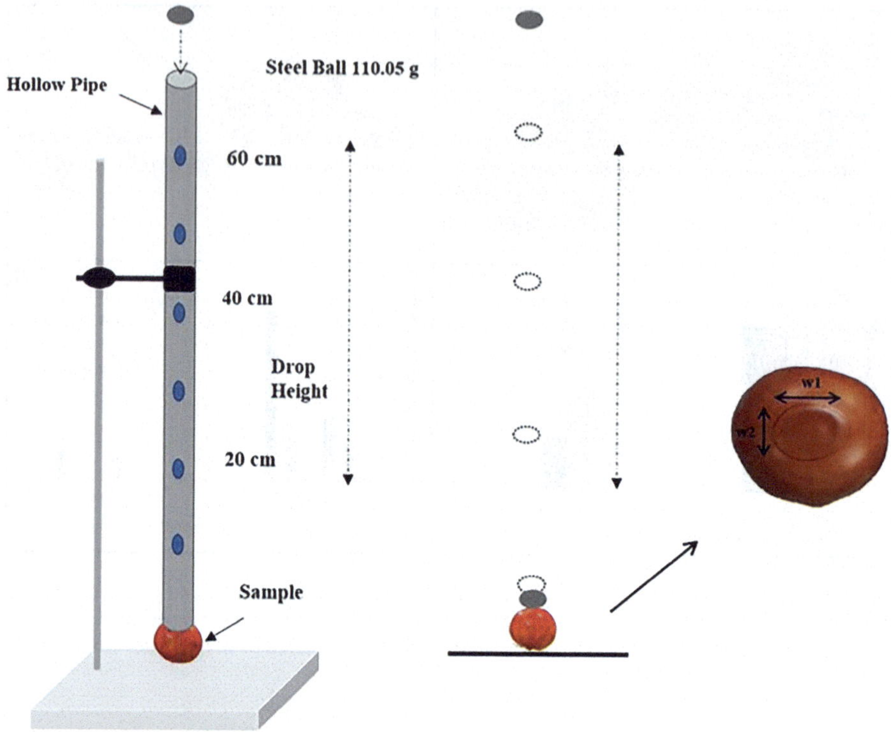

Fig. 9.3 Schematic experimental setup (drop impact test). *w*1 and *w*2 are major and minor widths (diameter) (Pathare & Al-Dairi, 2021a)

storage duration, and drop height (bruising) on the measured variables (bruise area, firmness, weight loss, total soluble solids, redness, and lycopene) at a 5% probability level with a determination coefficient (R^2) > 0.76.

Bruising intensity and quality attributes degradation could be decreased by improving handling practices and reducing the temperature during storage during the supply chain. This study can help tomato suppliers, transporters, and handlers to understand and increase awareness of the mechanism of bruising occurrence and how to reduce it.

9.3.2 Transportation

The reduction of fresh produce quality during transportation mostly occurs because of physical and biological damages/changes caused by vibration (Al-Dairi et al., 2022; Wu & Wang, 2014). As reported by Wei et al. (2019), vibration caused by transportation produced different internal and external damages to fresh produce. Internal damages are most difficult to visually identify. Walkowiak-Tomczak et al.

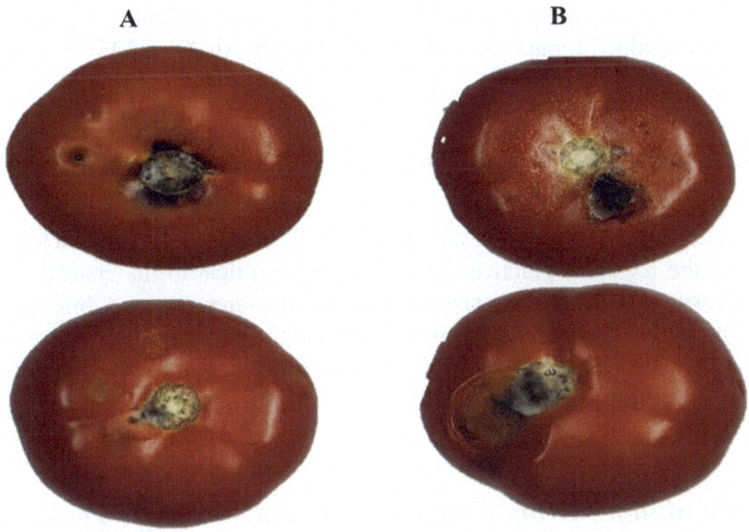

Fig. 9.4 Physiological damages on the vibrated tomatoes stored at (**a**) 10 °C and (**b**) 22 °C after10 days of storage (Pathare & Al-Dairi, 2021c)

(2021) noticed that the mechanical vibration of the simulated transport decreased the firmness of fresh horticultural produce after 14 days of storage. The influence of transport vibration has been studied on the quality of tomato fruit using transit (Al-Dairi et al., 2021b, c) and simulated vibration (Pathare & Al-Dairi, 2021c) experiments. Throughout fruit road transportation, the researchers aimed to determine the postharvest physical, chemical, and nutritional quality changes in tomato fruit as affected by transportation distance as well as storage conditions. Tomatoes were transported from a local farm to three distances, namely 100, 154, and 205 km. Once the samples arrived at the laboratory, tomatoes were stored at10 and 22 °C for 12 days. A three-Axis USB vibration/acceleration data logger was utilized during transportation to measure vibration data signals. Titratable acidity, weight loss, firmness, color, lycopene, and carotenoid contents were highly affected ($p < 0.05$) by transport distance, storage temperature, and storage duration. This is attributed to the repeated vibration stress that is occurred at the furthest transport distance (Al-Dairi et al., 2021d).

In the simulated transport vibration, the impact of vibration stress (generated from laboratory simulated transport using a vibration shaker) and storage on the quality of tomatoes for 10 days. They found that color lightness (L^*), redness (a^*), yellowness (b^*), hue°, and total color changes (ΔE^*) were dependent on the studied factors (vibration stress, storage temperature and duration). A great reduction in firmness, weight, L^*, b^*, O_2, and hue°, and an increment in TSS a^*, color index (CI), CO_2, and C_2H_4 content in the vibrated tomatoes at room temperature (22 °C). The percentage of damaged tomatoes from vibration was 38.80% at 22 °C, while it was 5.50% on the non-vibrated tomato group at the same temperature (Fig. 9.4). The results of this study showed that vibration stress during simulated transport and storage at ambient temperature expedited tomato degradation (Pathare & Al-Dairi, 2021c).

9.4 Bruise Detection and Analysis

The use of objective methods like environmental factors sensors and data loggers for monitoring the handling practices of fresh fruit is critical to understand the main causes of losses and developing several methods to overcome them. Information on tomato fruit susceptibility to bruising is very helpful for reducing and preventing bruise damage. Bruising in tomatoes is not directly visible, but it becomes apparent after postharvest handling and a certain amount of time (Cui et al., 2018).

9.4.1 Pendulum, Drop, Compression, and Vibration Tests

Laboratory experiments to determine the effect of applied various instruments have been designed to measure a particular impact and energy on individual fruit. The main measuring methods are the pendulum, drop (Sun et al., 2021), compression, and vibration tests (Opara & Pathare, 2014). The diverse and accurate data obtained from all techniques could provide a base or a comprehensive analysis of the bruise susceptibility of tomato fruit (Cui et al., 2018).

The instrument pendulum is presented as the most appropriate instrument to carry out a controlled impact experiment on tomatoes (Ahmadi & Abedi, 2013). A pendulum impactor is highly used for impact tests. It includes swinging a piece of individual products in a pendulum layout from different heights onto a particularly hard surface (Opara & Pathare, 2014). Van linden et al. (2006) used a half-spherical aluminum piece (radius of curvature: 25 mm) that best represents the shape of tomato fruit.

In another mentioned study, Pathare and Al-Dairi (2021a) dropped a uniform round stainless steel ball of a known mass from three different heights through a vertical hollow PVC pipe onto the tomato. The opposite sides of the pipe were perforated along the pipe length to reduce air resistance, which allows a free fall of the steel ball during impact testing (Fig. 9.3). Jiménez-Jiménez et al. (2013) designed an impact device to generate reproducible and controlled impact by dropping the fruit onto a metal plate and the impact parameters were recorded with a piezoelectric load cell attached to the metal plate.

Compression tests have also been employed by several researchers to study the potential of bruise damage on fresh produce. In this test, the individual fruit is compressed between two flat plates made of steel at a particular loading rate. The upper steel plate exerts compression force at the top of the produce and the lower plate bruised the bottom of the produce simultaneously (Opara & Pathare, 2014). Aktas et al. (2008) evaluated the bruise susceptibility and mechanical properties of tomatoes (Enda F1 cultivar) by using a compression test at 1 mm/min load velocity. For vibration bruising, the products are allowed to roll loosely in a container exposed to a particular vibration. Pathare and Al-Dairi (2021c) used a vibration shaker to identify the % of bruising damage to tomato fruit.

9.4.2 Non-destructive Methods (Image Processing)

Using machine learning has become a beneficial tool that allows for effective and quick assessment of food quality commodities based on empirical data. The non-destructive techniques have been increasingly implemented for effective fresh fruit quality assurance and evaluation (Nturambirwe & Opara, 2020).

Hyperspectral imaging, which combines imaging and spectral information, is a favorable non-destructive technique that allows the detection of external or internal quality attributes. This technique has been utilized for tomato quality detection like maturity, soluble solid contents, drought, cracks, etc. (Huang et al., 2018). Few studies based on hyperspectral imaging techniques have been done for tomato bruising detection. Sun et al. (2021) used hyperspectral imaging techniques to detect tomato bruising that was generated due to the falling technique. The influence of drop height, fruit size, and detection time point on hyperspectral characteristics was examined. Their results revealed that the spectra of bruised tomato fruit were lower than those of healthy control tomatoes. They also concluded that hyperspectral technology can categorize tomato fruit bruise damage over time (Fig. 9.5).

Visible and near-Infrared (Vis-NIR) spectroscopy used radiation in the range of 350–2500 nm wavelength region of the electromagnetic spectrum and has demonstrated accuracy in the detection of bruised surfaces. In NIR spectroscopy, the individual fruit is exposed to NIR radiation, and the transmitted and reflected radiation is measured. When the radiation penetrates the produce, its spectral attributes change across absorption processes and wavelength-dependent scattering.

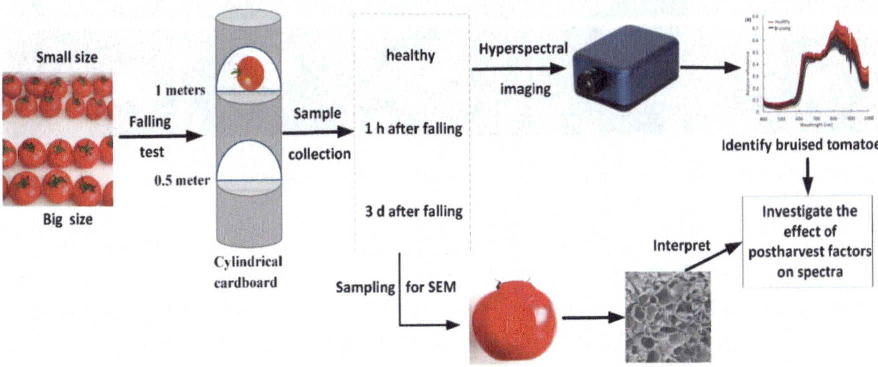

Fig. 9.5 Schematic overview of the analytical procedures of Sun et al. (2021)

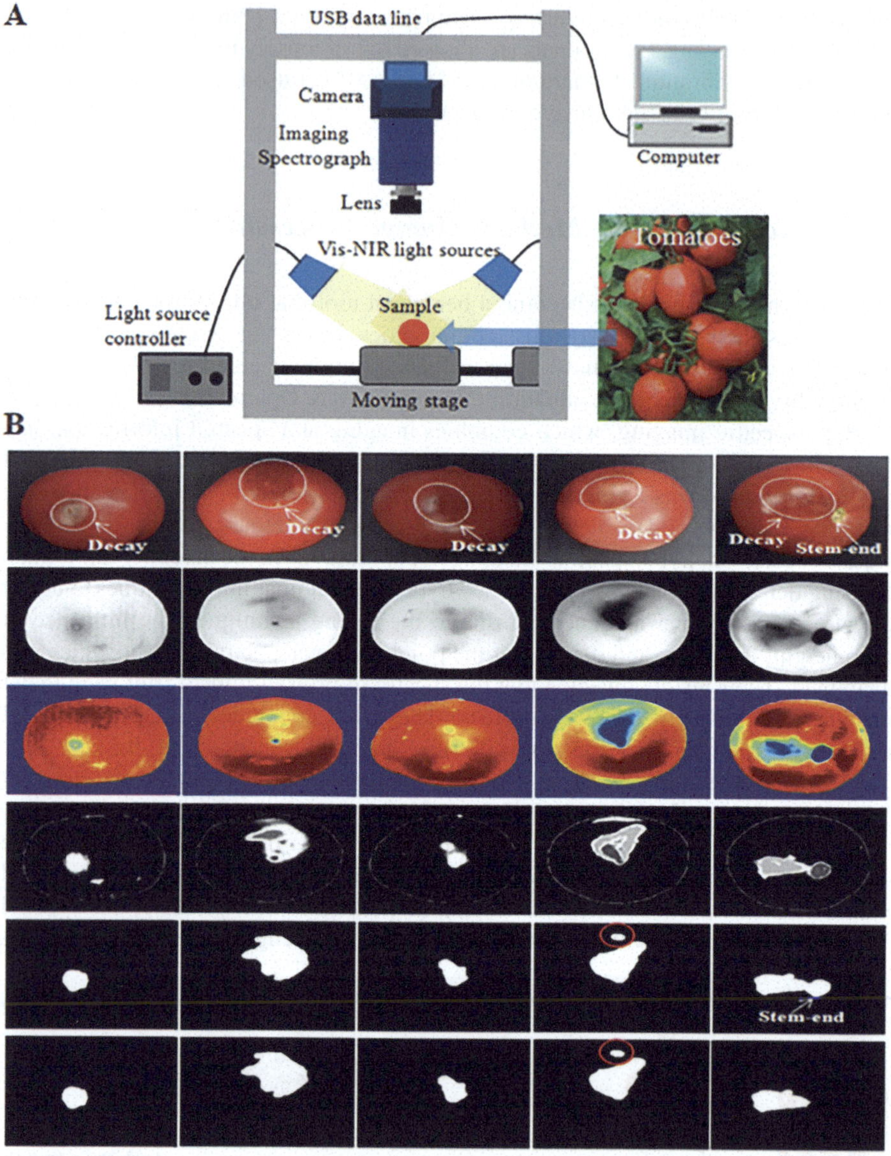

Fig. 9.6 Vis-NIR hyperspectral imaging (**a**) and segmentation results of typical decayed tomatoes samples (**b**) (Wang et al., 2021)

This change depends on the chemical composition of the product and on its light-scattering characteristics which are related to the microstructure (Opara & Pathare, 2014). The study of Wang et al. (2021) demonstrated that Vis-NIR hyperspectral imaging linked with image-spectrum merging processing technology could be a prospective approach for the detection of tomatoes with early decay and damage

(Fig. 9.6). Wu and Wang (2014) investigated bruise damage in tomatoes generated from simulated transport based on vis/NIR spectroscopy. Their study showed that Vis/NIR spectroscopy has the potential to predict tomato fruit damage resulting from vibration.

Thermal Imaging (TI) is a non-invasive and emerging process analytical method appropriate for the food industry. In TI the invisible heat radiation radiated by an object is transformed into a visible image (which is called a thermal image or thermogram) via infrared thermal cameras (Opara & Pathare, 2014). "ThermaCAM™ SC3000" (FLIR Systems AB, Sweden) was used to detect tomato bruising by Vereycken et al. (2002). With respect to bruise detection, they compared three temperature treatments, i.e., oven heating to 70 °C for 1 or 2 min, cooling the fruits for 90 min at 1 °C, and microwave heating for 7 or 15 s. Slight differences between intact and bruised tissue after a 15 s treatment were found and detected by means of microwave heating, reflecting cold circular spots of bruises on thermal images of tomato fruit surface.

Magnetic resonance imaging (MRI) is a contactless, non-invasive technique that is often applied to analyze the internal defects and internal structure of horticultural produce. Magnetic resonance (MR) imaging methods were implemented by Milczarek et al. (2009) to characterize the environment of water protons in bruised and non-bruised tomato tissues. MR images of tomatoes proved to be effective for predicting the conductivity score of pericarp tissue in the fruit for bruise detection.

9.4.3 Prediction Models

Employing models to predict bruising can provide valuable information about the effect of fruit attributes on bruise susceptibility, leading to delivering essential recommendations for fruit handling during postharvest handling. Many researchers have applied different analytical and statistical models to predict bruising in fruit (Opara & Pathare, 2014). A logistic regression function based on fruit parameters, precise impact data, and a sensory-based score for the subsequent bruise damage, proved to be proper to describe the bruise susceptibility of tomatoes (Van linden et al., 2006).

The damage assessment and multivariate analysis approach proposed by Cui et al. (2018) can be utilized to verify the bruise susceptibility of tomatoes non-destructively. Besides, multivariate models may be useful for examining the influence of different fresh fruit handling practices on the risk of developing bruises after mechanical damage and impact. In their study, they developed five estimation models combining quality factors and the existence of bruising resulting from peak contact forces. Artificial neural network (ANN) regression models, linear regression models, and a logistic regression model were also developed. Logistic regression proved a link between the impact force delivered to the tomatoes and the consequences of bruise damage.

Milczarek et al. (2009) developed an in-line technique to detect damaged pericarp tissue of tomato fruit utilizing multivariate analysis of NMR images. The finite element method (FEM) is also another numerical technique used to resolve indirect problems, like impact phenomena, in harvest and postharvest progressions (Rashvand et al., 2022). The mechanical damage of different forms of tissues in tomato fruit was predicted in the study of Li et al. (2013) by utilizing nonlinear multiscale finite element models. They concluded that the nonlinear multiscale finite element technique was able to predict the internal mechanical damage behavior of tomato fruit under different loading conditions. Besides, this technique can be expanded to examine the internal mechanical behavior of other fruits and vegetables.

Li et al. (2021) developed an extended finite element (XFEM) model for examining the fracture mechanical response of tomatoes under postharvest mechanical damage. The results of this study establish the XFEM application as a novel technique to understand how fruit is damaged under mechanical loading when the fruit mechanics differ under different conditions and the extent of damage propagation which are significant for developing or improving new mechanical handling technologies. As stated by Li et al. (2017), mathematical models of tomato fruit mechanical damage are vital to better postharvest classification handling systems for fruits or encourage the development of emerging methods to assess fruit quality.

9.5 Conclusion

Tomato is one of the most widely grown and important vegetable crops worldwide. Tomato is climacteric, sensitive, and one of the highly perishable fresh produce which encounters various problems during storage, transportation, and marketing. Bruising is the most common form of mechanical damage influencing fresh tomatoes and reduces quality to the consumer and income. The results of different studies discussed in this paper could be useful for the development of real-time monitoring of the quality characteristics of horticultural products during the postharvest supply processes. Identifying the impact conditions which can cause bruising is essential to improve harvesting, handling, sorting, and transporting equipment and procedure. Besides, the data can be of great help to designers of end-effector and processing equipment in reducing the mechanical damage resulting from mechanical impact and collision and providing products with high quality for consumers and processors.

Acknowledgement We would like to thank Sultan Qaboos University, Oman for their financial support under the project codes: : CL\SQU\QU\AGR\23\02 and IG/AGR/SWAE/19/03.

References

Addo, J., et al. (2015). Assessment of farmer level postharvest losses along the tomato value chain in three agro-ecological zones of Ghana. *International Journal, 2*(9), 2311–2476.

Ahmadi, E., & Abedi, G. (2013). Sensitivity of tomatoes to internal bruising induced by mechanical stress. *European Journal of Horticultural Science, 78*(5), 219–224.

Aktas, T., Polat, R., & Atay, U. (2008). *Mechanical properties and bruise susceptibility of tomatoes related to impact direction and drop-height.* Philippine Agricultural Scientist.

Al-Dairi, M., Pathare, P. B., & Al-Mahdouri, A. (2021a). Effect of storage conditions on postharvest quality of tomatoes: A case study at market-level. *Journal of Agricultural and Marine Sciences, 26*(1), 13–20.

Al-Dairi, M., Pathare, P. B., & Al-Yahyai, R. (2021b). Chemical and nutritional quality changes of tomato during postharvest transportation and storage. *Journal of the Saudi Society of Agricultural Sciences, 20*(6), 401–408.

Al-Dairi, M., Pathare, P. B., & Al-Yahyai, R. (2021c). Effect of postharvest transport and storage on color and firmness quality of tomato. *Horticulturae, 7*(7), 163.

Al-Dairi, M., Pathare, P. B., & Al-Yahyai, R. (2021d). Quality changes kinetic of tomato during postharvest transportation and storage. *Journal of Food Process Engineering, 44*, e13808. https://doi.org/10.1111/jfpe.13808

Al-Dairi, M., Pathare, P. B., Al-Yahyai, R., & Opara, U. L. (2022). Mechanical damage of fresh produce in postharvest transportation: Current status and future prospects. *Trends in Food Science & Technology, 124*(1), 195–207.

Allende, A., Desmet, M., Vanstreels, E., Verlinden, B. E., & Nicolaı̈, B.M. (2004). Micromechanical and geometrical properties of tomato skin related to differences in puncture injury susceptibility. *Postharvest Biology and Technology, 34*(2), 131–141.

Cherono, K., & Workneh, T. S. (2018). A review of the role of transportation on the quality changes of fresh tomatoes and their management in South Africa and other emerging markets. *International Food Research Journal, 25*(6), 2211–2228.

Costa, J. D. D., et al. (2021). Road transport vibration stress impact on 'Palmer'mangoes quality and shelflife. *Revista Brasileira de Fruticultura, 43*(2), e-641.

Cui, J., Yang, M., Son, D., Park, S., & Cho, S.-I. (2018). Estimation of tomato bruising by mechanical impact force using multivariate analysis. *Horticultural Science, 53*(9), 1352–1359.

Erba, D., et al. (2013). Nutritional value of tomatoes (Solanum lycopersicum L.) grown in greenhouse by different agronomic techniques. *Journal of Food Composition and Analysis, 31*(2), 245–251.

Huang, Y., Lu, R., & Chen, K. (2018). Assessment of tomato soluble solids content and pH by spatially-resolved and conventional Vis/NIR spectroscopy. *Journal of Food Engineering, 236*, 19–28.

Hussein, Z., Fawole, O. A., & Opara, U. L. (2017). Investigating bruise susceptibility of pomegranate cultivars during postharvest handling. *African Journal of Rural Development, 2*(1), 33–39.

Idah, P., Ajisegiri, E., & Yisa, M. (2007). An assessment of impact damage to fresh tomato fruits. *Assumption University Journal of Technology, 10*(4), 271–275.

Idah, P. I., Gana, Y. M., Ogbonnaya, C., & Morenikeji, O. O. (2012). Simulated transport damage study on fresh tomato (*Lycopersicon esculentum*) fruits. *Agricultural Engineering International: CIGR, 14*(2), 119–126.

Jiménez-Jiménez, F., et al. (2013). Table olive cultivar susceptibility to impact bruising. *Postharvest Biology and Technology, 86*, 100–106.

Li, Z., Li, P., Yang, H., & Liu, J. (2013). Internal mechanical damage prediction in tomato compression using multiscale finite element models. *Journal of Food Engineering, 116*(3), 639–647.

Li, Z., Andrews, J., & Wang, Y. (2017). Mathematical modelling of mechanical damage to tomato fruits. *Postharvest Biology and Technology, 126*, 50–56.

Li, D., Li, Z., & Tchuenbou-Magaia, F. (2021). An extended finite element model for fracture mechanical response of tomato fruit. *Postharvest Biology and Technology, 174*, 111468.

Lubes, G., & Goodarzi, M. (2017). Analysis of volatile compounds by advanced analytical techniques and multivariate chemometrics. *Chemical Reviews, 117*(9), 6399–6422.

Milczarek, R. R., Saltveit, M. E., Garvey, T. C., & McCarthy, M. J. (2009). Assessment of tomato pericarp mechanical damage using multivariate analysis of magnetic resonance images. *Postharvest Biology and Technology, 52*(2), 189–195.

Munhuewyi, K. (2012). *Postharvest losses and changes in quality of vegetables from retail to consumer: A case study of tomato, cabbage and carrot*. Master thesis, Stellenbosch University, Stellenbosch.

Nturambirwe, J. F. I., & Opara, U. L. (2020). Machine learning applications to non-destructive defect detection in horticultural products. *Biosystems Engineering, 189*, 60–83.

Opara, U. L., & Pathare, P. B. (2014). Bruise damage measurement and analysis of fresh horticultural produce—A review. *Postharvest Biology and Technology, 91*, 9–24.

Pathare, P. B., & Al-Dairi, M. (2021a). Bruise damage and quality changes in impact-bruised, stored tomatoes. *Horticulturae, 7*(5), 113.

Pathare, P. B., & Al-Dairi, M. (2021b). Bruise susceptibility and impact on quality parameters of pears during storage. *Frontiers in Sustainable Food Systems, 5*, 658132.

Pathare, P. B., & Al-Dairi, M. (2021c). Effect of simulated vibration and storage on quality of tomato. *Horticulturae, 7*(11), 417.

Rashvand, M., Altieri, G., Genovese, F., Li, Z., & Di Renzo, G. C. (2022). Numerical simulation as a tool for predicting mechanical damage in fresh fruit. *Postharvest Biology and Technology, 187*, 111875.

Sablani, S., Opara, L., & Al-Balushi, K. (2006). Influence of bruising and storage temperature on vitamin C content of tomato fruit. *Journal of Food Agriculture and Environment, 4*(1), 54.

Sarma, P. K. (2018). Postharvest losses of tomato: A value chain context of Bangladesh. *International Journal of Agricultural Education and Extension, 4*(1), 85–92.

Sibomana, C. I., Opiyo, A. M., & Aguyoh, J. N. (2015). Influence of soil moisture levels and packaging on postharvest qualities of tomato (Solanum lycopersicum). *African Journal of Agricultural Research, 10*(12), 1392–1400.

Sun, Y., Pessane, I., Pan, L., & Wang, X. (2021). Hyperspectral characteristics of bruised tomatoes as affected by drop height and fruit size. *LWT, 141*, 110863.

Tran, D. T., et al. (2017). Population modeling approach to optimize crop harvest strategy. The case of field tomato. *Frontiers in Plant Science, 8*, 608.

Van Linden, V., De Ketelaere, B., Desmet, M., & De Baerdemaeker, J. (2006). Determination of bruise susceptibility of tomato fruit by means of an instrumented pendulum. *Postharvest Biology and Technology, 40*(1), 7–14.

Vereycken, R., Bravo, C., Ramon, H., De Baerdemaeker, J., & Van Linden, V. (2002). Detection technique for tomato bruise damage by thermal imaging. *International Conference: Postharvest Unlimited, 599*, 389–394.

Walkowiak-Tomczak, D., Idaszewska, N., Łysiak, G. P., & Bieńczak, K. (2021). The effect of mechanical vibration during transport under model conditions on the shelf-life, quality and physico-chemical parameters of four apple cultivars. *Agronomy, 11*(1), 81.

Wang, H., Hu, R., Zhang, M., Zhai, Z., & Zhang, R. (2021). Identification of tomatoes with early decay using visible and near infrared hyperspectral imaging and image-spectrum merging technique. *Journal of Food Process Engineering, 44*(4), e13654.

Wei, X., et al. (2019). Excess water loss induced by simulated transport vibration in postharvest kiwifruit. *Scientia Horticulturae, 250*, 113–120.

Wu, G., & Wang, C. (2014). Investigating the effects of simulated transport vibration on tomato tissue damage based on vis/NIR spectroscopy. *Postharvest Biology and Technology, 98*, 41–47.

Xing, J., Ngadi, M., Wang, N., & De Baerdemaeker, J. (2006). Bruise detection on tomatoes based on the light scattering image. In *2006 ASAE Annual Meeting* (p. 1). American Society of Agricultural and Biological Engineers.

Chapter 10
Bruising of Avocado (*Persea americana* M.) Fruit

Muhammad Sohail Mazhar, Neil Tuttle, and Daryl Joyce

Abstract Avocado is prized as a fruit of high economic, nutritional, and medicinal value. Its worldwide production and trade are continuously rising. However, a major concern of avocado consumers is inconsistent value for money spent on purchase of the fruit. Despite decades of research, development, extension effort, and considerable financial investment, this perception persists. Flesh bruising and fruit rot are predominant among the quality issues that concern consumers. Flesh bruising is typically associated with physical impact and compression events. Skin abrasion bruising may result from vibration. Understanding the bruising of avocado fruit during handling through supply chains is important so that preventative measures can be implemented. This chapter reviews the causative factors of avocado fruit bruising, the occurrence and severity of bruising associated with supply chain handling practices, and technology options towards avoiding fruit bruising. Based on the present understanding, best practice recommendations for the avocado industry are presented.

Keywords Avocado · Bruising · Impact · Compression · Postharvest handling

M. S. Mazhar (✉)
Department of Industry, Tourism and Trade, Northern Territory Government of Australia, Darwin, NT, Australia
e-mail: MuhammadSohail.Mazhar@nt.gov.au

N. Tuttle
Griffith University, Nathan, Australia
e-mail: n.tuttle@griffith.edu.au

D. Joyce
Department of Agriculture and Fisheries, Queensland Government, Brisbane, Australia
e-mail: Daryl.Joyce@daf.qld.gov.au

P. B. Pathare, U. L. Opara (eds.), *Mechanical Damage in Fresh Horticultural Produce*, https://doi.org/10.1007/978-981-99-7096-4_10

10.1 Introduction

Avocado (*Persea americana* Mill) fruit are consumed worldwide for their high nutritional value and perceived health benefits. Avocados are a source of monounsaturated fat, soluble fibre, folate, vitamins B-6, C, E, and K, and minerals including copper, iron, and potassium. The fruit is consumed fresh as well as value-added (Dorantes et al., 2004). Fresh consumption includes as an accompaniment to main servings and in salad. Semi- and processed forms include guacamole, beverages, sauces, canned, frozen, dehydrated, and pureed. Value-added avocado products also include beauty, health, and sanitary lines.

In 2019, world avocado production was mostly from 22 countries between 27°N to 27°S (Food and Agriculture Organization (FAO), 2021, AltlasBig, 2021). World avocado production in terms of area under orchard, quantity produced, and international trade has increased steadily over the last 20 years (Fig. 10.1) (FAO, 2021). The fruit gained the reputation of a superfood fruit (International Trade Centre (ITC), 2018). In 2018, avocado ranked sixth globally among imported fruit (ITC, 2018). As estimated by ITC (2021), avocado has untapped export potential of about US$ 3.7 billion. Demand for avocado is expected to continue to increase worldwide due to increased understanding of benefits from consuming avocados, advances in production and supply chain logistics, and a trend in international trade arrangements to reduce tariff costs (NZ-Avocados, 2021). Consistent delivery of quality avocado fruit to consumers around the world should maintain growth in production, trade, and consumption.

However, and despite decades of research, development, and extension, avocado consumers are often disappointed by inconsistent quality, particularly with bruises and rots encountered at consumption (Hort-Innovation, 2017). Research has

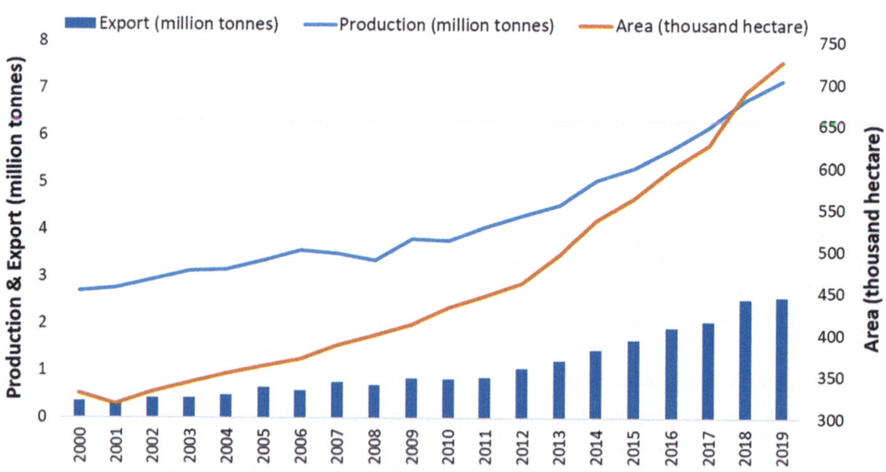

Fig. 10.1 World avocado production and export (million tonnes; LHS axis) and area (thousand hectares; RHS axis) for 2000–2019. (Data source: FAO, 2021)

explored many aspects of avocado bruising; e.g., Arpaia et al. (1987), Baryeh (2000), Gamble et al. (2010), Bill et al. (2014), Mazhar (2015). This chapter overviews bruising in avocado fruit in a whole of supply chain context. It covers causes, occurrence, and severity associated with handling practices across serial supply chain nodes. It also considers technological options for lessening bruising of avocado fruit. A set of best practice recommendations is offered for the avocado industry.

10.2 Overview of Bruising

Bruises in fruit tissue are expressions of mechanical damage events involving compression, impact, and abrasion injuries (MacLeod et al., 1976; Studman, 1995). Typical symptoms expressed upon bruising events are light to dark brown or grey discolouration of the affected tissue (Fig. 10.2). The discolouration results largely from enzymic (e.g., polyphenol oxidase) oxidation of soluble phenolic compounds (Lurie, 2009). Visual manifestation of oxidation also depends on tissue pH, fruit physiological characteristics (e.g., maturity, ripening stage) and handling conditions (e.g., temperature, duration) before and after the bruising event (Adato & Gazit, 1974; Yakoby et al., 2000; Mazhar, 2015). Water-soaked tissue is another form of bruise expression (Quintana & Paull, 1993). Bruises in avocado also express as hairline to coarse cracks (Hofman, 2003) (Fig. 10.2). Also, there may be no obvious expression at all (Pajuelo et al., 2003).

Avocado as a climacteric fruit (Biale et al., 1954) is typically harvested green mature and held until ripening is initiated postharvest (Arpaia et al., 1987). In this context, avocado fruit may not express symptoms of bruising until they soften during ripening (Bill et al., 2014). Historically, avocado was marketed green mature (Milne, 1998). However, consumer preferences for buying ready to use fruit (Lee & Coggins

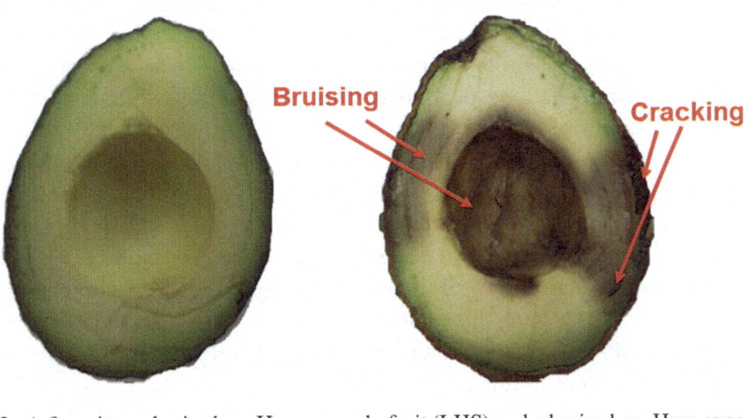

Fig. 10.2 A firm ripe unbruised cv. Hass avocado fruit (LHS) and a bruised cv. Hass avocado fruit (RHS) with visible bruising and cracking

Jr, 1982) encouraged supply chain stakeholders to supply fruit at various serial stages of ripeness. However, handling and marketing softening fruit increases the prospects of mesocarp bruising. Arpaia et al. (1987), Milne (1998), Baryeh (2000), and others have associated bruising in ripening avocado fruit with fruit maturity, firmness, and energy absorbed during the event.

Botanically, the avocado fruit is a berry (Schaffer et al., 2013). The mature unripe fruit has a cohesive pericarp comprised of a green epicarp, a light green to near white mesocarp, and a paper-thin endocarp surrounding the seed (Whiley et al., 2002). In green mature fruit, the skin and composite tissues are contiguous and so not easily separated. During ripening, the epicarp of certain varieties changes colour from green to purple/black due to synthesis of anthocyanins. At this stage, the epicarp may separate from the mesocarp and even form a 'shell' enclosing the underlying softening mesocarp (Fig. 10.2). Duly, the mesocarp can be easily separated from seed (Schaffer et al., 2013). Over the course of ripening, elastic deformability of individual cells and tissue under applied forces gradually decreases such that the mesocarp becomes increasing susceptible to plastic deformation which expresses as bruising (Studman, 1995). Upon mechanical damage, and modulated by elastic vs. plastic tissue characteristics, enzymic oxidation of phenolics manifests as light brown to black bruise symptoms (Linden & Baerdemaeker, 2005).

In a typical commercial supply chain, green mature avocado fruit are harvested, cooled, treated, packed, and dispatched from the farm. Thereafter, they are gas-ripened by wholesalers or dedicated ripeners (Hofman & Ledger, 2001a). Hofman and Ledger (2001b) considered that avocado fruit are susceptible to bruising at any stage in the supply chain. Nonetheless, evidence of bruising in fruit collected from the end of packing lines is relatively sparse. Fruit susceptibility to bruising increases particularly after the application of commercial ripening protocols (Hofman & Sandoval, 2002; Arpaia et al., 2005; Opara & Pathare, 2014). Moreover, bruising susceptibility increases as they pass through subsequent distribution channels to retail store handling and display and ultimately the point of consumption (Mazhar, 2015). Hofman et al. (2001) reported that up to 80% of avocados sampled from the retail shelf had more than 10% of the cut surface area of mesocarp affected by bruising. Moreover, up to 25% of fruit sampled had over 25% of fruit cut surface area affected by bruising. Embry (2009) reported that 63% of avocados sampled from retail store display had mesocarp disorders and that 29% of these were affected by more than 10% bruised mesocarp.

Avocado consumers have consistently reported their lack of confidence in the quality of fruit purchased off the shelf (Campbell, 1993; Hofman & Ledger, 1999; Harker et al., 2007). Smith et al. (1990) found that 40% of avocado consumers had bad experiences with avocado purchase and consumption. Dermody (1990) reported that 53% of participants surveyed complained that avocado fruit that appeared satisfactory at purchase were disappointing at the time of consumption. Gamble et al. (2010) determined that more than 10% mesocarp affected by disorders detracts from a consumer's intent to repeat purchase.

Better understanding the causative factors of mesocarp bruising in commercial supply chain handling operations, as outlined in subsequent sections, has returned positive outcomes. In Australia for example, quality concerns have reduced from 80% fruit with >10% of cut surface area damaged in 2001 (Hofman et al., 2001) to 20% in 2015 (Hort-Innovation, 2017). In 2015, Joyce reported that the proportion of consumers dissatisfied with avocado fruit quality was 18%. In Australia, the avocado industry committed to a target that "*by 2021, over 90 per cent of avocados received by consumers will meet or exceed their expectations of quality*" (Henderson, 2019).

10.3 Factors Contributing to Bruising

A range of physicochemical factors contribute to bruising and bruise expression in avocado fruit. Abrasion, compression, and impact events give rise to mechanical damage, and are the primary physical forces eliciting bruise development (MacLeod et al., 1976; Studman, 1995; Mazhar et al., 2012, 2013b).

10.3.1 Impact, Compression, Abrasion

Mechanical force transfers energy into fruit such as to create stress at cell and tissue levels. These anatomical structures experience strain manifesting as deformation in shape, size, and form. Depending on overall fruit characteristics, especially firmness, cells, and tissues may initially exhibit elastic deformation and then return to their original form once the energy dissipates. Alternatively, or as well, they may exhibit plastic deformation and display persistent deformation. From the onset of mechanical damage and depending on the fruit and handling conditions (Bill et al., 2014), bruising expresses with visual symptoms.

In ripening avocado, an impact of just ~0.38 kg m^2 s^{-2} (J) to fruit dropping 25 cm onto a solid metal surface induced bruising (Mazhar et al., 2012). Bruise severity expressed as volume of mesocarp affected varied across softening vs. firm ripe vs. soft ripe stages (White et al., 2009). Greater drop heights of ~0.81 and ~1.61 J impact energies fruit trended to increase bruise severity (Table 10.1).

In commercial supply chains, avocado fruit are handled in trays up to and including at the point of retail display. Firm ripe avocados in trays dropped

Table 10.1 Bruise severity (ml) ± standard deviation (SD) in avocado fruit (n = 10) impacted at different ripening stages and held at 20 °C for 48 h (Mazhar, 2015)

Energy absorbed (J)	Softening	Firm ripe	Soft ripe
~0.38	2.3 ± 0.7e	7.8 ± 2.3c	26.8 ± 5.9a
~0.81	3.5 ± 1.2de	16.1 ± 5.7b	29.8 ± 2.6a
~1.61	4.4 ± 1.3d	16.5 ± 3.8b	30.8 ± 4.6a

Values sharing the same letters are not significantly different at P = 0.05 from each other by Tukey's least significant difference (LSD)

Table 10.2 Bruise severity (ml) ± SD in firm ripe avocado fruit kept in trays ($n = 3$) impacted from different drop heights and angles and held at 20 °C for 48 h (Mazhar, 2015)

Tray drop height (cm)	Tray drop angle (degrees)	Impact force (*G*)	Bruise severity (ml)
15	0	83.8 ± 11.7bc	4.8 ± 0.8bc
	15	62.9 ± 23.4c	3.0 ± 0.3c
	30	64.6 ± 10.0c	3.0 ± 0.8c
25	0	133.5 ± 6.0a	10.2 ± 2.1a
	15	100.0 ± 1.9b	6.4 ± 2.7b
	30	88.6 ± 13.3b	5.3 ± 0.1b

Values sharing the same letters are not significantly different at $P = 0.05$ from each other by Tukey's LSD

Table 10.3 Bruise severity (ml) ± SD in firm ripe avocado fruit ($n = 3$) subjected to manual and mechanical compression force and held at 20 °C for 48 h (Mazhar et al., 2013b)

Manual compression		Mechanical compression	
Qualitative level	Bruise severity (ml)	Quantitative amount (N)	Bruise severity (ml)
Slight	1.7 ± 0.8c	~1 kg (~10 N)	0.7 ± 0.3d
Moderate	2.7 ± 0.6b	~2 kg (~20 N)	3.3 ± 0.8b
Firm	9.3 ± 1.2a	~3 kg (~30 N)	9.5 ± 0.5a

Values sharing the letters are not significantly different at $P = 0.05$ from each other by Tukey's LSD

Table 10.4 Bruise severity (ml) ± SD in firm ripe and soft ripe avocado fruit subjected to different levels of deformation and held at 20 °C for 48 h (Mazhar, 2015)

Ripening stage	1 mm ($n = 5$)	1.5 mm ($n = 5$)	3 mm ($n = 3$)
Firm ripe	0.1 ± 0.2d	0.4 ± 0.6c	6.2 ± 1.2b
Soft ripe	0.1 ± 0.2d	0.6 ± 1.1c	10.2 ± 0.6a

Values sharing the letters are not significantly different at $P = 0.05$ from each other by Tukey's LSD

horizontally from only 15 cm onto a hard surface suffered mesocarp bruising (Mazhar, 2015). Bruising severity increased with tray drop height and decreased with tray drop angle (Table 10.2).

For compression induced bruising, Mazhar et al. (2013b) showed that 1 kg force (9.8 N), which relates to 'slight' compression by thumb, was sufficient to cause bruising in firm ripe avocados. Greater compression forces of 2 and 3 kg force, corresponding to 'moderate' and 'firm' thumb pressures, respectively—comparable to the force required to depress a plunger on a hand sanitiser bottle—induced higher levels of bruise severity (Table 10.3).

Mazhar (2015) discerned permanent deformation and visible bruising in mesocarp tissue of firm and soft ripe avocados compressed to just 1 mm depth. Further compressions to 1.5 and 3 mm depths caused greater bruising (Table 10.4).

A range of devices are used in the industry to measure and monitor avocado fruit firmness (White et al., 1999). However, even the impact and compression energies from firmness measurement devices can bruise ripening avocado fruit (Mazhar et al., 2018b.) It is recommended that non-destructive and non-damaging firmness

Fig. 10.3 Skin damage upon abrasion of an avocado fruit. (Image Source: Roberto Marques, with permission)

measurement tools be sought and used to reduce impact and compression associated flesh bruising in avocado fruit (Landahl & Terry, 2020).

Pak et al. (2001) demonstrated the association on avocado fruit between abrasion energy and skin damage (Fig. 10.3). The likelihood of skin abrasion and ripe rot increased with greater protuberance of nodules (lenticels) on the fruit surface. Abrasion, compression, and impact energies can cause individually or in combinations contribute to mechanical damage, including mesocarp bruising.

10.3.2 Preharvest Fruit Characteristics

Popular avocado varieties cultivated globally include Hass, Shepard, Fuerte, and Pinkerton (Moore, 2001). New varieties are also being grown, including Gem, Nabal, and Reed (Dominguez et al., 2014). Characteristic fruit shapes and weights vary among varieties, such as from round to oblong and from about 0.25 to over 2 kg (Schaffer et al., 2013). Some varieties have smooth skin, others are rough with dents of varying dimensions (Whiley et al., 2002). Avocado fruit differ in their chemical composition at different stages of development and ripening (Golan et al., 1977). Hausch et al. (2020) reported varying aroma volatile profiles through the harvest season for cvs. Hass and Gem.

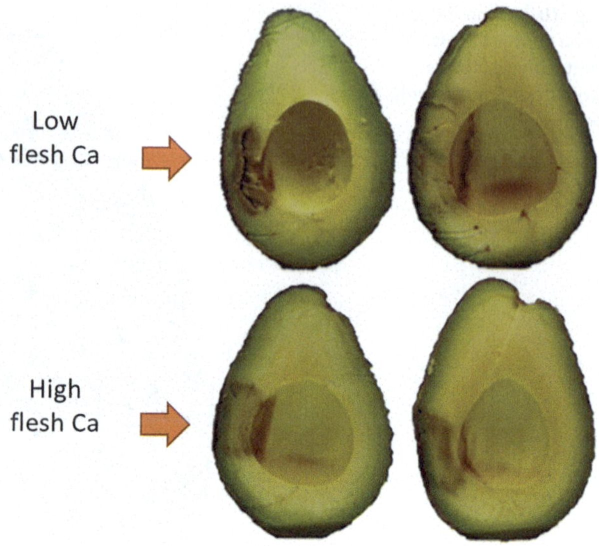

Fig. 10.4 Flesh bruising in firm ripe cv. Hass avocado fruit with relatively low (top) vs. high (bottom) Ca concentrations after impacts from ~40 cm height (~1 J). The impacted fruit were held at 20 °C for 48 h before cutting (Joyce, 2020)

Biale and Young (1971) reported that exocarp structure, particularly sclerenchyma tissue, can influence fruit resistance to mechanical damage. In comparing bruising susceptibility of cvs. Fuerte, Hass, and Pinkerton, Arpaia et al. (1987) observed a differential visible response in association with cultivar-specific phenolic substrates.

Avocado fruit quality, including bruise susceptibility, may be influenced by fruit mineral composition. Joyce (2020) observed that fruit with relatively higher calcium (Ca) and relatively lower nitrogen (N) concentrations are less susceptible to fruit quality issues. Fruit from vigorous trees with low flesh Ca expressed 43% greater flesh bruising as compared with fruit with high flesh Ca from non-vigorous trees (Fig. 10.4). Ratios among potassium (K) and magnesium (Mg) in addition to Ca and boron (B) levels influence fruit robustness with regards postharvest quality issues. Soil application of soluble Ca to avocado plants at critical stages of fruit development may improve fruit robustness (Joyce, 2020). Environmental factors, especially temperature and relative humidity, and rootstock may all also influence Ca accumulation in the fruit (Joyce, 2020).

Harvest maturity is suggested by Arpaia et al. (1987) to be inversely related with avocado fruit susceptibility to mesocarp bruising. Over a 4.5-month harvest window, fruit dry matter progressively increased and bruising severity concomitantly decreased in avocado fruit harvested fortnightly from the same trees (Fig. 10.5) (Mazhar et al., 2018a). With increasing fruit maturity, Ca concentration in the mesocarp increases (Bower & Cutting, 1988) and dry matter content also increases along with increased oil content (Lee et al., 1983). In contrast, water content in the fruit tissue decreases (Gupta & Kaur, 2000). The net effect mediated at cell and tissue level evidently relates to a shift from relatively more plastic deformation in

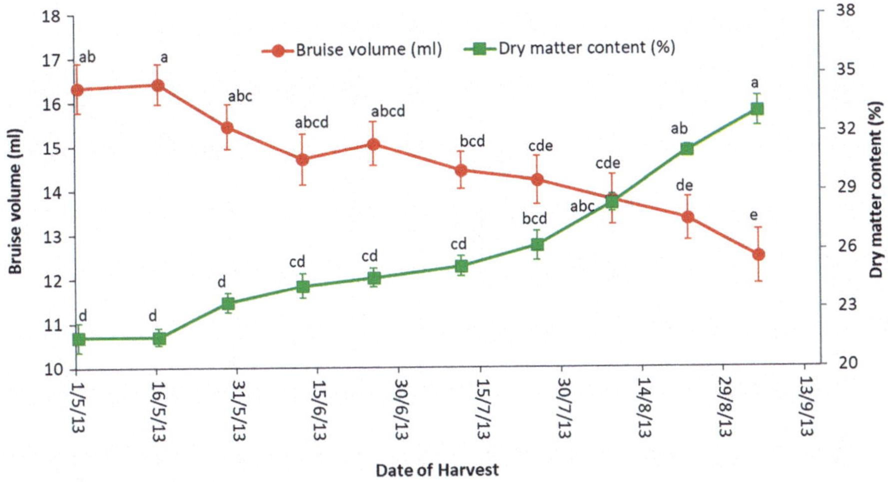

Fig. 10.5 Dry matter content (%) ($n = 5$) and bruise severity (ml) ($n = 20$) ± SD in avocado fruit harvested over ~4.5 months. Fruit for bruise severity assessment were impacted (energy ~0.9 J) at firm ripe and held at 20 °C for 48 h. Values with letters in common are not significantly different at $P = 0.05$ by Tukey's LSD (Mazhar et al., 2018a)

fruit with low dry matter concentration to comparatively more elastic deformation in fruit with higher dry matter concentration (Mazhar, 2015).

10.3.3 Postharvest Fruit Characteristics and Holding Conditions

Avocado fruit susceptibility to bruising increases as fruit ripening progresses from hard green mature fruit through rubbery, softening, firm ripe, and soft ripe hand firmness stages (Mazhar, 2015). Mazhar et al. (2013a) showed that hard green mature fruit having absorbed ~1.68 J energy (100 cm drop height) did not express bruising at the impact site until 3 days after event. Mazhar et al. (2014) showed that 7.3% of hard green mature fruit subjected to impacts from 50 cm and 100 cm drop heights expressed rot at the impact site from day 7 after impact. Rot expressed as the fruit started to soften upon ripening. In contrast, no impacted fruit expressed bruising until day 14 after impact. Better understanding the apparent association between bruising of hard green mature fruit and rot development is warranted.

Between firm and soft ripe avocado fruit compressed by 3 mm, soft ripe fruit expressed higher bruise severity (Table 10.4) (Mazhar, 2015). Mazhar et al. (2018a) showed that fruit exposed to similar impact events at softening, firm ripe, and soft ripe stages expressed serially increasing bruise severity (Table 10.1). Thus, avocado fruit at advancing stages of ripening should be protected from damaging compression and impact events.

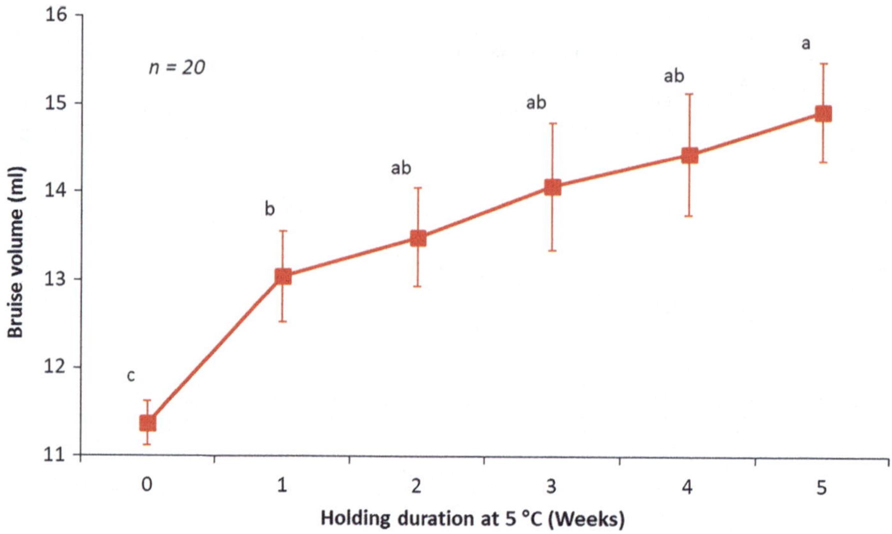

Fig. 10.6 Bruise severity (ml) ± SD in avocado fruit held at 5 °C pre-ripening, subsampled weekly and ripened, impacted (~1 J energy absorbed), and held at 20 °C for 48 h. Values sharing the letters are not significantly different at $P = 0.05$ from each other by Tukey's LSD (Mazhar, 2015)

Bower and Cutting (1988) reported that enzymes associated with bruise discolouration in avocado are latent in unripe fruit. The enzymes evidently activate when pH increases from 5.2 to 6.3 during fruit ripening (Yakoby et al., 2000). Increasing bruise susceptibility of avocado fruit with advancing ripening is associated with changing strength properties of fruit tissue at the cellular level (Baryeh, 2000). Cell wall elasticity and structure and mesocarp texture change during fruit ripening. The pectin cell wall component is degraded as reflected in disappearance of the middle lamella in concert with disorganization of sub-cellular organelles (Platt-Aloia & Thomson, 1981). Changes in rough endoplasmic reticulum (RER) are evident in parallel with the degradation of cell walls by cellulase and polygalacturonase such as to lessen tissue and cell elasticity. Cell wall bio-yield points drop in concert with mesocarp texture changes. Such changes have been observed by freeze fracture electron microscopy (Platt-Aloia & Thomson, 1981). Consequently, cells and tissues become more susceptible to mechanical damage upon bruising. A decline of Ca in the cell wall of mesocarp cells of avocado fruit during ripening may add to texture loss (Ferguson et al., 1994). It was suggested that the fall in Ca in cell wall is a consequence of breakdown of the pectin in cell walls, which may expose new binding sites and shift the equilibrium of the free and the bound Ca.

In general, increasing duration between harvest and ripening of avocado holding at low temperature is associated with increasing fruit susceptibility to bruising (Mazhar et al., 2018a). Mazhar (2015) showed that the bruise susceptibility of avocado fruit held at 5 °C for 5 weeks before ripening increased over weekly assessments (Fig. 10.6). Lee and Young (1984) considered that membrane lipid

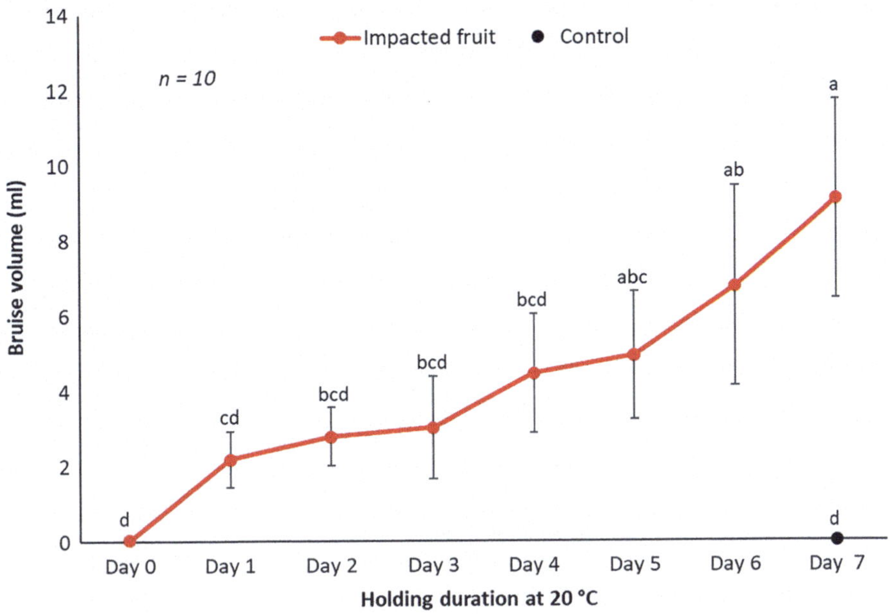

Fig. 10.7 Bruise severity (ml) ± SD in avocado fruit impacted (~0.8 J) at firm ripe and held at 20 °C for the assessment period. Control fruit received no impact and were assessed on day 7. Values sharing the letters are not significantly different at $P = 0.05$ from each other by Tukey's LSD (Mazhar, 2015)

phase changes during avocado storage at low temperature are coincident with increasing fruit susceptibility to mechanical damage and thence greater bruise expression. Platt-Aloia and Thomson (1987) established that while viscosity of lipids in avocado cell membrane is reversible to a degree, this reversibility is reduced once lipids gel. Van Lelyveld and Bower (1984) established that peroxidase and polyphenol oxidase activities were higher in avocado fruit held in storage such as to contribute as increased bruise expression.

Duration of fruit holding after a bruising event for ripening avocado fruit is also associated with increased bruise severity and intensity (viz., darkness) of the bruised mesocarp (Mazhar et al., 2018a). Upon impacting avocado fruit at softening, firm ripe, and soft ripe stages of hand firmness from different drop heights and conducting bruise assessment, Mazhar (2015) found that ripening avocado fruit did not express bruises until after 8 h from the impact event. However, most of the fruit expressed bruising within about 24 h. Bruise severity in firm ripe avocado fruit dropped from a height of 50 cm continued to increase over the assessment period of up to 7 days post-impact (Fig. 10.7). Bruise intensity measured as mesocarp tissue hue (colour) and chroma (saturation) progressively decreased over the assessment period (Table 10.5). Also observed from day 4 post-impact, the fruit started to rot at the impact site.

Table 10.5 Hue and Chroma ± SD in avocado fruit ($n = 10$) impacted (~0.8 J) at firm ripe and held at 20 °C for the assessment period (Mazhar, 2015)

Treatment/assessment	Hue	Chroma
Day 0	102.8 ± 2.6a	41.2 ± 4.8a
Day 1	102.6 ± 2.5a	36.6 ± 3.7ab
Day 2	93.6 ± 8.2b	32.4 ± 11.0bc
Day 3	89.0 ± 14.2bc	29.5 ± 9.5bcd
Day 4	88.8 ± 7.3bc	28.4 ± 7.8cd
Day 5	88.2 ± 9.6bc	26.6 ± 8.4cd
Day 6	87. 3 ± 7.8bc	24.0 ± 8.8d
Day 7	84.2 ± 8.8c	24.2 ± 9.5d
Control (no impact, assessed on day 7)	94.2 ± 1.3b	36.8 ± 3.4ab

Values with letters in common are not significantly different at $P = 0.05$ from each other by Tukey's LSD

Table 10.6 Bruise severity (ml) ± SD in avocado fruit ($n = 20$) impacted at firm ripe stage, energy absorbed ~0.8 J, and held at treatment specific temperature for assessment (Mazhar, 2015)

Treatment	Bruise severity (ml)
Impacted and held at 5 °C	0.8 ± 1.7c
Impacted at 5 °C, held at 10 °C	11.2 ± 3.4b
Impacted at 5 °C, held at 20 °C	13.7 ± 3.4a
Impacted at 20 °C, held at 5 °C	1.4 ± 2.5c
Impacted at 20 °C, held at 10 °C	11.3 ± 1.8b
Impacted and held at 20 °C	15.0 ± 3.4a

Values sharing the letters are not significantly different at $P = 0.05$ from each other by Tukey's LSD

Yakoby et al. (2000) reported that, depending on pH of the mesocarp, bruise expression due to the oxidative activity of enzymes can take from minutes to hours. Once triggered, oxidative enzymes activity continues provided that substrate is accessible (Lurie, 2009).

Consistent with Dixon et al. (2003), Mazhar (2015) also observed that the incidence and severity of body rots in avocado fruit increased with storage duration. As a matter of general practice, Kader (1999) advised to market ripened fruit as quickly as possible.

Temperature management is critical in the avocado supply chain (Kassim et al., 2013). Temperature mediates chemical and physical, including sensory, fruit attributes (Bill et al., 2014). Kader (1999) advocated maintaining low temperature in the supply chain for ripening avocados up to the point of fruit display at retail. Considering results across related trials, Mazhar et al. (2018a) concluded that fruit holding temperature at the time of a mechanical injury event affects bruise susceptibility and expression in ripening avocado. For example, bruise severity was relatively low for fruit maintained at 5 °C pre- and post-impact event compared with other combinations of higher fruit holding temperatures pre- and post-event (Table 10.6). Various factors may account for the influence of temperature on bruise expression in avocado. They encompass temperature regulated physicochemical ripening processes (Hopkirk et al., 1994), such as softening associated substrate availability for bruise expression (Yakoby et al., 2000). Temperature can also mediate the rate of oxidative bruise manifestation (Lurie, 2009).

Physicochemical factors contribute individually and interactively to determine the incidence and severity of bruising expression at different stages of the supply chain. Fruit characteristics of genotype, mineral composition, harvest maturity, and firmness confer differential responses to energies applied by abrasion, compression, and/or impact. Bruise severity and expression will be greater with longer storage durations pre-ripening and/or post-bruising event and with higher holding temperatures (Mazhar, 2015).

10.4 Incidence and Severity of Bruising in the Supply Chain

In the course of postharvest life, avocado fruit can be moved over considerable time and distance (Bill et al., 2014). In Australia, ~65% of avocado fruit are shipped by road and rail, and retailed through supermarket chains, with the balance being sold through independents (McGrath, 2008). Export of avocado fruit by air and sea is increasing. The harvested fruit pass through various operations and handling practices in their supply chains (Fig. 10.8). Thereby, they are potentially subject to mechanical damage, as discussed above, that can lead to bruising.

10.4.1 Pre-Ripening

Avocado fruit are usually harvested at commercial maturity assessed on the basis of dry matter content (%) (Chen et al., 1993). Fruit harvested early in the season with

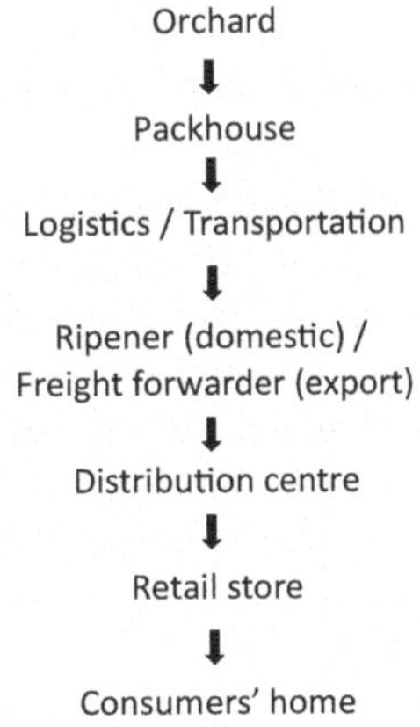

Fig. 10.8 Serial stages of domestic and export avocado supply chains (Mazhar, 2015)

low dry matter content can be more susceptible to bruising than those harvested later (Mazhar et al., 2018a).

Upon harvest, avocado fruit are generally placed into a field bin and then moved to the packhouse where they pass through packing line operations before being accumulated in a cool store to fill a transport container. Thereafter, they are delivered to the wholesaler and/or ripener in refrigerated trucks or containers. Fruit that experience longer time in supply chains before ripening can be relatively more susceptible to bruising (Mazhar et al., 2018a) and potentially also to lenticel damage (Mazhar, 2015).

Up to the ripener, avocado fruit are handled hard green mature and are relatively less susceptible to bruising (Hofman & Sandoval, 2002; Arpaia et al., 2005). Mazhar (2015) affirmed the relative tolerance of fruit at this stage to bruising. In that work, non-ethylene treated fruit were acquired from a ripener and impacted from 100 cm. Fruit held at 20 °C post-impact did not show any visible symptoms of bruising over 4 subsequent days of sacrificial assessment (Mazhar et al., 2013a). Mazhar et al. (2014) also found that hard green mature avocado fruit impacted from 50 to 100 cm drop heights immediately after harvest and then held at 20 °C did not express bruising over 15 days of assessment. Nonetheless, some fruit impacted from either height did express rot at the impact site as they started to soften from day 7 post-impact. Non-impacted control fruit did not show rot development throughout the 15 days assessment period. While hard green mature avocado fruit are relatively resistant to bruising, there is potential for rot development at the impact site. Overall, observations suggest that staff at the packhouse and ripener should handle fruit carefully to mitigate bruise expression and prospective rot development.

10.4.2 Ripener to Retail Store

Depending on market demands, avocados are dispatched from ripener's facilities at various stages of ripening (Bill et al., 2014). Fruit susceptibility to bruising continues to increase as ripening progresses (Baryeh, 2000; Mazhar et al., 2018a) while they travel further along the supply chain (Glogoski, 1995; Hofman et al., 2001). Joyce et al. (2015) conducted multi-year serial sampling of avocado fruit at ripener, distribution centre, and retail store stages and affirmed that each of the serial stages contribute to the bruise severity that consumers discern (Fig. 10.9).

At distribution centre or later in retail store, palletisation gives way to fruit handling in individual trays. Handling fruit in smaller tray units can potentially increase the risk of bruising (Mazhar, 2015). Thereafter on retail display, fruit may be handled individually by store staff for shelf stacking. Working with fruit free from damage and handled carefully, Joyce et al. (2015) found that when the fruit were collected from retail display and assessed, they had higher bruising severity compared with control fruit that had not been subject to handling by retail store staff (Table 10.7).

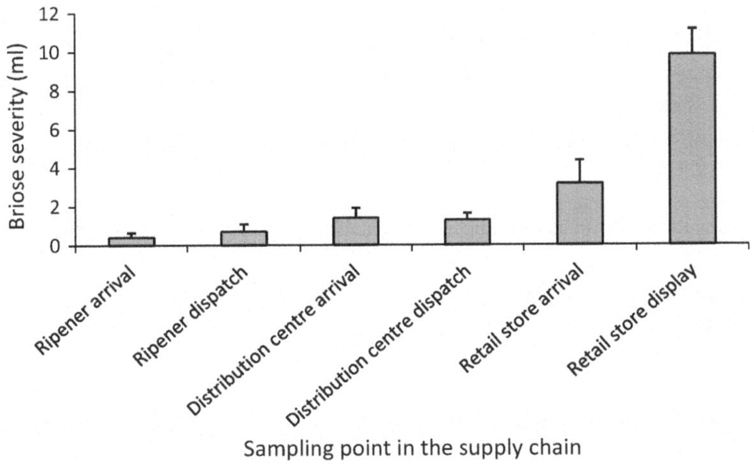

Fig. 10.9 Bruise severity (ml) in avocado fruit ($n = 20$) randomly sampled from serial stages of the supply chain. Perpendicular lines on the bars present the SD (Mazhar, 2015)

Table 10.7 Bruise severity (ml) ± SD in firm ripe avocado fruit ($n = 20$) subjected to normal store staff handling practices in two supply chains and held at 20 °C for 48 h (Joyce et al., 2015)

Supply chain 1		Supply chain 2	
Control fruit	Fruit handled by store staff	Control fruit	Fruit handled by store staff
0.1 ± 0.2c	3.3 ± 1.8a	0.1 ± 0.2c	2.1 ± 1.9b

Values sharing the letters are not significantly different at $P = 0.05$ from each other by Tukey's LSD

10.4.3 Handling by Shoppers

Avocado fruits on retail display are subject to random handling by shoppers. Shopper's contributing to avocado bruising has been considered by Ledger (1994), Milne (1998), and Embry (2007). Avocado fruits on retail display are often at various stages of ripening. Such variation allows shoppers to purchase fruit at the desired stage for use; e.g., firmer for salad, softer for guacamole (Redgwell & Fischer, 2002). Firmness judged by hand (White et al., 2009) is the usual method of assessment. In a study by Gamble et al. (2010), 76% of 107 participants considered that squeezing is their most reliable method for ripening assessment. In a survey conducted in-store and online, Mazhar (2019) found that 97% of participants considered hand firmness reliable for fruit firmness assessment. Mazhar (2015) ascertained that shoppers interacted with at least three fruit for each fruit purchased. Every handling of a ripening avocado fruit event by shoppers can potentially cause bruising (Rogers, 2018).

Joyce et al. (2015) reported further evidence of shoppers' contributing to avocado fruit bruising. Firm ripe avocado fruit subjected to single, multiple, and random handling by shoppers all showed greater bruise severity than did control fruit (Table 10.8). The bruising caused by shoppers' handling fruit on retail display was

Table 10.8 Bruise severity (ml) ± SD in firm ripe avocado fruit (n = 20) subjected to shoppers handling and held at 20 °C for 48 h (Joyce, 2015)

Single handling		Multiple handling		Random handling	
Control	Shopper's handling	Control	Shopper's handling	Control	Shopper's handling
0.1 ± 0.1d	0.9 ± 0.1c	0.1 ± 0.3d	6.3 ± 1.9a	0.0 ± 0.1d	1.9 ± 0.6b

Values sharing the letters are not significantly different at $P = 0.05$ from each other by Tukey's LSD

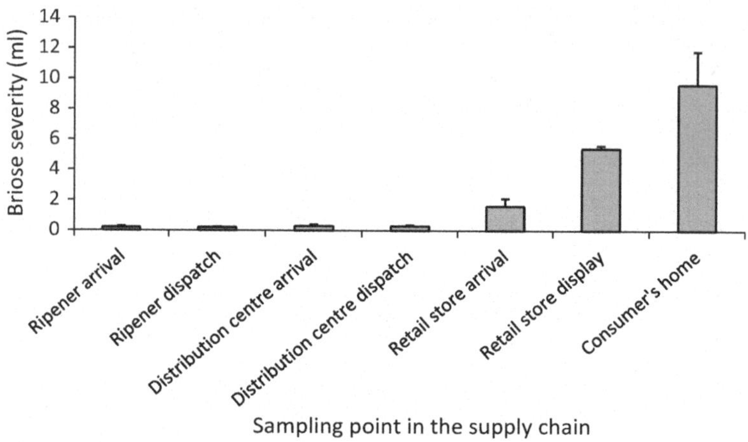

Fig. 10.10 Bruise severity (ml) in avocado fruit (n = 24) randomly sampled from serial stages of the supply chain. Perpendicular lines on the bars represent the SD (Mazhar, 2015)

the greatest incremental contributor to bruising discerned by the consumer at their home (Fig. 10.10).

In that study (Joyce, 2015), a pressure sensor (Force sensing resistor (FSR) 406, Interlink Electronics®, Camarillo, CA, USA, https://www.interlinkelectronics.com/fsr-406) and a Grip™ sensor system (Tekscan®, South Boston, MA, USA, https://www.tekscan.com/products-solutions/systems/grip-system) were utilized in in-store trials. The common aim was to elaborate putative relationships between compression forces applied by shoppers and consequent mesocarp bruising as well as to map pressures as applied across shopper's hands (Fig. 10.11). It was determined that a compression force above 7.7 N can potentially cause bruising and above 12.7 N consistently elicits mesocarp bruising in ripening fruit. Over 74% shoppers used either their thumb alone or in combination with index or middle fingers. Mazhar et al. (2013b) reported that ~10 N compression force (roughly twice the force required to click a retractable pen) was perceived by shoppers as slight thumb pressure however consistently caused bruising in ripening avocados. Only a small number of consumers used our recommended method of 'palming' the fruit which resulted in less pressure and thereby less bruising.

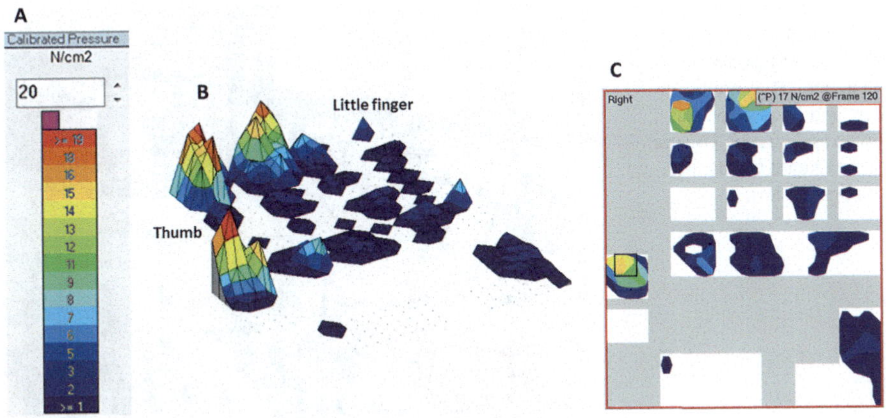

Fig. 10.11 Illustrative pressure map for the right hand of a shopper assessing firmness of a softening avocado. (**a**) Calibration scale (N cm^{-2}) recorded by the Grip™ sensor. (**b**) 3D plot of pressure distribution as applied by the shopper's hand. (**c**) Equivalent 2D plot of pressure across the pressure sensor pads (Mazhar, 2015)

On reducing fruit squeezing by consumers in store, Mission Avocados (2016) and Rogers (2018) reported that a combination of in-store intervention by way of separate displays of ripe vs. unripe fruit and consumer education can reduce avocado fruit squeezing towards increased sales (Fig. 10.12).

10.4.4 Handling by Consumers

Avocado consumer supply chain research has considered expectations (Dermody, 1990), fruit quality, and satisfaction from the decision to purchase perspective (Harker et al., 2007) and the intent to repeat purchase based on their consumption experience (Embry, 2009; Gamble et al., 2008). Lee and Coggins Jr (1982) reported a consumer preference for ripened fruit to avoid the vagaries of fruit ripening in home, including an attendant risk of rot development. However, purchasing ripened fruit elevates the risk of bruising (Campbell, 1993). Harker et al. (2007) and Gamble et al. (2010) established that consumer's intent to repeat purchase avocados fruit is negatively affected at >10% of a fruit's mesocarp volume showing internal disorders.

Recognition of the potential contribution of consumers to mesocarp bruising (Campbell et al., 2009) led towards work on quantifying consumer's contributions to mesocarp bruising experienced at the time of consumption (Joyce, 2015). From serial fruit samplings from ripener's premises to consumer's home, bruising escalated at every step (Fig. 10.10). For avocados handled carefully to avoid bruising and duly offered to consumers upon exiting the retail check-out point and collected later from them at their homes after 48 h, handled fruit had higher bruise severity than

Fig. 10.12 R store display with avocados sorted into ripening and ripe fruit labelled as 'Buy now, Eat later' and 'Buy now, Eat now', respectively. (Image source: Adam Goldwater, with permission)

Table 10.9 Bruise severity (ml) ± SD in firm ripe avocado fruit (n = 20) subjected to normal consumer handling practices and held at 20 °C for 48 h (Joyce, 2015)

Control fruit	Fruit handled by consumers
0.0 ± 0.1b	0.9 ± 0.1a

Values sharing the letters are not significantly different at $P = 0.05$ from each other by Tukey's LSD

control fruit (Table 10.9). Thus, in addition to in-store fruit handling by shoppers (Table 10.8), consumer's handling, including keeping fruit among other grocery items, during handling and transport to home, and upon placements on benchtops and in cupboards, refrigerators, and fruit baskets contributes to the overall level of bruising experienced at the time of consumption.

10.5 Monitoring Bruising in Supply Chains

Researchers around the world work to support industry towards monitoring and management of mesocarp bruising through commercial supply chains. In a value chain context (Mazhar, 2015), the overarching aim is that value for money spent on avocado purchases is maintained and improved (Campbell, 1993). Thereby, the industry can continue to grow in overall volume and value (Hort-Innovation, 2017).

10.5.1 Monitoring Methods

Milne (1998) described avocado fruit characteristics, their production, and postharvest management, and in-market handling. Adoption was suggested of preventive measures to maintain fruit quality based on the principles of hazard analysis at critical control points (HACCP) at each step from production to consumption. Avocado bruising can result from fruit mishandling at any supply chain stage (Blackburn & Scudder, 2009). Establishing fruit quality monitoring system along chain stages can inform reduction of disorders, like bruising. Where economic threshold levels are discerned for bruising, interventions are needed to avoid loss of investment, including in subsequent supply chain stages.

Destructive sacrificial bruise assessment is typically practiced on avocado fruit sampled randomly from batches (White et al., 2009). Fruit collected at certain points in the supply chain are quartered longitudinally and peeled. Visible bruises on exposed fruit surfaces are collectively and subjectively scored on a scale like: 0 = 0%, 1 = 10%, 2 = 25%, and 3 = 50%. A destructive bruise assessment method reported by Rashidi et al. (2007) for kiwifruit was adapted by Mazhar (2015) for avocado bruise severity assessment. It involves scooping-out the bruised mesocarp volume from individual fruit and determining volume by water displacement in measuring cylinders. Mazhar (2015) also used a colour meter (CR 400, Minolta Ltd. Osaka, Japan) for determining bruise intensity. Mathematical models have been applied for bruise severity assessment in commodities like pome (Bollen et al., 1999) and tomato (Van Linden et al., 2006) fruit. They are typically based on measurements of bruise shape, including thickness. Each method has pros and cons, and all are destructive (Kabas, 2010).

Pak et al. (2005) used library or retention trays for monitoring fruit quality in supply chains. Retaining these reveals fruit quality issues, including those relating to preharvest environment and management factors and postharvest treatments and handling practices, including conditions. Avocado fruits are typically sampled randomly at the end of the packing line. They are considered as representative of the commercial consignment. They are duly subject to conventional storage and ripening regimes and assessed for external and internal fruit quality issues, such as are described in best practice guides. Lagadec and Campbell (2011) extended the library tray approach by exposing avocado fruit to simulated handling conditions of the commercial supply chain. Quality assessment (White et al., 2009) of library tray fruit potentially best reflects fruit quality experienced in market.

Sensing fruit quality in real time with temperature, gas, colour, etc. sensors (e.g., ripening; Kuswandi, 2020) can potentially enhance supply chain efficiencies for supplying best possible quality produce to consumers. Advances in traceability, like forward and reverse traceability in avocado fruit supply chains (Bill et al., 2014) and conventional vs. sparse partial least squares-discriminant analysis (Jiménez-Carvelo et al., 2021), can also inform improvement of fruit quality. Other, non-destructive methods have also been used to evaluate the development of bruising in the supply chain.

10.5.2 Non-invasive Technologies

Relatively sophisticated non-destructive methods of bruise assessment have been applied in studies on avocado fruit. They include acoustic resonance (Mizrach & Flitsanov, 1999), proton magnetic resonance imaging (^{1}H-MRI; Sanches et al., 2003), near infra-red spectroscopy (NIR; Wedding et al., 2012), ultrasound (Mizrach et al., 1999), and X-ray (Sharma et al., 2020).

Among these, Mazhar et al. (2013a) employed ^{1}H-MRI towards understanding avocado fruit tissue responses to impact. ^{1}H-MRI discerned morphological fruit features and mesocarp damage at the impact site. For firm and soft ripe bruised fruit, relatively hyperintense (viz., bright) regions were apparent on the day of impact at points of contact of 25 and 50 cm drops (~0.38 and ~0.81 J), respectively. Thereafter, bruise volume became relatively hypointense (viz., dull) over the 3-day assessment period. Bruise progression was confirmed by destructive assessments conducted in parallel. For hard green mature fruit that had absorbed ~1.68 J energy (100 cm drop height), image intensity at the impact site was likewise hyperintense on the day of impact and decreased thereafter. Hard green mature fruit impacted from 100 cm drop height did not express visible bruise symptoms over the 6-day assessment period. ^{1}H-MRI images revealed an atypical region in one softening fruit that became hyperintense (Fig. 10.13). Upon destructive assessment on day 6, it was evident that pathogen activity accounted for the atypical contrast.

Avocado supply chain operators remain abreast of technological solutions for improving the fruit robustness and operational efficiency (Freshplaza, 2021). With advances in the non-invasive technologies and reduction in their cost, sophisticated diagnostics may become commercially practical for early detection of bruising and potentially decay. However, costs of infrastructure and application, and speed, would need to markedly reduce and increase, respectively.

10.5.3 Impact Event Recording Devices

Technological aids to inform avocado supply chains by recording for mapping any rough handling events during postharvest treatment and handling are commercially available. Tennes et al. (1990) used an impact recording device (IRD) to associate apple bruising with impact events. Timm and Brown (1991) used an instrumented sphere (IS) to record impacts and vibrations that can result in quality issues from packhouse operations for avocados.

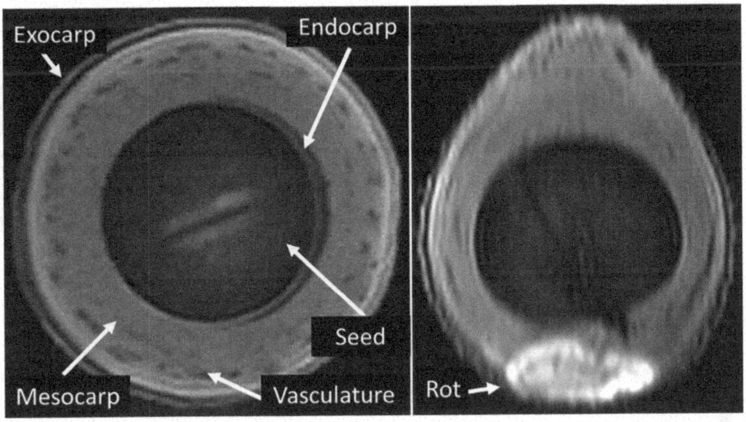

Fig. 10.13 Left—Transverse ^{1}H-MRI image of a ripe avocado cv. 'Hass' fruit. The internal morphology of the avocado fruit included exocarp (skin), mesocarp, vasculature, endocarp (seed coat), and seed. Right—Longitudinal ^{1}H-MRI image of a ripe avocado cv. 'Hass' fruit showing hyperintense area that proved upon destructive assessment to be a rot (Mazhar et al., 2013a)

In laboratory and commercial avocado supply chain settings, Joyce et al. (2015) trialled an Impact Recording Device® (IRD) (SN Techmark, Inc., Lansing, MI, USA, https://www.techmark-inc.com/assets/ird_brochure.pdf; Yu et al., 2011) and a tri-axial ShockLog (SL298 Shockwatch®, Sydney, NSW, Australia, https://www.shockwatch.com.au/product/shocklog-298-impact-recorders/; Bollen, 2006). In the laboratory, impact energies absorbed by individual fruit and by those in trays subjected to impact by drops from various heights were quantified and resultant mesocarp bruising severity was assessed. In the supply chain, both devices corroborated destructively assessed data on bruise incidence and severity arising from impact events during handling operations (Fig. 10.14). Associations between the severity of impact events and resultant bruise development in commercial supply chains can inform prediction and management of fruit bruising.

10.5.4 Fruit Firmness Assessment

Fruit firmness assessment is an integral practice in commercial avocado supply chain operations (Kader, 1999). Avocado fruit firmness informs decisions around packing, fruit storage, ripening, and marketing (Bill et al., 2014). Ideally, it is important to adopt 'industry wide' practises for firmness assessment to facilitate common understanding. Hand firmness assessment (White et al., 2009) is the traditional method. However, it relies on individual assessor's subjective ability to squeeze fruit and determine firmness stage without exceeding bio-yield point, such as to inadvertently cause mesocarp bruising. As described above, manual fruit firmness gauging by

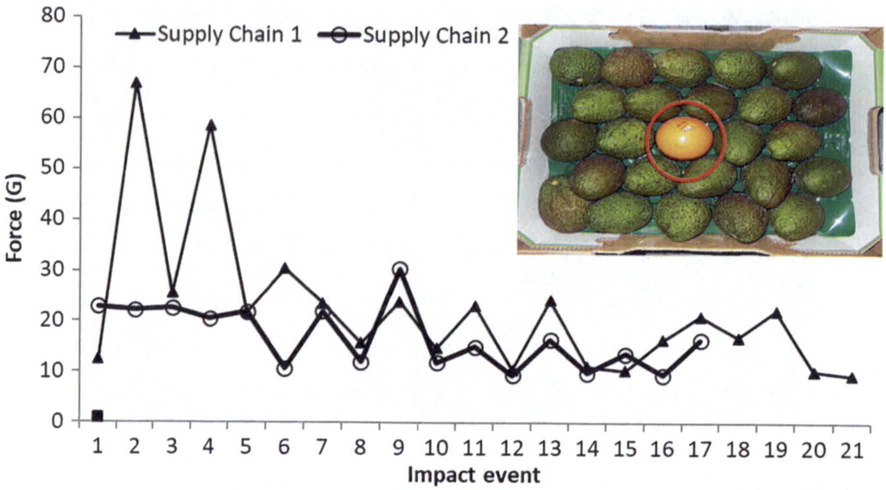

Fig. 10.14 Magnitudes of impact events (*G*-force) recorded with an Impact Recording Device® (IRD) for two commercial supply chains. The IRD (circled in red) was placed in a fruit tray prior to dispatch from ripener and retrieved on arrival at retail store for two different supermarkets (Joyce et al., 2015)

shoppers also and probably more so involves risk of mesocarp bruising (Opara & Pathare, 2014).

Researchers have explored objective means for avocado firmness assessment. Macnish et al. (1997) described a non-destructive firmness assessment device based on mechanical mesocarp displacement under a resting mass applied for a prescribed duration. White et al. (1999) compared various purportedly non-destructive firmness assessment methods; viz., hand firmness, Anderson Firmometer (Anderson Manufacturing and Toolmaking, Arataki, New Zealand) and an Instron material testing machine model 4301 (Instron, Norwood, Massachusetts, USA, https://www.instron.com/en/) on the basis of relative efficiency and reliability for purpose. Mazhar et al. (2018b) also comparatively evaluated commercially available allegedly non-destructive firmness assessment devices; viz., Sinclair™ Internal Quality Firmness Tester (Sinclair™ International Ltd., supplied by J Tech Systems, Albury Australia) and Fruitfirm Meter (Turoni Instruments, Forlì, Italy, https://www.trturoni.com/en) with Digital Firmness Meter (Macnish et al., 1997) and Force Sensor Decision Aid Tool (DAT) (Joyce, 2015). The firmness assessment devices that involved transmission of impact or compression energy to softening avocado fruit beyond the bio-yield point could induce bruising.

Understanding associations between forces and pressures applied by shopper's hands can potentially inform development of non-damaging point of sale aids for avocado shoppers deciding which fruit to buy. Novel haptic technologies involving applied forces, vibrations, or motions are being explored (Sreekanth & Megalingam, 2018). Such technologies can align soft robotics and human perception (Campbell, 2018). Rapid advancements in the internet of things (IoT), artificial intelligence,

Fig. 10.15 The Readycado™ prototype developed for evaluation of prospective commercial acceptance as an in-store decision aid tool for firmness assessment by retail store staff and / or shoppers. Upon shoppers assessing fruit firmness with the pressure sensor between a fruit and their thumb, green LED strip lights on the device serially illuminate to show fruit firmness relative to the depicted intended purpose of consumption. A red LED warning or stop light indicates that a force applied by shopper is potentially damaging to the fruit (Mazhar et al., 2020)

robotics, and sensors, soft grippers, and robotics hold promise for more efficient and effective handling of fresh produce (Navas et al., 2021). For example, a haptic sensor based robotic hand (Biotac®) is available 'off the shelf' for operational evaluation (Narang et al., 2021). Haptic technology offers rapidly evolving utility of application in avocado fruit handling (A. Pandey, personal communication). Landahl and Terry (2020) present a case for further development and ultimately commercial adoption of a safe objective non-destructive firmness assessment tool for use by commercial supply chain operators as well as retailers and shoppers.

By way of example, a device named Readycado™ (Fig. 10.15) was developed as a potential point-of-sale DAT with which avocado shoppers might assess fruit firmness without damaging the intended purchase (Joyce, 2015). It addressed the proposition that most bruising experienced at consumption result from avocado shoppers squeezing fruit on retail display when they are assessing firmness. The prototype was comprised of a single zone FSR (402 FSR, Interlink Electronics®, Westlake Village, CA USA, https://www.interlinkelectronics.com/fsr-402), a force sensor input connector, a DC power in jack (9–12 V, centre positive), a LCD display, three potentiometers, a micro-SD card slot, LED lights (green, amber, and red), and a console box (Joyce, 2015). Utility and acceptance were evaluated in laboratory trials, stakeholder consultations, a retailer survey, and in-store and on-line shopper surveys (Muriro, 2015; Mazhar et al., 2018b; Mazhar, 2019). Mazhar et al. (2020) proposed prototype refinement with alternative and/or improved sensors such as are described above.

10.6 Conclusion

Avocado fruit production, trade, and consumption are growing worldwide. However, consumers are often disappointed by fruit quality. In this context, it is important for supply chain players to appreciate inherent fruit properties and preharvest, harvest, and postharvest handling factors that influence skin abrasion and mesocarp bruising. Ideally, recognition, support, and response systems will be established across all supply chain stages towards mitigating the issue. Recommendations in the literature and pertinent research were reviewed herein. Recommendations related to factors that occur prior to arrival at the retail outlet include harvesting fruit at specified dry matter contents, handing individual fruit and fruit trays carefully to avoid bruise-causing impact events, passing fruit through supply chains as efficiently as possible, and maintaining optimum cool chain conditions. Up to 80% of the bruising appears to be related to what occurs after arrival at the retail store including handling by store staff as well as by consumers in store and after purchase. Damage at retail outlets may be able to be reduced by pre-sorting fruit employing technological aids and innovations and educating retailers to reduce and improve handling. Damage by shoppers, and consumers can be reduced by pre-sorting or packaging to reducing the need to handle fruit and educating consumers on how to avoid bruising when testing and transporting the fruit. Continuous improvement of avocado supply chain handling operations will contribute to maintaining and increasing of avocado consumer's satisfaction to increase repeat purchasing and to further growth in avocado production and supply.

Acknowledgements The authors gratefully acknowledge support from Alan Lisle, Anicet Muriro, Brett Jahnke, Gary Cowin, Madan Gupta, Melinda Perkins, Nga Vu, Qiu Xiaohang, Ray Collins, and Ximing Sun of The University of Queensland; Andrew Macnish, Leanne Taylor, Noel Ainsworth, Peter Hofman, and Roberto Marques of the Department of Agriculture and Fisheries (Queensland); Gary Poole of Murray Brothers; Matt Hall of Department of Industry, Tourism and Trade of the Northern Territory Government; Paul Gapes of Pacific Data Systems; Priya Joyce of Sugar Research Australia; Ali Asghari of Gorgan University of Agricultural Sciences and Natural Resources; and staff of Avocados Australia Limited and Hort Innovation towards the work reported herein.

References

Adato, I., & Gazit, S. (1974). Postharvest response of avocado fruits of different maturity to delayed ethylene treatments. *Plant Physiology, 53*, 899–902.

Altlasbig. (2021). *World avocado production by country*. Retrieved August 4, 2021, from https://www.atlasbig.com/en-au/countries-by-avocado-production.

Arpaia, M. L., Mitchell, F. G., Katz, P. M., & Mayer, G. (1987). Susceptibility of avocado fruit to mechanical damage as influenced by variety, maturity and stage of ripeness. *South African Avocado Growers' Association Yearbook, 10*, 149–151.

Arpaia, M. L., Requejo, C., Woolf, A., White, A., Thompson, J. F., & Slaughter, D. S. (2005). Avocado postharvest quality. In *Proceedings of the California Avocado Research Symposium*, 29 October 2005 (pp. 120–133). University of California.

Baryeh, E. A. (2000). Strength properties of avocado pear. *Journal of Agricultural Engineering Research, 76*, 389–397.

Biale, J. B., & Young, R. (1971). The avocado. In AC, H. (Ed.), *Biochemistry of fruits and their products*. Academic Press.

Biale, J. B., Young, R. E., & Olmstead, A. J. (1954). Fruit respiration and ethylene production. *Plant Physiology, 29*, 168.

Bill, M., Sivakumar, D., Thompson, A. K., & Korsten, L. (2014). Avocado fruit quality management during the postharvest supply chain. *Food Reviews International, 30*, 169–202.

Blackburn, J., & Scudder, G. (2009). Supply chain strategies for perishable products: the case of fresh produce. *Production and Operations Management, 18*, 129–137.

Bollen, A. F. (2006). Technological innovations in sensors for assessment of postharvest mechanical handling systems. *International Journal of Postharvest Technology and Innovation, 1*, 16–31.

Bollen, A. F., Nguyen, H. X., & Rue, B. T. (1999). Comparison of methods for estimating the bruise volume of apples. *Journal of Agricultural Engineering Research, 74*, 325–330.

Bower, J. P., & Cutting, J. G. (1988). Avocado fruit development and ripening physiology. *Horticultural Reviews, 10*, 229–271.

Campbell, T. (1993). To market, to market, to buy avocado. Home again, home again; with our bravado. *Talking Avocados, 4*, 17.

Campbell, S. (2018). The robotics revolution will be soft: soft robotics proliferate-along with their sources of inspiration. *IEEE Pulse, 9*, 19–24.

Campbell, R. L., Smith, B. G., Jaeger, S. R., & Harker, F. R. (2009). Deterioration and disposal of fruit in the home: consumer interviews and fruit quality assessments. *Journal of the Science of Food and Agriculture, 89*, 24–32.

Chen, P., Mccarthy, M. J., Kauten, R., Sarig, Y., & Han, S. (1993). Maturity evaluation of avocados by NMR methods. *Journal of Agricultural Engineering Research, 55*, 177–187.

Dermody, J. (1990). Quality in the marketplace. In *Proceedings of the Australian Avocado Growers Federation Conference 'Profit through Quality', Gold Coast, Australia* (pp. 14–23).

Dixon, J., Pak, H., Smith, D., Elsmly, T., & Cutting, J. (2003). New Zealand avocado fruit quality: The impact of storage temperature and maturity. In *Proceedings V World Avocado Congress* (pp. 647–652).

Dominguez, P., Araus, K., Bonert, P., Sánchez, F., San Miguel, G., & Toledo, M. (2014). The avocado and its waste: an approach of fuel potential/application. In *Environment, energy and climate change II*. Springer.

Dorantes, L., Parada, L., & Ortiz, A. (2004). *Avocado: Postharvest operation*. Food and Agriculture Organization of the United Nations.

Embry, J. (2007). How are the retailers performing? *Talking Avocados, 17*, 26.

Embry, J. (2009). Avocado retail quality surveys. In *4th Australian and New Zealand Avocado Growers Conference*, Cairns, 21–24 July 2009.

FAO. (2021). *Statistics* [Online]. Food and Agriculture Organization. Retrieved April 24, 2021, from http://www.fao.org/faostat/en/#data.

Ferguson, I., Volz, R., Harker, F., Watkins, C., & Brookfield, P. (1994). Regulation of postharvest fruit physiology by calcium. *Acta Horticulturae, 398*, 23–30.

FreshPlaza. (2021). *Device scans avocados ripeness within seconds*. Retrieved August 22, 2021, from https://www.freshplaza.com/article/9346111/device-scans-avocados-ripeness-within-seconds/.

Gamble, J., Wohlers, M., & Jaeger, S. R. (2008). *A survey of Australian avocado consumers: Product experiences, health benefit awareness, and impact of defect on purchase intentions*. Horticulture Australia Limited.

Gamble, J., Harker, F. R., Jaeger, S. R., White, A., Bava, C., Beresford, M., Stubbings, B., Wohlers, M., Hofman, P. J., Marques, R., & Woolf, A. (2010). The impact of dry matter, ripeness and internal defects on consumer perceptions of avocado quality and intentions to purchase. *Postharvest Biology and Technology, 57*, 35–43.

Glogoski, A. (1995). The future market. *Talking Avocados, 6*, 18–20.

Golan, A., Kahn, V., & Sadovski, A. (1977). Relationship between polyphenols and browning in avocado mesocarp. Comparison between the Fuerte and Lerman cultivars. *Journal of Agricultural and Food Chemistry, 25*, 1253–1260.

Gupta, A. K., & Kaur, N. (2000). *Carbohydrate reserves in plants-synthesis and regulation*. Elsevier.

Harker, F., Jaeger, S., Hofman, P., Bava, C., Thompson, M., Stubbings, B., White, A., Wohlers, M., Heffer, M., & Lund, C. (2007). *Australian consumers' perceptions and preferences for 'Hass' avocado*. Horticulture Australia Limited.

Hausch, B. J., Arpaia, M. L., Kawagoe, Z., Walse, S., & Obenland, D. (2020). Chemical characterization of two california-grown avocado varieties (Persea americana mill.) over the harvest season with an emphasis on sensory-directed flavor analysis. *Journal of Agricultural and Food Chemistry, 68*, 15301–15310.

Henderson, T. (2019). *Industry-specific impact assessment program: Avocado*. Horticulture Australia Limited.

Hofman, P. J. (2003). *Bruising of 'Hass' avocado from harvest to the packhouse. Final report AV02015*. Horticulture Australia Limited.

Hofman, P., & Ledger, S. N. (1999). Retail surveys show little quality improvement. *Talking Avocados, 10*, 22–23.

Hofman, P., & Ledger, S. (2001a). Rots and bruising main quality problems. *Talking Avocados, 12*, 20–22.

Hofman, P. J., & Ledger, S. N. (2001b). *Reducing avocado defects at retail level. Final report AV99007*. Horticulture Australia Limited.

Hofman, P. J., & Sandoval, G. (2002). AvoCare and bruising in avocados. *Talking Avocados, 13*, 18–19.

Hofman, P., Ledger, S., Stubbings, B., Adkins, M., Cook, J., & Barker, J. (2001). Avocado quality on the retail shelf: We all have a role to play. In *Australian and New Zealand Avocado Growers Conference*, Bundaberg, Australia.

Hopkirk, G., White, A., Beever, D., & Forbes, S. (1994). Influence of postharvest temperatures and the rate of fruit ripening on internal postharvest rots and disorders of New Zealand 'Hass' avocado fruit. *New Zealand Journal of Crop and Horticultural Science, 22*, 305–311.

Hort-Innovation. (2017). *Avocado Strategic Investment Plan 2017-2021*. Hort Innovation.

ITC. (2018). *Avocado: global trade's golden fruit* [Online]. Retrieved from https://www.intracen.org/layouts/2coltemplate.aspx?pageid=47244640256&id=47244673702.

ITC. (2021). *Spot export opportunities for trade development* [Online]. Retrieved from https://exportpotential.intracen.org/en/?type=product&code=080440.

Jiménez-Carvelo, A. M., Martín-Torres, S., Ortega-Gavilán, F., & Camacho, J. (2021). PLS-DA vs sparse PLS-DA in food traceability. A case study: Authentication of avocado samples. *Talanta, 224*, 121904.

Joyce, D. (2015). *Understanding and managing avocado flesh bruising. Final report AV12009*. Horticulture Innovation Australia.

Joyce, D. (2020). *Improved fruit quality and robustness in avocado supply chains (mineral nutrition)*. Hort Innovation.

Joyce, D., Mazhar, S., & Hofman, P. (2015). *Reducing flesh bruising and skin spotting in Hass avocado*. Hort Innovation.

Kabas, O. (2010). Methods of measuring bruise volume of pear (*Pyrus communis* L.). *International Journal of Food Properties, 13*, 1178–1186.

Kader, A. A. (1999). Fruit maturity, ripening, and quality relationships. *Acta Horticulturae, 485*, 203–208.

Kassim, A., Workneh, T., & Bezuidenhout, C. (2013). A review on postharvest handling of avocado fruit. *African Journal of Agricultural Research, 8*, 2385–2402.

Kuswandi, B. (2020). Real-time quality assessment of fruits and vegetables: Sensing approaches. In: B. Kuswandi, & M.W. Siddiqui (Eds.), *Sensor-based quality assessment systems for fruits and vegetables.* Google.

Lagadec, D. L., & Campbell, T. (2011). *Development of an avocado rapid library tray system for Hass and Shepard.* Hort Innovation.

Landahl, S., & Terry, L. A. (2020). Non-destructive discrimination of avocado fruit ripeness using laser Doppler vibrometry. *Biosystems Engineering, 194*, 251–260.

Ledger, S. (1994). 'Hass' mask problems but consumers find them. *Talking Avocados, 5*, 14–15.

Lee, S., & Coggins, C., Jr. (1982). Feasibility of marketing soft avocado fruit. *California Avocado Society Yearbook, 66*, 57–62.

Lee, S., & Young, R. (1984). Temperature sensitivity of avocado fruit in relation to C_2H_4 treatment. *Journal of the American Society for Horticultural Science, 109*, 689–692.

Lee, S., Young, R., Schiffman, P., & Coggins, C. (1983). Maturity studies of avocado fruit based on picking dates and dry weight. *Journal of the American Society for Horticultural Science, 108*, 390–394.

Linden, V. V., & Baerdemaeker, J. (2005). The phenomenon of tomato bruising: where biomechanics and biochemistry meet. *Acta Horticulturae, 682*, 925–930.

Lurie, S. (2009). Stress physiology and latent damage. In W. J. Florkowski, R. L. Shewfelt, B. Brueckner, & S. E. Prussia (Eds.), *Postharvest handling* (2nd ed.). Academic Press.

Macleod, R. F., Kader, A. A., & Morris, L. L. (1976). Damage to fresh tomatoes can be reduced. *California Agriculture, 30*, 11.

Macnish, A. J., Joyce, D. C., & Shorter, A. J. (1997). A simple non-destructive method for laboratory evaluation of fruit firmness. *Australian Journal of Experimental Agriculture, 37*, 709–713.

Mazhar, M. (2015). *Bruising in avocado (Persea americana M.) cv. 'Hass' supply chains: from the ripener to the consumer.* PhD, The University of Queensland.

Mazhar, M. (2019). *Decision aid tool to help avocado fruit shoppers with firmness assessment.* Department of Agriculture and Fisheries Queensland.

Mazhar, M., Joyce, D., Hofman, P., Collins, R., & Gupta, M. (2012). Impact induced bruising in ripening 'Hass' avocado fruit. *Talking Avocados, 22*, 34–37.

Mazhar, M., Joyce, D., Cowin, G., Hofman, P., Brereton, I., & Collins, R. (2013a). MRI as a non-invasive research tool for internal quality assessment of 'Hass' avocado fruit. *Talking Avocados, 23*, 22–25.

Mazhar, M., Joyce, D., Tuttle, N., Jahnke, B., Nolan, C., Gupta, M., Hofman, P., & Collins, R. (2013b). Bruising by hand in softening avocado fruit. *Talking Avocados, 24*, 29–31.

Mazhar, M., Joyce, D., Hofman, P., & Collins, R. (2014). Bruising and rots development in impacted hard green mature avocado cv. 'Hass' fruit. *Talking Avocados, 25*, 30–33.

Mazhar, M., Joyce, D., Hofman, P., & Vu, N. (2018a). Factors contributing to increased bruise expression in avocado (Persea americana M.) cv.'Hass' fruit. *Postharvest Biology and Technology, 143*, 58–67.

Mazhar, M., Perkins, M., & Joyce, D. (2018b). *Comparison of devices for 'non-destructive' firmness measurement of avocado fruit.* University of Queensland, Australia.

Mazhar, M., Joyce, D., & Faircloth, A. (2020). *Progressing Readycado^{TM} readiness through further development and stakeholder engagement. H&FS development proposal final report - Confidential.* Department of Agriculture and Fisheries.

Mcgrath, C. (2008). *Avocado market in the European Union.* Department of Primary Industries and Fisheries.

Milne, D. (1998). Avocado quality assurance; who?, where?, when?, how? *Talking Avocados, 9*, 24–27.

Mission Avocados. (2016). *Avoca Dos and Don'ts - Avocado best practice guide for grocery retailers.* Mission Avocados. Retrieved August 4, 2021, from https://mission-produce-static.

sfo2.cdn.digitaloceanspaces.com/craft/download-center/pdf/Mission-AvocaDOs-and-DONTs_Retail.pdf.
Mizrach, A., & Flitsanov, U. (1999). Nondestructive ultrasonic determination of avocado softening process. *Journal of Food Engineering, 40*, 139–144.
Mizrach, A., Flitsanov, U., El-Batsri, R., & Degani, C. (1999). Determination of avocado maturity by ultrasonic attenuation measurements. *Scientia Horticulturae, 80*, 173–180.
Moore, F. (2001). Australian avocado industry update for 2001 - The history and development of the Australian avocado industry. *California Avocado Society Yearbook, 85*, 65–75.
Muriro, A. (2015). *Shoppers perceptions on the utility of a decision aid tool for in-store assessment of avocado fruit firmness.* Master of Agriculture Science, The University of Queensland, Australia.
Narang, Y. S., Sundaralingam, B., Van Wyk, K., Mousavian, A., & Fox, D. (2021). Interpreting and predicting tactile signals for the SynTouch Biotac. *arXiv preprint arXiv:2101.05452*.
Navas, E., Fernández, R., Sepúlveda, D., Armada, M., & Gonzalez-de-Santos, P. (2021). Soft grippers for automatic crop harvesting: a review. *Sensors, 21*, 2689.
NZ-AVOCADOS. (2021). *World avocado market* [Online]. Retrieved from https://industry.nzavocado.co.nz/world-avocado-market/.
Opara, U. L., & Pathare, P. B. (2014). Bruise damage measurement and analysis of fresh horticultural produce - A review. *Postharvest Biology and Technology, 91*, 9–24.
Pajuelo, M., Baldwin, G., Rabal, H., Cap, N., Arizaga, R., & Trivi, M. (2003). Bio-speckle assessment of bruising in fruits. *Optics and Lasers in Engineering, 40*, 13–24.
Pak, H., Bettesworth, D., & Dawes, H. (2001). *The role of surface ridging and protuberances on avocado fruit in the development of ripe rots. NZ Avocado Growers Association Annual Research Report, 1* (pp. 20–25).
Pak, H., Dixon, J., & Cutting, J. (2005). Library trays - a valuable tool for managing fruit quality. In *New Zealand and Australia Avocado Grower's Conference,* Tauranga, New Zealand.
Platt-Aloia, K., & Thomson, W. (1981). Ultrastructure of the mesocarp of mature avocado fruit and changes associated with ripening. *Annals of Botany, 48*, 451–466.
Platt-Aloia, K., & Thomson, W. (1987). Freeze fracture evidence for lateral phase separations in the plasmalemma of chilling-injured avocado fruit. *Protoplasma, 136*, 71–80.
Quintana, M. E. G., & Paull, R. E. (1993). Mechanical injury during postharvest handling of solo' papaya fruit. *Journal of the American Society for Horticultural Science, 118*, 618–622.
Rashidi, M., Seyfi, K., & Gholami, M. (2007). Determination of kiwifruit volume using image processing. *Journal of Agricultural and Biological Science, 2*, 17–22.
Redgwell, R. J., & Fischer, M. (2002). Fruit texture, cell wall metabolism and consumer perceptions. In M. Knee (Ed.), *Fruit quality and its biological basis.* CRC Press.
Rogers, G. (2018). *Supply chain quality improvement–retailer point of purchase improvements.* Hort Innovation.
Sanches, J., Biscegli, C. I., Durigan, J. F., Sim Es, M. L., & Da Silva, W. T. L. (2003). Diagnosis of mechanical injuries in avocados by magnetic resonance imaging. In *V World Avocado Congress*, Granada, 19–24 October 2003 (pp. 695–700).
Schaffer, B. A., Wolstenholme, B. N., & Whiley, A. W. (2013). *The avocado: Botany, production and uses.* CABI.
Sharma, R., Reddy, S. V. R., & Gajanan, G. (2020). X-ray imaging for quality detection in fruits and vegetables. In B. Kuswandi & M. Wasim (Eds.), *Sensor-based quality assessment systems for fruits and vegetables.* Google Books.
Smith, K. L. C., Turner, S., & Tak, M. (1990). *A strategic marketing plan for the Queensland avocado industry.* Queensland Depertment of Primary Industries.
Sreekanth, M., & Megalingam, R. K. (2018). Haptic solutions for agriculture. In *2018 International conference on computing, power and communication technologies (GUCON)* (pp. 416–420). IEEE.
Studman, C. (1995). A model of fruit bruising. In *2nd Australasian Postharvest Conference.* Monash University Melbourne.

Tennes, B. R., Zapp, H. R., Marshall, D. E., & Armstrong, P. R. (1990). Apple handling impact data acquisition and analysis with an instrumented sphere. *Journal of Agricultural Engineering Research, 47*, 269–276.

Timm, E., & Brown, G. (1991). Impacts recorded on avocado, papaya and pineapple packing lines. *Applied Engineering in Agriculture, 7*, 418–422.

Van Lelyveld, L., & Bower, J. (1984). Enzyme reactions leading to avocado fruit mesocarp discoloration. *Journal of Horticultural Science, 59*, 257–263.

Van Linden, V., De Ketelaere, B., Desmet, M., & De Baerdemaeker, J. (2006). Determination of bruise susceptibility of tomato fruit by means of an instrumented pendulum. *Postharvest Biology and Technology, 40*, 7–14.

Wedding, B. B., Wright, C., Grauf, S., & White, R. D. (2012). The application of near infrared spectroscopy for the assessment of avocado quality attributes. In T. Theophanides (Ed.), *Infrared spectroscopy: Life and biomedical sciences*. IntechOpen.com.

Whiley, A. W., Schaffer, B., & Wolstenholme, B. N. (2002). *The avocado: Botany, production and uses*. CABI.

White, A., Woolf, A., Harker, F., & Davy, M. (1999). Measuring avocado firmness: Assessment of various methods. In *IV World Avocado Congress*, Uruapan, Mexico (pp. 389–392).

White, A., Woolf, A., Hofman, P. J., & Arpaia, M. L. (2009). *The international avocado quality manual*. UC Davis Press.

Yakoby, N., Kobiler, I., Dinoor, A., & Prusky, D. (2000). pH regulation of pectate lyase secretion modulates the attack of *Colletotrichum gloeosporioides* on avocado fruits. *Applied and Environmental Microbiology, 66*, 1026–1030.

Yu, P., Li, C., Rains, G., & Hamrita, T. (2011). Development of the berry impact recording device sensing system: Hardware design and calibration. *Computers and Electronics in Agriculture, 79*, 103–111.

Chapter 11
Bruise Damage Susceptibility and Assessment of Guava

Saowapa Chaiwong and Rattapon Saengrayap

Abstract Guava (*Psidium guajava*) is an economical fruit crop widely grown in Asian countries. This soft thin-skinned fruit is very susceptible to postharvest mechanical damage caused by impact, vibration, and compression bruising, resulting in significant fruit browning, leading to consumer rejection and monetary loss. Impact and vibration play important forces for guava bruising throughout the guava supply chain, particularly during transportation. Incidence and severity of guava fruit impact bruising were impacted by drop heights, number of drops, and storage temperature causing moderate to severe browning on either peel or flesh. Vibration bruising as scattered brown spots on the peel reflected transport acceleration, frequency, and duration. Cushioning was applied to reduce bruising by covering the fruit with foam netting and paper-based materials as bagging during fruit growth development and after harvesting during transportation. Digital image analysis identified the characteristics of guava bruising damage with accurate results. Size and shape features were used to assess the impact bruising. Color feature can describe physiological changes of guava bruising during storage. Textual feature extraction through fractal dimension has the capability to evaluate scattered bruising spots due to vibration damage as well as bruise susceptibility due to impact damage. Image analysis with feature extraction was determined as a feasible technique to assess postharvest guava damage and quality.

Keywords Digital image analysis · Fruit quality · Impact bruising · Mechanical damage · Vibration bruising

---

S. Chaiwong (✉) · R. Saengrayap
School of Agro-Industry, Mae Fah Luang University, Chiang Rai, Thailand

Integrated Agri-Tech Ecosystem Research Group, Mae Fah Luang University, Chiang Rai, Thailand
e-mail: saowapa@mfu.ac.th; rattapon.sae@mfu.ac.th

P. B. Pathare, U. L. Opara (eds.), *Mechanical Damage in Fresh Horticultural Produce*, https://doi.org/10.1007/978-981-99-7096-4_11

11.1 Introduction

Guava (*Psidium guajava* L.) is commercially cultivated in several tropical and sub-tropical countries (Singh, 2011). The top ten ranked global producers of guava, mango, and mangosteen are India, Indonesia, China, Mexico, Pakistan, Malawi, Brazil, Thailand, Egypt, and Bangladesh, respectively (FAO, 2021). This soft thin-skinned fruit is very susceptible to postharvest mechanical damage caused by impact, vibration, and compression bruising, resulting in significant fruit browning leading to consumer rejection and monetary loss (Singh, 2011). Impact and vibration damages are the main causes of guava bruising throughout the supply chain, particularly during transportation. Incidence and severity of guava fruit impact bruising were impacted by drop heights, number of drops and storage temperature causing moderate to severe browning incidence on flesh with less incidence on skin (Htike et al., 2023). Vibration bruising as scattered brown spots on skin reflected transport acceleration, frequency, and time duration (Chaiwong et al., 2021). Limited research has assessed mechanical damages in guava fruit under simulated impact and vibration tests, with limited scientific understanding of environmental factors causing damages, especially compression damage. To ameliorate mechanical damages throughout the guava supply chain, cushioning was applied to reduce bruising by covering the fruit with foam netting and paper-based materials in both fruit growth development as bagging and postharvest operation.

Image analysis has been applied to quantify bruises caused by impact and vibrational forces and overcome any disadvantages of conventional manual inspections. Features such as color, size, shape, and texture are extracted by capturing images of fruit using image analysis software (Zheng et al., 2016). This information is used to describe bruising characteristics. Color, size, and shape features have successfully been used to quantify both impact and vibrational damages to guava caused by simulated conditions (Htike et al., 2023; Chaiwong et al., 2021). Color features can describe physiological changes of guava bruising during storage, while size features can describe severity of bruise occurrence. In terms of textural features, fractal dimension (FD) values can evaluate scattered bruising spots due to vibration damage as well as bruise susceptibility from impact damage. Therefore, image analysis with suitable extraction features was determined as a feasible technique to assess postharvest guava damage and quality.

11.2 Global Guava Production

In 2019, worldwide guava production was 55.85 million tons (FAO, 2021). India was the top guava producer at 45.9% of global production and around 10 times higher than second-ranked Indonesia. Guava production is increasing for all the top producers except China and Thailand, as given in Table 11.1. In 2020, the total

Table 11.1 Top 10 producers of guava in 2019 (FAO, 2021)

Rank	Country	Production quantity (million tons)	5-year growth in quantity (2014–2019) %
1	India	25.63	+39.06
2	Indonesia	3.29	+20.53
3	China	2.42	−6.52
4	Mexico	2.40	+36.59
5	Pakistan	2.27	+5.37
6	Malawi	2.08	+61.23
7	Brazil	2.00	+33.97
8	Thailand	1.63	−54.74
9	Egypt	1.47	+58.90
10	Bangladesh	1.46	+46.76

exported value of guava was 3.88 billion USD. Thailand ranked as the number one exporter at 18.91% of the total exported value, with Mexico (11.82%) and the Netherlands (10.72%) as second and third. The top export flow in 2020 was from Thailand to China with a value of 595.92 million USD. On the other hand, China, the USA, and the Netherlands were the top three guava importers.

11.3 Guava Fruit Characteristics

Guava (*Psidium guajava* L.) has year-round production and belongs to the family Myrtaceae (Pommer & Murakami, 2009). Guava is a climacteric fruit with a respiration rate of 8-60 and 18-130 mgCO_2 kg^{-1} h^{-1} at storage temperature (10 and 20 °C), while ethylene (C_2H_4) production was 1–20 μL kg^{-1} h^{-1} at 20 °C during ripening (Paull & Chen, 2016). Guava has a berry-type fruit with medium to large size, average diameter of 5–10 cm and weight of 100–250 g. Variation of peel, surface, fruit shape, and color depends on the cultivar. The surface of guava is rough to smooth with different spherical, ovoid or pyriform shapes. The dark green color of immature fruit changes to yellowish-green, pale yellow, and yellow at the ripening stage. Guava flesh color depends on the cultivar as white, pink, and salmon-red (Singh, 2011). For example, Thai guava ("Glom Sali" cultivar) is spherical and light green in color with a semi-smooth peel surface, while "Gim Ju" has the same shape and color with a smooth peel surface (Fig. 11.1) (Table 11.2) (Chaiwong et al., 2021).

Guava fruit skin is thin and prone to mechanical injuries such as cuts, punctures, vibration, impact, and compression bruising (Singh, 2011). The skin is composed of the cuticle as a barrier to transpiration, covering layers of polyhedral parenchyma cells with scatterings of well-formed sclereid cells in vascular bundles and crystals. The primary role of tissue mechanical strength is associated with highly developed sclereid cells in bundles with thick secondary walls (Oliveira et al., 2014). The radius

Fig. 11.1 Characteristics of Thai guava fruit, "Glom Sali" (**a**) and "Gim Ju" (**b**)

Table 11.2 Fruit characteristics and some chemical properties of "Glom Sali" guava (Chaiwong et al., 2021)

Fruit characteristic	Mean ± SE
Weight (g)	251.81 ± 1.52
Volume (mL)	293.79 ± 9.87
Density (g mL^{-1})	0.87 ± 0.03
Horizontal diameter (cm)	7.39 ± 0.05
Height (cm)	8.72 ± 0.12
Radius of curvature (mm)	51.85 ± 3.00
Firmness (N)	6.07 ± 0.37
Total soluble solids (TSS) (%)	10.37 ± 0.26
Total acidity (TA) (%)	0.10 ± 0.00
Dry matter (%)	13.35 ± 0.15

Values are means ± SE ($n = 15$)

of curvature and fruit shape are associated with bruising. A small radius of curvature caused higher absorbed energy and resulted in increased impact bruising damage in kiwifruit (Ahmadi, 2012). In pomegranate fruit, the region of peel (stem shoulder, cheek, and calyx shoulder) had a higher relationship to fruit firmness than peel thickness, while an increase in radius of curvature increased the volume and area of impact bruising (Mohammad Shafie et al., 2015). In fruit-to-fruit impact bruising of apple, the top and middle regions exhibited higher bruising sensitivity than the bottom region (Fu et al., 2020). In case of vibration bruising, a larger bruise volume of apple resulted in a larger radius of curvature of apple fruit with a larger contact area (Rostampour & Motlagh, 2018). In vibration bruising of guava, the middle region at the cheek point exhibited the highest vibration bruising (Chaiwong et al., 2021). Thus, the middle region as a cheek with a large radius of curvature and contact area should be protected to reduce mechanical damages.

11.4 ASEAN and Thailand Quality Standards of Guava

Quality standards of guava for fresh consumption are based on fruit appearance, i.e., adequate shape, color, and size, as well as free from bruises, rotting or any damage caused by pest or microbial deterioration. Figure 11.2 shows the examples of common defects found in the packinghouse. Guava classification standards (TAS,

Fig. 11.2 Defects of guava (TAS, 2010). (**a**) Healthy fruit. (**b**) Damage caused by sunburn (acceptable). (**c**) Damage caused by sunburn (unacceptable). (**d**) Damage caused by Anthracnose disease. (**e**) Damage caused by insects. (**f**) Damage caused by plant pathogen

Table 11.3 Classification criteria of guava

Class	Parameter
"Extra" class	1. Stem attached
	2. Free of defects
Class I	1. Slight defects in shape and color
	2. Slight defects due to sunburn and blemishes (<5% of total surface area)
Class II	1. Defects in shape and color
	2. Defects due to sunburn and blemishes (5–10% of total surface area)

Table 11.4 Standard of guava sizing (TAS, 2010; ASEAN Stan7, 2008)

	ASEAN standard		Thai standard
Size code	Weight (g)	Diameter (mm)	Weight (g)
1	>450	>100	>450
2	351–450	96–100	351–450
3	251–350	86–95	251–350
4	201–250	76–85	201–250
5	151–200	66–75	151–200
6	101–150	54–65	101–150
7	61–100	43–53	–
8	35–60	30–42	–
9	<35	<30	–

2010; ASEAN Stan7, 2008) are used in the field and in packinghouses to assess fruit quality (Table 11.3).

Weight and fruit diameter are used to determine the size of guava, as given in Table 11.4. Commercial classification depends on trade requirements or seasonal limitations.

11.5 Postharvest Management of Guava

11.5.1 Postharvest Losses of Guava

Guava mechanical damages caused by rough handling result in bruising and wounding as browning incidence on skin and flesh that are susceptible to microbial spoilage (Singh, 2011; Kamsiati, 2013) with short shelf life during storage and transport (Reyes & Paull, 1995; Tandon et al., 1989). Postharvest shriveling from mass loss of guava is also a major problem together with mechanical injury. Desiccation resulted in a dull yellow and wrinkled skin (Singh, 2011). Thus, good and proper practices in postharvest handling are required to maintain the quality of guava (Kamsiati, 2013). In India, postharvest losses of guava in farm operation (14%) were higher than postharvest losses of storage (4%) (Narayana et al., 2014). In Ethiopia, postharvest losses of guava were 16% (Bantayehu et al., 2017). In Brazil

(MA) estimated total postharvest losses of guava were 17% for groceries stores and 21% in supermarkets, with 10.5% from mechanical damages (Mendes et al., 2019). Recently, mechanical damages from guava handling in Kenya accounted for 24% of total deterioration (Katumbi et al., 2021). Therefore, overall average postharvest losses of guava were 14–20% with high mechanical damages.

11.5.2 Postharvest Handling of Guava

Fruit bagging in guava is commonly applied to young fruit around 6–9 weeks before harvesting or 25 weeks from fruit set to improve fruit appearance, protect from fruit fly and guava weevil damage and reduce black spot and anthracnose diseases (Martins et al., 2007; Bilck et al., 2011). Various bagging materials are applied including waxed paper, nylon fabric, telephone book paper (Morera-Montoya et al., 2010), and nonwoven material (Sharma & Nagaraja, 2016). In Thailand, commercial guava bagging is mostly applied using a plastic bag as the inner layer and paper as the outer layer (Fig. 11.3a). Covering with a foam net as the inner layer and a plastic bag with two holes at the bottom corner (Fig. 11.3c) at 6–8 weeks after fruit set is also practiced for fruit size of 2.5 cm. Plastic bagging of guava fruit reduces scratching (abrasion) and the paper layer is removed before sorting and grading (Fig. 11.3b). Bagging with a foam net improves efficiency to protect against shock

Fig. 11.3 "Gim Ju" guava bagging using a thin plastic bag (inner layer) and paper (outer layer) (**a**), removal of paper after harvesting (**b**), "Glom Sali" guava bagging with a foam net (inner layer) and thin plastic bag (outer layer) (**c**), removal of plastic bag after harvesting (**d**)

Fig. 11.4 Postharvest handling of guava from farm to retail store in Thailand

and vibration damages in farm handling and transportation. The plastic bag (outer layer) is removed after harvesting (Fig. 11.3d).

Figure 11.4 shows a flow chart of postharvest handling of guava in Thailand from harvesting, collecting, sorting/grading, packaging and transportation to storage/retail. Before harvesting, guava maturity was observed by color from dark green to light green, combined with other maturity indices such as size, skin, and specific gravity (<1.0) (Cavalini et al. 2006). In Thailand, guava maturity is generally determined by counting for 6–8 weeks after fruit bagging. Hand harvesting is done by simply removing the fruit from the plant and placing it in a suitable container. Care should be taken to prevent impact damage by dropping the fruit (Bakshi, 2015). At sorting/grading, the standard guava criteria are applied (see Sect. 11.4). Defects on the guava fruit must be less than 10% of the total area following Thai and ASEAN guava fruit standards (Thai Agricultural Standard TAS 16-2010, 2010; AEAN Stan7:2008, 2008). For distribution of guava in Thailand, various packaging types are commonly used as a trapezoid-shaped plastic crate or perforated polyethylene bag for packaging distribution in fresh food markets and wholesalers or used corrugated boxes in supermarkets (Fig. 11.3b, d). A trapezoid-shaped crate is made of high density polyethylene (HDPE) or polypropylene (PP) with attached steel handles for stacking (Chonhenchob et al., 2011). Cushioning material as a foam net is widely used to protect against mechanical bruising of guava, as discussed in Sect. 11.7 (Cushioning) (Fig. 11.3c). The optimal storage temperature for guava is 8–10 °C with 90% relative humidity (RH), giving a shelf life of 7 days at 20 °C (Paull & Chen, 2016). During shelf life of 6–8 days, mass loss of guava was 10–20% depending on temperature and RH conditions (Singh, 2011).

11.6 Mechanical Damage of Guava

Three types of mechanical damage of guava were reviewed as impact, vibration, and compression forces. Impact damage is a common occurrence throughout supply chain management from harvesting, postharvest handling to transportation, and distribution. Impact damage has been identified as the most significant cause of mechanical fruit damage as fruit-to-fruit impact or impact against the package surface (Li & Thomas, 2014). Vibration damage results from fruit movement inside the package during transport (Dubey & Mishra, 2018). Compression damage is caused by static weight or pressure acting on the fruit. Long duration and static loads involved in compression damage frequently result in a creeping phenomenon that deforms tissue over time (Miller & Knee 2002). Few studies have addressed specific mechanical damages in guava. Apple, peach, and pomegranate were selected as similar fruit based on their spherical shape and size. External factors involved each mechanical damage were summarized to show how experimental settings impact fruit damage.

11.6.1 Impact Damage

Three methods often used for impact bruising comprise two simulated impact tests and a shock test. The two impact tests involve dropping fruit from a specific height onto the impact surface or mounting the fruit with a pendulum impactor. When dropping fruit from a specific height, the impact energy (E) of the impact test is calculated by the equation $E = mgh$, where m is the fruit mass, g is the acceleration of gravity (9.81 m s^{-2}), and h is the drop height (m) (Ruiz-Altisent & Moreda, 2011). The impact bruising shock test uses a shock machine to select shock programming and types of shock pulse (Goodwin & Young, 2011). Here, guava bruising focused on the impact method using a drop test with three factors such as drop height, number of drops, and storage temperature. The experimental design focused on these three factors. Many published studies have compared or described impact bruising from different drop heights followed by varied storage temperatures, while a few

Table 11.5 Height, number of drops, and storage temperature affecting impact bruising of guava, apple, pomegranate, and peach

No.	Commodity	Height (m)	No. of drops	Temp (°C)	Results	References
Guava						
1	Guava	0.20, 0.50	1	RT	Bruise depth of 0.2 m was smaller than that at 0.5 m.	Nila et al. (2017)
2	"Gim Ju" guava	0.30, 0.60	1, 5	20	Highest bruise susceptibility was height of 0.3 and 0.6 m from 5 drops.	Htike et al. (2023)
Apple						
3	"Pink Lady" Apple	0.25, 0.50, 0.75	1	0, 20	• Increase of drop height and storage period affected impact bruising. • Bruising area measurement was more accurate than bruise volume measurement.	Celik et al. (2021)
Pomegranate						
4	Pomegranate	0.15, 0.30, 0.60	1	5, 15, 25	• Impact energy and impact region played major roles in bruising susceptibility of pomegranate fruit. • Pomegranates stored at 5 °C suffered more bruise damage than at 25 °C.	Shafie et al. (2015)
5	Pomegranate	0.20, 0.40, 0.60	2	5, 20	• Highest browning was observed on fruit bruised at 0.4 and 0.6 m impact levels after 48 h incubation at both 5 and 20 °C. • Increase of browning intensity was associated with both temperature and time after impact.	Hussein et al. (2019a)
6	Pomegranate	0.20, 0.40, 0.60	2	5, 20	• Bruise area of fruit increased from 0.2 to 0.6 m. • Pomegranate fruit stored at 5 ° C had higher bruise volume and bruise susceptibility than at 20 ° C.	Hussein et al. (2019b)

Note: n/a means not available. RT means room temperature

proposed different numbers of drops. However, no study has investigated three-factor combinations with different levels. The three factors affected impact bruising of guava, apple, pomegranate, and peach. Higher drop height and number of drops influenced impact bruise susceptibility. Maximum drop height of an individual fruit for the impact test was 0.75 m, with 5 drops at storage temperature of 25 °C (Table 11.5).

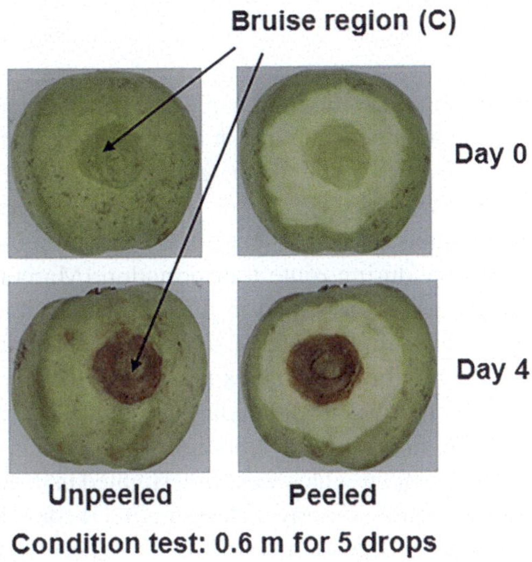

Fig. 11.5 Browning incidence from impact bruising in guava fruit appeared significantly on flesh tissue after drop test from a height of 0.6 m for 5 drops and storage at 25 °C under 70% RH for 4 days (Htike et al., 2023)

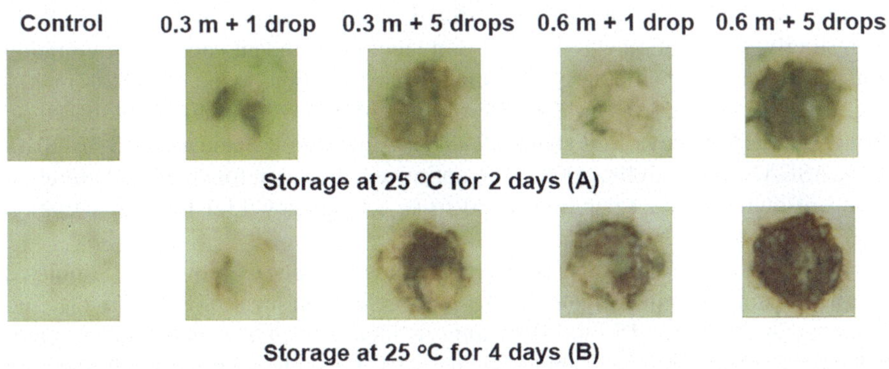

Fig. 11.6 Impact bruising of "Gim Ju" guava fruit after drop test with different drop heights and number of drops and storage at 25 °C under 70% RH for 2 days (**a**) and 4 days (**b**) (Htike et al., 2023)

Two impact bruising studies were performed in guava. Drop heights of 0.2 and 0.5 m for simulated impact testing were investigated. Results showed that bruising depth at a drop height of 0.5 m was higher than for 0.2 m at room temperature (Nila et al., 2017). Recently, Htike et al. (2023) studied drop height levels (0.3 and 0.6 m) and number of drops (1 and 5) in impact bruising of "Gim Ju" guava. The impact bruise as browning incidence appeared on flesh, not skin (Fig. 11.5) on the initial day. Five drops from a height of 0.3 (impact energy of 3678.75 J) and 0.6 m (impact energy of 7357.50 J) exhibited highest impact bruising susceptibility with the lowest color features (L^*, b^*, and C^*). These results indicated that the number of drops greatly influenced impact bruising of guava (Fig. 11.6).

11.6.2 Vibration Damage

Most vibration studies in apple and peach investigated either simulated transport or a combination between transit and simulated transport as transit-simulation (Fernando et al., 2018). Three main factors in actual transport research focused on power spectrum density (PSD) (G^2 Hz^{-1}) converted from acceleration (m s^{-2}) and time (s) domains, and frequency (Hz) indicating a strong or weak peak PSD at each frequency during route transportation (Marcondes, 2009). Thus, three independent variables were assigned in the simulated vibration study. Conditions of simulated vibration testing for guava, apple ,and peach are summarized in Table 11.6 Previous simulated vibration studies mostly focused on one independent variable or a combination of variables including frequency, acceleration, and time duration. In guava fruit, the first study of vibration bruising was reported by Chaiwong (2021). Three independent variables were investigated using response surface methodology (RSM) with a central composite design (CCD) for vibration testing of "Glom Sali" guava. The three independent variables were acceleration (2.942, 5.884, and 8.826 m s^{-2}), frequency (7, 13.5, and 20 Hz), and vibration duration (15, 30, and 45 min). After the vibration test, all fruit samples were stored at 20 °C under 70% RH for 48 h. Interestingly, acceleration was the most dominant independent variable for vibration bruising in guava, particularly the highest acceleration level of 8.826 m s^{-2} (Fig. 11.7). The fixed target value of bruise area (10% of total area) was based on Thai and ASEAN guava fruit standards (Thai Agricultural Standard TAS 16-2010, 2010; ASEAN Stan 7:2008, 2008). The optimized condition for simulated vibration testing of guava was acceleration 6.570 m s^{-2}, frequency 13.5 Hz, and vibration duration 30 min.

The period after testing also affected the development of browning. Simulated vibration testing was investigated under an acceleration level of 8.826 m s^{-2}, and frequency at 20 Hz for 15 min. Browning incidence from vibration bruising gradually increased and obviously appeared on guava skin with no bruising in flesh tissue within 12 h under storage at 25 °C with 70% RH (Fig. 11.8).

11.6.3 Compression Damage

Compression damage caused by stacked fruit received the greatest load from above, and was mostly exhibited at the bottom layers of the stack (Dubey & Mishra, 2018). Compression testing was performed by a rigid plate-probe using uniaxial compression. A fruit was compressed by the rigid plate-probe at a constant velocity. Fruit cells at the contact area were sheared, with bending between adjacent cells (Li et al., 2017). Very few studies have investigated compression bruising of fruit compared with impact and vibration bruising. Mosa (2019) investigated "Mamora" guava at three maturity stages (immature, medium, and mature). Maximum rupture forces of

Table 11.6 Simulated vibration conditions with varied acceleration, frequency, and time duration levels for guava, apple, and peach

No.	Commodity	Acceleration ($m\ s^{-2}$)	Frequency (Hz)	Time duration (min)	Result	References
1	Guava	2.942, 5.884, 8.826	7, 13.5, 20	15, 30, 45	Optimization condition was 6.570 $m\ s^{-2}$ with 13.5 Hz for 30 min at 10% bruising area.	Chaiwong et al. (2021)
2	Apple	3.236, 6.178	8.19, 12.59	10, 15, 20	Highest bruising index increased with increasing acceleration, frequency and time duration.	Vursavuş and Özgüven (2004)
3	Apple	2.942, 4.903, 6.864	7.5, 10. 13	n/a	Higher acceleration (>4.903 $m\ s^{-2}$) with frequency (>13 Hz) increased bruising volume	Rostampour and Motlagh (2018)
4	Apple	4.903	3	0, 180, 360, 540, 720	Vibration treatment decreased firmness and soluble solids compared to the control (untreated vibration)	Lu et al. (2019)
5	"Gala," "Idared," "Topaz," "Red Prince" Apple	n/a	28	360	Total color difference varied in skin color with only small effects in flesh color	Walkowiak-Tomczak et al. (2021)
6	Peach	8.826	5	60–120	Vibration affected PPO activity in peach pulp	Dantas et al. (2013)
7	Peach	n/a	5, 10, 15	120	Highest levels of damage and mold growth were exhibited at 15 Hz followed by 10 and 5 Hz	Dagdelen and Aday (2021)

Note: n/a means not available

Fig. 11.7 Vibration bruising of "Glom Sali" guava before testing (**a**, **c**) on the first day, and after vibration testing and storage at 20 °C under 70% RH for 48 h (**b**, **d**) (Chaiwong et al., 2021)

Fig. 11.8 Vibration bruising of "Gim Ju" guava after testing (8.826 m s^{-2} with frequency at 20 Hz for 15 min) developed within 0–12 h

immature fruit were higher than mature fruit by 1.84 (longitudinal position), 1.7 (radial position), and 1.39 folds (minimum axis).

11.7 Cushioning for Guava

Cushioning materials for perishable produce reduce mechanical damages from impact/shock, vibration, and compression forces (Dubey & Mishra, 2018). Cushioning material as fruit covering and cushioning sheets also reduces free space inside the package (Kitazawa, 2018). Foam nets as a commercial cushioning material are widely used for fruit covering. An alternative as paper-based cushioning material was applied to compare with foam net for reducing impact, shock, and vibration studies in apple and pear (Jarimopas et al., 2005; Zhou et al., 2008; Kitazawa et al., 2018). For the impact bruising study, foam net was compared to various paper-based cushioning materials. Single face with corrugated medium outside provided low bruise volume with high absorbed energy and low impact energy in "Fuji" apple.

The foam net had a higher bruise volume than single face with corrugated medium outside due to low absorbed energy and medium impact energy (Jarimopas et al., 2005). Two layers of pulp-molded trays also provided lower peak acceleration and velocity changes in "Fuji" apple than the foam net cushion, resulting in lower bruise area (Kitazawa et al., 2018). Reduction of acceleration value using foam net (1.618 m s^{-2}) caused the lowest vibration bruising of pear followed by paper-wrap packages (1.755 m s^{-2}) and no cushioning material (1.795 m s^{-2}) (Zhou et al., 2008). However, foam-based material has recently shown highest performance in reducing vibration bruising of soft texture fruit. Cushioning polyurethane (PU) + corrugated fiberboard boxes (CFB) gave better results than expandable polyethylene (EPE) + CFB and CFB formats to reduce vibration bruising of ripe peaches (Lin et al., 2020). Thus, paper-based material showed potential in terms of reducing mechanical damages as well as an alternative environmentally friendly cushion.

Guava packed with styrofoam net and placed in a cardboard box had lower damage (39.59%) than without filler material (84.03%) (Rolle, 2008). Different cushioning or wrapping was investigated in guava fruit. For cushioning, no significant differences in total soluble solids, total acidity, and ascorbic acid contents were shown among neem leaves, rice straw, bamboo leaves, and the control (Chandra & Kumar, 2012). Shrink and cling wrapping in an individual guava fruit preserved freshness with weight loss (<3.5%) and delayed the ripening process after storage at 7 °C for 21 days (Rana et al., 2015). Cling wrapping and shrink wrapping for guava also provided the greatest covering and increased shelf life by 4 days at 37 °C compared to vacuum and modified atmosphere packaging (MAP) (Rana & Siddiqui, 2018). Recently, guava fruit packed in a corrugated fiberboard box lined with newspaper cuttings showed higher overall chemical quality parameters compared to poly bags and a wooden box for road transportation of 200 km in India (Avesh et al., 2019). Most previous studies investigated material wrapping or MAP condition, and no cushioning study has been conducted on mechanical damages of guava.

In Thailand, foam net is a popular commercial cushioning material for guava fruit during distribution and transportation packed in corrugated boxes and plastic crates (see Sect. 11.5.2) (Fig. 11.4). To reduce mechanical damages, foam net is used preharvest as fruit bagging and postharvest as a cushion before packaging (Fig. 11.3c). However, plastic bags or paper materials are usually used to protect fruit abrasion during domestic transportation (Fig. 11.3b). Initially, two cushioning materials (foam net and honeycomb paper) were investigated for different frequency levels, time durations, and acceleration levels, and compared with the control (no cushioning). As shown in Fig. 11.9, guava samples of the three treatments were vibrated under a condition of 7 Hz + 45 min + 8.826 m s^{-2} and stored at 20 °C under 70% RH for 48 h. Observations illustrated that foam net protected vibration bruising better than honeycomb and no cushioning.

Fig. 11.9 Vibration bruising of "Glom Sali" guava in control (no cushioning) (**a**), foam net (**b**) and honeycomb paper (**c**) under 7 Hz + 45 min + 8.826 m s^{-2} and stored at 20 °C under 70% RH for 48 h

11.8 Assessment of Guava Bruise Damage

In small packinghouses, visible defect testing for damages caused by sunburn, pests, and postharvest disease is commonly performed manually. However, manual inspection cannot detect mechanical damages if the bruise has not fully developed. The efficacy of manual inspection also depends on the experience of the inspector and may lead to inconsistent inspecting results. For visible damages, image analysis is superior to manual inspection, with higher performance and greater accuracy (Zhang et al., 2014; Du et al., 2020). Since the development of computers and analytical techniques, machine vision has reduced sorting time and quantified visible defects accurately. However, bruises caused by mechanical damage require time to develop, so simple image analysis techniques are not suitable to inspect defects instantly. Many researchers compared different novel techniques to detect fresh bruises such as X-ray, fluorescence, hyperspectral imaging, and thermal imaging (Opara & Pathare, 2014; Diels et al., 2017; Li et al., 2018; Du et al., 2020).

For guava, manual inspection is suitable for visible defects on the surface (Fig. 11.2). Conventional manual measurement of bruising was reported by many researchers to describe the dimension of bruise damage corresponding to the area and volume (Opara & Pathare, 2014). This technique is suitable for a single bruising area caused by impact or compression forces. Bruise severity is reported as diameter, area, and volume. Different methods for quantifying and describing characteristic bruise volume were summarized by Opara and Pathare (2014) such as (a) bruise thickness method, (b) full depth method, (c) enclosed volume method, (d) unbruised volume removal method, and (e) ellipsoid method. Selection of the appropriate shape and related calculating formulae of bruise volume is important since bruise shape varies with different factors such as type of fruit, maturity, impactor, and energy applied (Opara & Pathare, 2014). Htike et al. (2023) estimated bruise severity of guava using area and volume as bruise damage parameters caused by impact force. Bruise area (BA) was calculated by assuming an elliptical shape (Lewis et al.,

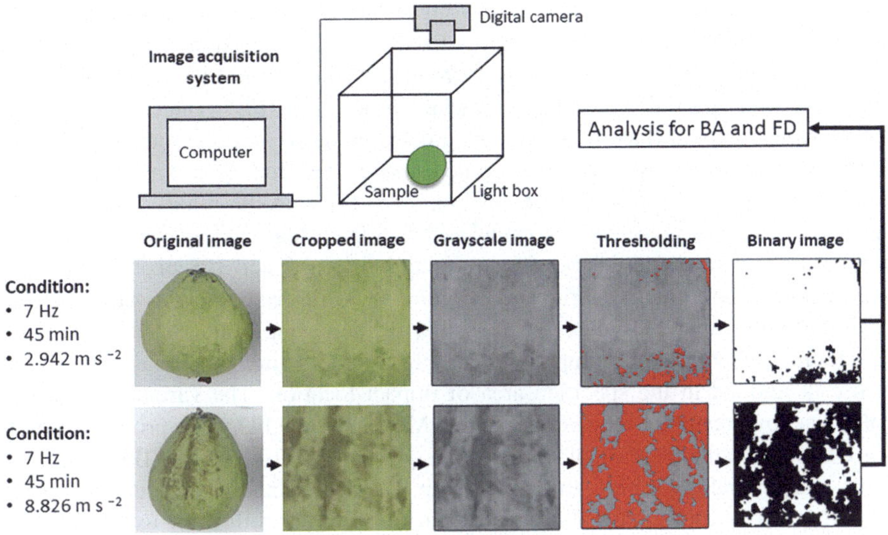

Fig. 11.10 Image analysis procedure for vibrational bruising damage (Chaiwong et al., 2021)

2007). $BA = \frac{\pi}{4}ab$, where a is the major diameter and b is the minor diameter. Bruise volume (BV) was calculated by assuming an elliptical shape with full depth method. $BV = \frac{\pi d}{24}\left(3ab + 4d^2\right)$, where d is full depth of bruise. Results showed that both BA and BV gradually increased as higher impact energy was applied. Thus, for guava, estimation of BA and BV by assuming the bruise as elliptical shape showed a good result.

Unlike impact and compression bruising, vibrational bruising may induce tiny bruising scattered through the fruit surface (Fig. 11.7); thus, the conventional method for determining bruise damage is not suitable. Visual inspection may be a better choice for inspecting defects and making decisions to accept or reject products (Opara & Pathare, 2014). Image analysis showed better results for the quantification of bruise damage due to vibrational force (Chaiwong et al., 2021). Images of guava fruit were quantified for size and severity of damage using image analysis software (Fig. 11.10). Image features (color, size, shape, and texture) were extracted by various image analysis software techniques. This basic information was used to determine bruise characteristics or used to calculate further parameters (Zheng et al., 2016). Important information extracted from captured images used to quantify vibrational bruises were the size and textural features. Image analysis software computed the area of the bruise by counting the black pixels of binary images and then reporting the result as percentage of bruising area (%BA). On the other hand, textural features represent the variation of pixel intensity compared with neighboring pixels (Zheng et al., 2016).

Textural features were successfully used to describe defects on the fruit surface such as browning occurrence in fresh-cut produces (Quevedo et al., 2016; Cáez Ramírez et al., 2017), senescence spots (Quevedo et al., 2008; Zheng & Lu, 2012), and bruises (Zheng et al., 2011; Chaiwong et al., 2021). Chaiwong et al. (2021) quantified vibrational bruises as normalized fractal dimension different values ($\Delta FD/FD_0$). Higher non-homogeneous browning color caused by vibration on the guava surface provided a higher fractal dimension (FD) value, which also represented higher surface roughness or greater variation in pixel intensity (Nayak et al., 2018). FD value was determined using the Box Counting method by intercepting an N number of different box sizes (r) with the binary image. Thus, the fractal dimension value was calculated as $FD = \log(Nr)/\log(1/r)$, where N is number of box size, and r is box size ranging between 2 and 128 pixels. FD value was reported to quantify bruise damage due to impact energy. Htike et al. (2023) reported that FD values decreased with increasing bruise severity. For the developing phenomenon of impact bruise (Fig. 11.5), the homogeneous color of the bruise developed in the specified area of impact location. The variation of bruise color was low, resulting in lower FD value. Moreover, the FD value provided a high correlation with the increase of a bruise area (BA) and bruise volume (BV) in guava. Thus, image analysis with feature extraction was determined as an accurate simple tool to quantify mechanical damages of guava, with capability for use with other fruit.

References

Ahmadi, E. (2012). Bruise susceptibilities of kiwifruits as affected by impact and fruit properties. *Research in Agricultural Engineering, 58*(3), 107–113.

ASEAN Stan 7:2008. (2008). *ASEAN standard for guava*. Retrieved from https://asean.org/storage/2012/05/7-ASEAN-STANDARD-FOR-GUAVA-2008.pdf

Avesh, Singh, S., Rana, M. K., & Nain, S. (2019). Effect of packaging materials on quality parameters of guava (Psidium guajava L.) cv. Hisar Safeda during storage after transportation. *International Journal of Chemical Studies* 7(3), 2880–2886.

Bakshi, P. (2015). Maturity indices of guava. *Journal of Agricultural Science,* 18.

Bantayehu, M., Alemayehu, M., Abera, M., & Bizuayehu, S. (2017). Postharvest losses assessment of tropical fruits in the market chain of North Western Ethiopia. *Food Science and Quality Management, 66*, 13–24.

Bilck, P. A., Roberto, S. R., Maria, V. E. G., & Yamashita, F. (2011). Efficacy of some biodegradable films as pre-harvest covering material for guava. *Scientia Horticulturae, 130*(1), 341–343.

Cáez Ramírez, G., Téllez-Medina, D. I., García-Armenta, E., & Gutiérrez-López, G. F. (2017). Digital image analysis and fractal metrics as potential tools to monitor colour changes in fresh-cut papaya (*Carica papaya* L.). *International Journal of Food Properties, 20*(Suppl 1), S177–S189.

Cavalini, F. C., Jacomino, A. P., Lochoski, M. A., Kluge, R. A., & Ortega, E. M. M. (2006). Maturity indexes for 'Kumagai' and 'Paluma' guava. *Revista Brasileira de Fruticultura, 28*(2), 176–179.

Celik, H. K., Ustun, H., Erkan, M., Rennie, A. E. W., & Akinci, I. (2021). Effects of bruising of 'Pink Lady' apple under impact loading in drop test on firmness, colour and gas exchange of fruit during long term storage. *Postharvest Biology and Technology, 179*, 111561.

Chaiwong, S., Yoythaisong, P., Arwatchananukul, S., Aunsri, N., Tontiwattanakul, K., Trongsatitkul, T., Kitazawa, H., & Saengrayap, R. (2021). Vibration damage in guava during simulated transportation assessed by digital image analysis using response surface methodology. *Postharvest Biology and Technology, 181*, 111641.

Chandra, D., & Kumar, R. (2012). Qualitative effect of wrapping and cushioning materials on guava fruits during storage. *HortFlora Research Spectrum, 1*(4), 318–322.

Chonhenchob, V., Chariyachotilert, S., & Singh, S. P. (2011). Packaging in fresh produce supply chains in Thailand. In *Packaging in fresh produce supply chains in Southeast Asia*. RAP PUBLICATION 2011/20. Retrieved from http://www.fao.org/3/ba0135e/ba0135e00.htm

Dagdelen, G., & Aday, M. S. (2021). The effect of simulated vibration frequency on the physico-mechanical and physicochemical properties of peach during transportation. *LWT - Food Science and Technology, 137*, 110497.

Dantas, T. B. H., Ferraz, A. C. D. O., Honório, S. L., & Lima, G. P. P. (2013). Polyphenoloxidase activity in peaches after vibration. *Engenharia Agrícola, 33*(2), 312–318.

Diels, E., van Dael, M., Keresztes, J., Vanmaercke, S., Verboven, P., Nicolai, B., & Smeets, B. (2017). Assessment of bruise volumes in apples using X-ray computed tomography. *Postharvest Biology and Technology, 128*, 24–32.

Du, Z., Zeng, X., Li, X., Ding, X., Cao, J., & Jiang, W. (2020). Recent advances in imaging techniques for bruise detection in fruits and vegetables. *Trends in Food Science and Technology, 99*, 133–141.

Dubey, N., & Mishra, V. (2018). Cushioning materials for fruits, vegetables and flowers. In M. W. Siddiqui, M. S. Rahman, & A. A. Wani (Eds.), *Innovative packaging of fruits and vegetables: Strategies for safety and quality maintenance* (pp. 313–316). Apple Academic Press.

FAO. (2021). *Food and agricultural commodities production: Guava, mango, and mangosteen*. Retrieved from http://www.fao.org/faostst/en/#data/QC

Fernando, I., Fei, J., Stanley, R., & Enshaei, H. (2018). Measurement and evaluation of the effect of vibration on fruits in transit—Review. *Packaging Technology and Science, 31*(11), 723–738.

Fu, H., Karkee, M., He, L., Duan, J., Li, J., & Zhang, Q. (2020). Bruise patterns of fresh market apples caused by fruit-to-fruit impact. *Agronomy, 10*(1), 59.

Goodwin, D., & Young, D. (2011). *Protective packaging for distribution*. Destech Publications.

Htike, T., Saengrayap, R., Kitazawa, H., & Chaiwong, S. (2023). Fractal image analysis and bruise susceptibility evaluation of impact damage in guava. *Information Processing in Agriculture*. https://doi.org/10.1016/j.inpa.2023.02.004.

Hussein, Z., Fawole, O., & Opara, U. (2019a). Determination of physical, biochemical and microstructural changes in impact-bruise damaged pomegranate fruit. *Journal of Food Measurement and Characterization, 13*(3), 2177–2189.

Hussein, Z., Fawole, O. A., & Opara, U. L. (2019b). Bruise damage susceptibility of pomegranates (*Punica granatum,* L.) and impact on fruit physiological response during short term storage. *Scientia Horticulturae, 246*, 664–674.

Jarimopas, B., Singh, S. P., & Saengnil, W. (2005). Measurement and analysis of truck transport vibration levels and damage to packaged tangerines during transit. *Packaging Technology and Science, 18*, 179–188.

Kamsiati, E. (2013). Postharvest handling practices in maintaining quality and shelf life of guava (*Psidium guajava*). In *Proceeding of the International Conference on Agricultural Postharvest Handling, and Processing* (pp. 363–373).

Katumbi, J. N., Imungi, J. K., Abong, G. O., Gachuiri, C. K., Mwang'ombe, A. W., Omayio, D. G., & Owade, J. O. (2021). Harvesting, postharvest handling, hygiene knowledge and practices of guava fruit farmers: A comparative study of two counties of Kenya. *African Journal of Food Science, 15*(5), 177–189.

Kitazawa, H. (2018). Cushioning packaging for fresh produce. In *Reference module in food science* (pp. 1–8). Elsevier.

Kitazawa, H., Li, L., Hasegawa, N., Rattanakaran, J., & Saengrayap, R. (2018). Evaluation of shock-proof performance of new cushioning system for portable packaging of apples. *Environmental Control in Biology, 56*(4), 167–172. https://doi.org/10.2525/ECB.56.167.

Lewis, R., Yoxall, A., Canty, L. A., & Romo, E. R. (2007). Development of engineering design tools to help reduce apple bruising. *Journal of Food Engineering, 83*(3), 356–365.

Li, Z., & Thomas, C. (2014). Quantitative evaluation of mechanical damage to fresh fruits. *Trends in Food Science and Technology, 35*(2), 138–150.

Li, Z., Miao, F., & Andrews, J. (2017). Mechanical models of compression and impact on fresh fruits. *Comprehensive Reviews in Food Science and Food Safety, 16*(6), 1296–1312.

Li, J., Chen, L., & Huang, W. (2018). Detection of early bruises on peaches (*Amygdalus persica* L.) using hyperspectral imaging coupled with improved watershed segmentation algorithm. *Postharvest Biology and Technology, 135*, 104–113.

Lin, M., Chen, J., Chen, F., Zhu, C., Wu, D., Wang, J., & Chen, K. (2020). Effects of cushioning materials and temperature on quality damage of ripe peaches according to the vibration test. *Food Packaging and Shelf Life, 25*, 100518.

Lu, F., Xu, F., Li, Z., Liu, Y., Wang, J., & Zhang, L. (2019). Effect of vibration on storage quality and ethylene biosynthesis-related enzyme genes expression in harvested apple fruit. *Scientia Horticulturae, 249*, 1–6.

Marcondes, J. (2009). Vibration. In K. L. Yam (Ed.), *The Wiley encyclopedia of packaging technology* (3rd ed., pp. 1266–1270). John Wiley & Sons.

Martins, M. C., Lilian, A., Lourenço, S. A., Gutierrez, A. S., & Watanabe, H. S. (2007). Incidence of postharvest damages in guavas at the wholesale market of São Paulo and its relationship to preharvest bagging. *Revista Brasileira de Fruticultura, 29*(2), 245–248.

Mendes, M. S., Chaves, F. M. S., Nascimento, S. S., Pires, I. C. G., Almeida, E. I. B., Sousa, A. N. S., & Ferrão, G. E. (2019). Postharvest losses of guava in the retail trade at Chapadinha (MA). *Scientific Electronic Archives, 12*(5), 57–61.

Miller, A. R., & Knee, M. (2002). Mechanical injury. In M. Knee (Ed.), *Fruit quality and its biological basis*. Sheffield Academic Press.

Mohammad Shafie, M., Rajabipour, A., Castro-García, S., Jiménez-Jiménez, F., & Mobli, H. (2015). Effect of fruit properties on pomegranate bruising. *International Journal of Food Properties, 18*(8), 1837–1846.

Morera-Montoya, R., Blanco-Metzler, H., & Luis-Loria, C. R. (2010). Evaluation of different bagging materials for the control of the fruit fly *Anastrepha* sp. (Diptera: Tephritidae) and fruit pathogens in Taiwanese guava fruits (*Psidium guajava* L.). *Acta Horticulturae, 849*, 283–292.

Mosa, M. M. A. (2019). Effect of mechanical loads on guava quality at handling. *Journal of Soil Sciences and Agricultural Engineering, 10*(4), 253–258.

Narayana, C. K., Bengaluru, I., Pandey, B. K., Lucknow, C., Malhotra, S. K., & Pandey, V. (2014). *Postharvest losses in selected fruits and vegetables in India*. Indian Institute of Horticultural Research.

Nayak, S. R., Mishra, J., Khandual, A., & Palai, G. (2018). Fractal dimension of RGB color images. *Optik, 162*, 196–205.

Nila, I. R., Saputro, A. H., & Imawan, C. (2017). The prediction system of bruising depth of guava (*Psidium guajava* L.) based on Vis-NIR imaging. In *2017 International Conference on Sustainable Information Engineering and Technology (SIET)* (pp. 420–424).

Oliveira, R. F., Medes, V. D., Milina, S., & Clemente, E. (2014). Study post-harvest about impact and compression mechanical in the cell quality of guava fruit (cv. Paluma). *International Journal of Sciences, 3*(8), 30–34.

Opara, U. L., & Pathare, P. B. (2014). Bruise damage measurement and analysis of fresh horticultural produce—A review. *Postharvest Biology and Technology, 91*, 9–24.

Paull, R. E., & Chen, C. C. (2016). *Guava*. Retrieved from https://www.ars.usda.gov/arsuserfiles/oc/np/commercialstorage/commercialstorage.pdf

Pommer, C. V., & Murakami, K. R. N. (2009). Breeding guava (*Psidium guajava* L.). In S. M. Jain & P. M. Priyadarshan (Eds.), *Breeding plantation tree crops: Tropical species* (pp. 83–120). Springer.

Quevedo, R., Mendoza, F., Aguilera, J. M., Chanona, J., & Gutiérrez-López, G. (2008). Determination of senescent spotting in banana (*Musa Cavendish*) using fractal texture Fourier image. *Journal of Food Engineering, 84*(4), 509–515.

Quevedo, R., Pedreschi, F., Bastias, J. M., & Díaz, O. (2016). Correlation of the fractal enzymatic browning rate with the temperature in mushroom, pear and apple slices. *LWT - Food Science and Technology, 65*, 406–413.

Rana, S., & Siddiqui, S. (2018). Comparative effect of different individual wrappings on shelf life of guava (*Psidium guajava*). *Journal of Food Science and Technology, 55*(8), 2935–2944.

Rana, S., Siddiqui, S., & Goyal, A. (2015). Extension of the shelf life of guava by individual packaging with cling and shrink films. *Journal of Food Science and Technology, 52*(12), 8148–8155.

Reyes, M. U., & Paull, R. E. (1995). Effect of storage temperature and ethylene treatment on guava (*Psidium guajava* L.) fruit ripening. *Postharvest Biology and Technology, 6*(3), 357–365.

Rolle, R. (2008). *Good practice for assuring the post-harvest quality of exotic tree fruit crops produced in Jamaica*. Retrieved from http://www.fao.org/3/ak832e/ak832e.pdf

Rostampour, V., & Motlagh, A. M. (2018). Evaluation of the bruising susceptibility of apple in transport conditions. *Bulgarian Journal of Agricultural Science, 24*(5), 902–908.

Ruiz-Altisent, M., & Moreda, G. P. (2011). Fruits, mechanical properties and bruise susceptibility. In J. Gliński, J. Horabik, & J. Lipiec (Eds.), *Encyclopedia of agrophysics* (pp. 318–321). Springer.

Shafie, M., Rajabipour, A., Castro-Garcia, S., Jiménez-Jiménez, F., & Mobli, H. (2015). Effect of fruit properties on pomegranate bruising. *International Journal of Food Properties, 18*, 1837–1846.

Sharma, R. R., & Nagaraja, A. (2016). Bagging guava fruits to reduce fruit fly incidence. *Indian Council of Agricultural Research, 22*(2), 23.

Singh. (2011). Guava (*Psidium guajava* L.). In E. M. Yahia (Ed.), *Postharvest biology and technology of tropical and subtropical fruits. Volume 3: Cocona to mango*. Woodhead Publishing.

Tandon, D. K., Singh, B. P., & Kalra, S. K. (1989). Storage behavior of specific-gravity-graded guava fruits. *Scientia Horticulturae, 41*(1), 35–41.

Thai Agricultural Standard TAS 16-2010. (2010). *Thai agricultural standard for guava*. Retrieved from https://www.acfs.go.th/standard/download/eng/guava.pdf

Vursavuş, K., & Özgüven, F. (2004). Determining the effects of vibration parameters and packaging method on mechanical damage in Golden Delicious apples. *Turkish Journal of Agriculture and Forestry, 28*(5), 311–320.

Walkowiak-Tomczak, D., Idaszewska, N., Łysiak, G. P., & Bieńczak, K. (2021). The effect of mechanical vibration during transport under model conditions on the shelf-life, quality and physico-chemical parameters of four apple cultivars. *Agronomy, 11*(1), 81.

Zhang, B., Huang, W., Li, J., Zhao, C., Fan, S., Wu, J., & Liu, C. (2014). Principles, developments and applications of computer vision for external quality inspection of fruits and vegetables: A review. *Food Research International, 62*, 326–343.

Zheng, H., & Lu, H. (2012). A least-squares support vector machine (LS-SVM) based on fractal analysis and CIELab parameters for the detection of browning degree on mango (*Mangifera indica* L.). *Computers and Electronics in Agriculture, 83*, 47–51.

Zheng, H., Jiang, B., & Lu, H. (2011). An adaptive neural-fuzzy inference system (ANFIS) for detection of bruises on Chinese bayberry (*Myrica rubra*) based on fractal dimension and RGB intensity color. *Journal of Food Engineering, 104(4)*, 663–667.

Zheng, C., He, H.-J., & Sun, D.-W. (2016). Chapter 3—Object measurement methods. In *Computer vision technology for food quality evaluation* (pp. 65–85). Elsevier.

Zhou, R., Su, S., & Li, Y. (2008). Effects of cushioning materials on the firmness of Huanghua pears (*Pyrus pyrifolia* Nakai cv. Huanghua) during distribution and storage. *Packaging Technology and Science, 21*(1), 1–11.

Chapter 12
Bruise Damage Susceptibility of Blueberry and Strawberry

Piyush Sharma and Arun Prasath Venugopal

Abstract Bruising is the damage to the skin and subcutaneous tissues of the fruits that reduces their commercial potential and makes them more susceptible to microbial damage. Increased mechanization in the harvesting and post-harvesting process for soft-tissue fruits has increased the incidences of damage. Different harvest and post-harvest factors are responsible for the increased susceptibility to bruising which is discussed in detail. Most of the bruising that occurs inside the fruits makes it more challenging to determine in mass processing. Different novel techniques for the detection of bruising such as hyperspectral, impedance and thermographic imaging are briefly elaborated. The level and quantification of bruising are essential to sort the fruits based on the extent of the damage for better handling and preventing the quality changes in them. Different models like the bruise thickness model, full depth model, enclosed volume model, spherical model, unbruised volume remove model, semi-oblate spheroid model, and indices such as bruise index, and bruise harmonic index are described briefly. Different quality changes resulting due to bruising and increased susceptivity of fruits to microbial damage along with strategies to prevent them are included in the discussion. The use of non-destructive bruising determination techniques along with proper harvesting planning and development of a resilient cold chain might serve as a better strategy to minimize the occurrence of bruising in blueberries and strawberries.

Keywords Blueberry · Strawberry · Bruising · Bruise thickness model · Non-destructive techniques

P. Sharma · A. P. Venugopal (✉)
Department of Food Process Engineering, National Institute of Technology Rourkela, Rourkela, Odisha, India
e-mail: varun@nitrkl.ac.in

P. B. Pathare, U. L. Opara (eds.), *Mechanical Damage in Fresh Horticultural Produce*, https://doi.org/10.1007/978-981-99-7096-4_12

12.1 Introduction

Production and trade in blueberries and strawberries have seen a major boost in recent years due to the increased facilities and avenues for processing, product development, pigment/colour extraction and improved storage facilities. With an annual production of about 8.86 million tonnes for strawberries and 0.86 million tonnes for blueberries (FAOSTAT, n.d.), they become one of the most grown fruits. Due to the increased availability of processing facilities, its consumption went beyond fresh to powdered, puree, etc. forms. It is usually consumed for its flavour and health benefits. With demand propping for new usages as food ingredients and additives it has gained the attention of export industries. The estimated export is totalled $2.1 billion in 2019 (FAS, USDA) and expected to cross $3 billion by 2025. The USA is the largest producer of blueberry and Canada covers a share of 80%. In strawberries, China leads in the production by producing around 38% of the world's total. In 2020, net production from China was 33.4 lakh tonnes with the US being at 10.59 lakh tonnes (FAOSTAT, n.d.).

An increase in demand and consumption of blueberries and strawberries has called for an increase in their production and a reduction in post-harvest losses. In an attempt at this many cultivators and production houses have started shifting towards increased mechanization in the pre- and post-harvest handling of the produce. Therefore, extensive mechanical operations in the collection, handling, transportation, packing and storage make the blueberries more susceptible to mechanical damage. Involvement and application of static and dynamic forces lead to fruit tissue damage, tearing of skin or peel, crushing and bruising. This not only degrades the quality of the fruits but also makes them prone to microbial spoilage. Bruising (Figs. 12.1 and 12.2) being the most common among all types of mechanical damage involves the application of large force on one rigid surface against another surface which might be a packaging material or any other fruit surface (Hussein et al., 2020a). This leads to the injury to the subcutaneous tissue cells, some of which rupture leading to discolouration marking the spot indicating damage. The extent of damage might vary with the extent of force applied and it might take up to 12 h to show up the bruising spot (Opara & Pathare, 2014). Internally inside the tissues when the stress over the cells exceeds their threshold value then breakage of cellular membranes occurs resulting in the release of cellular cytoplasmic enzymes into the free matrix thereby allowing them to act on other cells and vacuolar contents. Bruising hastens the deterioration of fruit quality and reduces its cosmetic appearance. It also results in several other changes which include changes in metabolic activities of fruits such as transpiration, respiration, ethylene production, etc. Bruises and mechanical damage over the skin disturb the natural barrier of the fruit, i.e., its peel which accelerates the microbial spoilage in fruits. Pathogens once entered the fruit may render it inconsumable in a very short period. This all makes fruit bruising one of the major factors for post-harvest losses.

As bruises are present internally inside the fruits, it might be challenging to detect them. On the outer surface, the fruit might appear normal and damage-free, internally

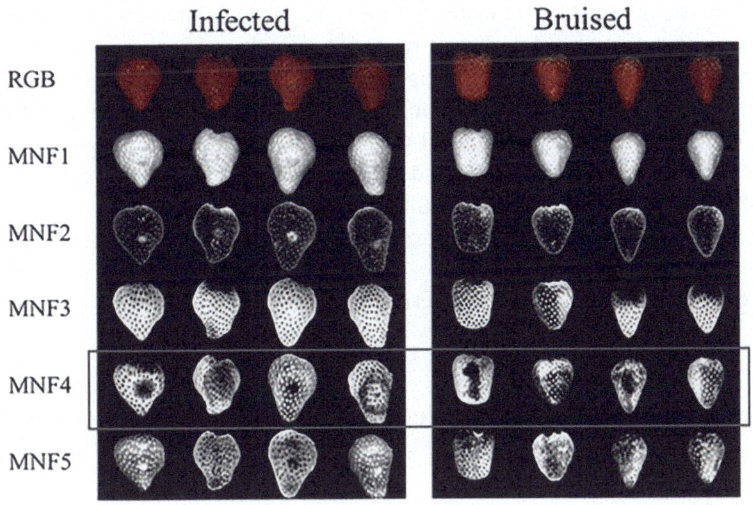

Fig. 12.1 Different bruising locations observed in the strawberry after image processing using MNF (Minimum Noise fraction) method (Liu et al., 2018)

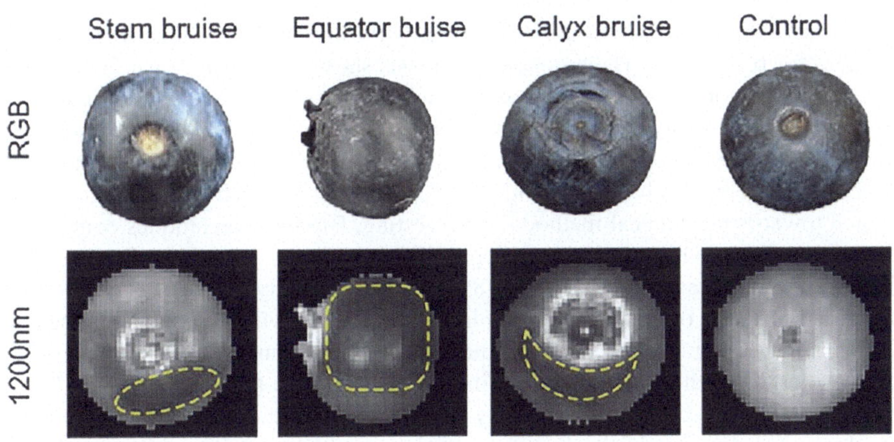

Fig. 12.2 Bruising in blueberries after 12 h of impact (Lobos et al., 2018)

it might carry ruptured and damaged tissues which is unacceptable in export and consumer markets. Therefore, to sort out such fruits from the bulk mass various non-destructive techniques are devised such as reflectance, transmittance and interactance spectroscopy which uses the near-infrared waves to detect the bruising inside the fruits. Various other food properties such as electrical impedance, conductivity and resistance which changes upon bruising or mechanical damage are also being used in techniques such as the electrical impedance method. Similarly, in pulsed thermographic imaging, the thermal properties of food such as conductivity, diffusivity, etc. are being used. These non-destructive techniques have immense potential in being utilized industrially on a large scale thereby allowing better segregation of defected and damaged fruits from fresh and high-quality ones.

Quantifying the mechanical damage or bruise percentage is also essential to determine the severity of the damage. Various models such as bruise thickness model, enclosed volume model, full depth model, ellipsoidal model, unbruised volume removal model, spherical model, semi-oblate model, etc. are developed to accurately and precisely determine the extent of damage in fruit tissues. These models along with several indices such as bruising index, bruise harmonic index and bruise susceptibility indicate the quality of the fruit and the efficiency and effectiveness of the fruit handling operations.

12.2 Factors Affecting Mechanical Properties and Bruise Susceptibility

12.2.1 Harvest Factors

12.2.1.1 Method of Harvesting

With growing mechanization in horticulture, harvesting methods have seen a lot of changes for blueberries. The ergonomic analysis shows that for the year 2016 more than 30% of work away days in private industry are related to musculoskeletal disorders related to work (MSDs) (Kim et al., 2018). Long harvesting times that stretch up to 1500 labour hours/hectare/year, and unavailability of labour result in a loss for blueberry producers (Brondino et al., 2021). All such factors route us to switch towards mechanical methods of harvesting. But these techniques come with their cons among which major is susceptibility to bruising. A study by Brown et al. (1996) reported severe bruises in 78% of mechanically harvested blueberries whereas only 23% of hand-harvested blueberries were having detectable bruises.

Bruising during harvest happens due to the impact and compression stresses during picking of the fruits, field packing, dumping of the fruits in collection bags, etc. These stresses develop in the fully mature berries which are soft and delicate resulting in the tearing of the skin. Different harvesting methods subject berries to different types of compressive stress and impact. A low bruising index is considered good that also reflects the least damage to fruits while handling them. In a study on bruising in strawberries due to various forces, it was found that slow compression results in greater tissue bruising than sudden impacts due to its impact-absorbing characteristic (Ferreira et al., 2008). Finding finds usage in the way of collection handling during harvesting and commercial packaging of the strawberries. Table 12.1 represents various studies carried out in this area.

Manual harvesting is the one carried through hands using fingers and thumb; picking the fruits gently to avoid any damage. It can also be assisted with clippers, picking shears or secateurs. Properly trained and skilled manpower can handle blueberries/strawberries with minimum damage. Tissue damage in the fruits occurs

Table 12.1 Effect of harvest and post-harvest factors on bruising susceptibility

Study	Main finding
Effect of harvest factors on bruising	
Method of harvesting	
Takeda et al. (2017)	Less than 8% bruise area in fruits harvested with hands and semi-mechanical system. Pneumatic shakers harvested 3–15 times more fruit than done by hand but the non-bruised fruits were 90% of the total harvested
Gallardo and Zilberman (2016)	Mechanical harvesting reduced 20% yield losses and 29% in quality losses
DeVetter et al. (2019)	The highest firmness was observed among mechanically harvested berries with soft catch surfaces in comparison to hard catch surfaces
Sargent et al. (2021)	Hand-harvested fruits did not show bruises, this increased up to 2% after 22 days. Mechanically harvested fruits have 3% bruising that increased to 22% upon storing
Maturity at harvest	
Lobos et al. (2018)	Additional development of blueberries for 6 days transforms them from soft to very soft making them more susceptible to bruising
Hung and Prussia (1989)	Mature fruits are more susceptible to mechanical damage than least and medium matured one
Tsukamoto (1979)	Reported the degree of correlation between damage and maturity of fruits but did not find a relation between the weight of fruit and maturity
Effect of post-harvest factors on bruising	
Precooling of produce	
Robbins and Moore (1992)	Weight loss of greater in fruits with an increase in delay of precooling and suggested for immediate precooling of the produce
Nunes et al. (1995b)	A cooling delay of 6–8 h makes the strawberries softer and more shrivelled. Rapid cooling and storage at 1 °C were necessary
Ishii and Shinbori (1988)	A differential pressure cooling system was found to be more efficient and effective than forced air cooling. Cooling of 27 to 4 °C was achieved within 2.5 h
Zhao et al. (2018)	Treatment with 1-MCP or no delay in precooling can prevent the peel yellowing and extend the storage life
Vigneault et al. (2001)	Delayed precooling has a regressive impact on the overall quality of the colour, odour and other indices
Mode of cooling	
Elansari et al. (2019)	Different types of precooling modes are available such as hydro cooling, forced air cooling, mobile cooling, vacuum cooling, ice water mix
Ferreira et al. (1994)	Hydro cooling was found to be better than immediately forced cooling
Liu et al. (2020)	Sub-zero temperature cooling is very useful for fruit and vegetable preservation as it is capable of inhibiting respiratory rate, deterioration reactions and microbial activity
Effect of temperature and humidity	
Van Linden et al. (2006)	Suggested that lowering temperature slows down the metabolic activity and has a direct impact on bruise susceptibility
Crisosto et al. (1993)	Impact and bruising were maximum when the fruit's temperatures were less than 10 °C and suggested that berries should be kept at

(continued)

Table 12.1 (continued)

Study	Main finding
	temperatures between 10 and 20 °C during handling to minimize bruising
Ferreira et al. (2009)	Strawberries pulp with temperatures of 24 °C showed higher resistance to bruising and mechanical damage. If low temperatures of 1 °C are used then a severe impact was observed
Effect of packing	
Singh and Xu (1993)	Four different packaging materials like polystyrene, corrugated boards and paper pulp trays were evaluated for apples where results indicated that polystyrene trays are more effective in preventing bruising and damage
Opara and Fadiji (2018)	Bruising was less in packages with a higher length-to-height ratio and have high vent area which indicated that packaging design affects bruising susceptibility
Xia et al. (2020)	Three different materials made up of wooden box, polystyrene and HDPE were tested for minimum weight loss and observed that minimum weight loss was observed in fruits packed in EPS boxes
Effect of transportation	
Van Zeebroeck et al. (2007a)	Factors responsible for bruising during road transport via trucks are road profile, suspension and tyre characteristics and acceleration. He also found out that the top layer is more at risk for the development of bruises than other layers
Lewis et al. (2008)	Hypothesized the failure mechanism that myofibrils fail due to straightening and slipping out under quasi-static and dynamic load
Van Zeebroeck et al. (2007a)	Hypothesized that myofibrils fail to slip up under quasi load and snap under dynamic load

when the human grasp force exceeds that of the fruit's threshold (Hussein et al., 2020a). In this case, the mass rupturing of cells happen known as bruising. This rupturing makes fruits more prone to pathogenic spoilage. The extent of bruising might be insignificant if the fruit further goes into processing operations where its quality can be managed or improved. Bruising incidences during manual harvest for blueberries had reported a range of results with least as 2% (Yu et al., 2014a) and as high as 23% (Brown et al., 1996). It indicates that to minimize the bruising losses for manual harvesting a standard harvesting protocol should be designed where training is given on an artificial picking simulator equipped with a blueberry impact recording device (BIRD) (Yu et al., 2014b).

Mechanical harvesting of blueberries and strawberries is gaining demand due to a variety of reasons some of them are efficiency, less harvesting time, low availability of labour, etc. Studies were done to understand and find the harvesting mechanism that yields the least damage to fruits. In a study by Yu et al. (2014b) which quantified and evaluated the mechanical stress and impact developed at the picking point in fruit with the help of a sensor mounted on the blueberry bushes; it was found that out of a variety of mechanical harvesting methods that can be classified into rotatory, sway away or slapper mechanisms of operation the least stress was recorded in rotatory type in comparison with latter. The entire distribution of 90% of impact in the fruit showed 190 g for rotary than 250 g in sway away and slapper. The rotary mechanism involves agitation movement with the help of horizontal shaking rods mounted on a vertical spindle. It mimics manual harvesting by rotating the fruit in the clockwise and anticlockwise directions. Slapper involves hitting bushes with the help of beater bars from both sides. Sway away on other hand involves the same side swinging of beaters (Mainland, 1993; Peterson et al., 1997; Peterson & Brown, 1996). Attempts to development of mechanical harvesters were done by researchers (Peterson et al., 1997; Takeda et al., 2008). Due to slow speed at ground (<1 km/h) in comparison to 1.6 km/h at present and the inability to be used for tall bushes (>2 m), mechanical models developed were unpopular among cultivators (Yu et al., 2014b). A study by Hu et al. (2017) which compared different harvesting methods with different varieties of blueberries noted that around 8–25% difference in bruising was there with higher bruising with a mechanical harvester. They had also pointed out that internal bruising was there with mechanical harvesting which gradually soften in cold storage. A reduction in yield up to 16% was observed with mechanical harvesters in comparison to hand harvesters. Harvesting of green fruits was 4% and 0.35%, respectively (van Dalfsen & Gaye, 1999). Similar results were obtained by Ali (2016). Sargent et al. (2021) noted that mechanical harvesters modified to have soft catch surfaces has minimal bruising and significantly improves harvest quality.

Robotic harvesting is an emerging technology being researched all over the world to develop a mechanical arm capable of mimicking the human arm that collects the harvest fastly and efficiently with minimum bruises. A tendon-driven robot was designed by Gunderman et al. (2021), which grips berries with as low force as 0.5 N thereby harvesting berries in 4.8 s. Moreover, the force applied can be controlled with a feedback loop that enables harvesting fruits with minimum possible damage. Another system for automatic detection of the height of blueberry bushes was designed using LiDAR technology (Sun & Li, 2016). Similar research is growing in the agro-robotics to create a new order where these mechanical devices perform the harvesting functions with greater precision.

12.2.1.2 Maturity at the Time of Harvest

Ripening brings various textural and compositional changes in the fruit that affects their bruises susceptibility. Studies (Table 12.1) have reported that firmness changes to a greater extent even after a short time once the fruit gets matured. Fruit firmness is considered a good index for the measurement of bruising damage (Valero et al., 2007). Lobos et al. (2018) reported that blueberries transformed from soft to very soft within 6 days. They suggested that frequent harvest maintains the firmness and hence bruising during collection. Macrae et al. (1989) reported that with an increase in soluble solids concentration the firmness of fruits decreases, and they become more susceptible to bruising damage. During ripening the sucrose level increases while that of starch decreases which is reflected in the differences between the core and flesh firmness; it is greater in immature fruit than in mature fruit (Macrae et al., 1989).

Different methods were used to measure firmness in the fruits popular being penetrometer or compression test (Duprat et al., 1997) but recent advances have shifted to non-destructive acoustic modulus methods, the first one being patented by Clark and Walter (1942). Installing these non-destructive firmness measurement devices upon a robotic harvester might ensure the adjustment of forces based on the maturity and firmness of fruit which ultimately will help us to control bruising in fruits.

12.2.1.3 Field Time Elapsed After Harvest

A longer time of production in the field after harvesting makes it susceptible to faster senescence and firmness loss which further impacts the bruising susceptibility of the fruits (Hussein et al., 2020a). In a study of bruising on strawberries it was found that with high pulp temperature (around 24 °C), fruits are more susceptible to greater compression and impact damage (Ferreira et al., 2009). They also suggested that strawberry growers should take into consideration the fruit temperatures before harvest scheduling. Similar results have been reported in another study by Pyrotis et al. (2012).

12.2.2 Post-harvest Factors

12.2.2.1 Effect of Precooling of Produce After Harvest

Blueberries are climacteric fruits. It respires and produces heat after harvest that deteriorates the quality of fruits. Cooling of produce after harvest, therefore, becomes very important to remove the field heat. The activity of various enzymes like pectin lyase, methyl esterases polygalacturonases and galactosidases needs to subside to preserve the integrity of the fruits (Duan et al., 2020). A study on three different varieties of strawberries, i.e., 'Chandler', 'Sweet Charlie' and 'Oso Grande' to understand the effect of delay in precooling has resulted in more than 50% greater water loss and shrivelling of fruits (Nunes et al., 1995b). Firmness in tissue was lowered by about 14–22%. Pelletier et al. (2011) reported that delay in precooling makes the produce unacceptable in markets due to inferior quality. The presence of bruises caused due to mechanical damage to shrivelled and softened strawberries accelerates the decay too. Park and Hwang (2010) study suggested storing strawberries at 4 °C after precooling at 2 °C to maintain freshness. Similar results for different seasons of storage of strawberries and blueberries (Kenny, 1979; Kozos et al., 2014). Precooling fruits within 2 h of harvest in the morning has been known to result in the least weight loss compared to harvesting in the afternoon and precooling it.

12.2.2.2 Effect of Mode of Cooling

Different types of cooling methods (Table 12.1) are available which include forced air cooling, hydro cooling, vacuum cooling, ice-liquid water mix, etc. (Elansari et al., 2019). In a study conducted by Ferreira et al. (1994) on various varieties of strawberries, they found that hydro cooling was better for immediate cooling than forced cooling (FC). Immediate and instant cooling of fruits reduces the extent of bruising (Hussein et al., 2020a). Non-uniform cooling effects were reported in FC with a difference of over 8 °C at different locations within the tunnel (Mercier et al., 2019). Evaporative cooling is also a known technique for reducing temperatures which involves adiabatic cooling through the evaporation of water (Samuel & Beera, 2013). Vacuum cooling technology is a quick and effective method for precooling cherries compared to conventional freezing (Lomeiko et al., 2019). Attempts were done to improve the effectiveness of the cooling system by creating discrete and uniformly distributed air flow inside the packages and were found to increase efficiency (Nalbandi et al., 2016).

12.2.2.3 Effects of Temperature and Humidity During Storage

Storage conditions of the fruit mainly temperature and humidity play a major role in the preservation and maintenance of the quality of blueberries. Temperature affects the tissue susceptibility to bruising by affecting the hydration of cells which impacts the cell turgor and increases the elastic modulus and stiffness (Hussein et al., 2020a, b; Lee, 2005). An increase in the internal cell pressure leads to an increase in susceptibility to tissue damage (Hussein et al., 2020a; Bajema et al., 1998).

Nunes et al. (1995a) evaluated the variation in the quality with storage temperatures. He noted that storage at too low temperatures also degrades the quality. For blueberries stored a 10 or 15 °C darkening was the major limiting factor. Though respiration rate gets lowered at low temperatures the shelf life estimated through the Q_{10} factor was more than the actual shelf life. Decay in blueberries got increased significantly when stored above 1.1 °C. The lowest weight loss was observed at 0 °C among all temperatures analysed. The quality of blueberries was highest at 0–1 °C. A study have reported that blueberries stored at 4 °C exhibit lesser decay and weight loss than one stored at 20 °C (Chun et al., 2013). They have also considered their anthocyanin levels which are better maintained and showed the least damage at 4 °C. Ścibisz and Mitek (2007) evaluated the long-term frozen storage at −18 and −35 °C and its impact on quality in comparison with the fresh ones. He found that storage at such low temperatures preserves the blueberries for up to 6 months and there was insignificant quality damage. Gonzalez et al. (2003) reported similar results when −24 °C was used as storage temperature. The strawberries stored at lower temperatures also reduced the increase in antioxidant content (Jin et al., 2011). Treatment of strawberries with essential oils along with their storage at low temperatures improved their shelf life (Shehata et al., 2020). Storage of strawberries coated with aloe vera gel at 6 °C and 50% RH does not show quality degradation for up to 15 days (Nasrin et al., 2017). The textural properties of strawberries were maintained upon thawing when carried.

12.2.2.4 Effect of Mode of Transportation

Blueberries and strawberries are susceptible to mechanical damage due to impacts during transportation and packing. Major forces are vibrational or impact (Jarimopas et al., 2007). During transportation, these forces result from the impact of single fruits on one another, drop impact of fruit container and transport vibration which results in increased impact forces. The presence of packaging material imparts a cushioning effect on the fruits which reduces the possibility of bruising the fruits. Studies were done to correlate various factors of transportation such as acceleration, the motion of vehicle tray, suspension characteristics, etc. Movement in trucks has been found to increase bruising (Van Zeebroeck et al., 2007a). Fruits, when loaded in many layers, cause collisions amongst them which leads to damage to fruits. Some studies have used the ultrasonic technique to study the point of impact and found that maximum stress of value 0.5 MPa happens at the centre for all loads (Lewis et al., 2008). Failure mechanisms were hypothesized as microfibrils fail due to straightening out or slipping for both quasi-static and dynamic loading (Lewis et al., 2008). Another hypothesis was as under quasi-static loading fibrils will slip and under dynamic loading will snap after getting straightened up (Van Zeebroeck et al., 2007a). Packaging with cardboard boxes was found to be more effective in comparison to other packaging materials to reduce the chances of bruising at 16 Hz and 2.5 m s^{-2} acceleration (Bülent & Alayunt, 2010). In a study, it was reported that bruising was least in fruits with high internal dampening compared to fruits with low internal dampening (Azam et al., 2022). Bruising and mechanical damage is found to be significantly increased by raising acceleration and stack height (Tabatabaekoloor et al., 2013). It was reported that using corrugated or clap board for packaging before transportation can reduce the bruising by 20% (Lu et al., 2009). It was also found that packages with a low length-to-height ratio are less susceptible to damage than packages with a high length-to-height ratio. Moreover irrespective of packaging design the fruits at the top layer are more prone to damage in comparison to other layers (Fadiji et al., 2016).

12.2.2.5 Effect of Packing

Packaging materials and arrangement inside the package affect the extent of bruising or mechanical damage in fruits (Table 12.1). A package of fruits gives the desired cushioning effect required to reduce the impact of collision amongst the fruits. In a comparison study between two ventilated corrugated packaging boards (MK4 and MK6) it was found that the latter one has higher packaging transmissibility ranging from 100% to 250% (Fadiji et al., 2016). It was also found that fruits in the top layer were more susceptible than other layers. In yet another study considering the packaging materials made up of polystyrene or foam trays, pulp trays and paperboard trays it was found that foam tray was best for packaging followed by paperboard partitions (Singh & Xu, 1993). Ventilated corrugated boards were often used in industries' supply chain and storage dealing with strawberries and blueberries to maintain air circulation while minimizing the bruising during handling. In a CFD study using the finite element method to analyse heat transfer flow and cooling rates inside the package, it was found that face ventilation area up to 7–8% can reasonably increase the cooling rates while the increase in the area above 10% results in loss of strength (Mukama et al., 2020). Modifying the atmosphere in the package can still reduce the quality and firmness loss in fruits. In a study, it was found that MAP packaged fruits are firmer and have less decay irrespective of precooling (Magazin et al., 2015).

12.2.2.6 Effect of Storage Time

The bruising susceptibility of fruits is known to decrease with the duration of storage (Hussein et al., 2020a; Klein, 1987). The lowered susceptibility of the fruits to mechanical damage and bruising is attributed to the changes in the turgidity of the fruits. According to the study by García et al. (1995) it is found that flaccid fruits became more resistant to the damage than turgid fruits. It is also stated that with long storage duration can increase the fruit's skin resistance and variation in the fruit texture can increase the resistance to mechanical impact (Hussein et al., 2020a, b). Brusewitz and Bartsch (1989) also reported a reduction in bruise volume per unit of impact energy with the storage time. However, some of the studies have reported different results. They observed that upon storing the fruits at lower temperatures fruit softening and bruising get induced in it (Lippert & Blanke, 2004). The changes might be associated with the physiological and structural changes in the cells which include solubilisation of pectin, cellulosic and hemicellulosic cell wall that decreases the fruit firmness (Hussein et al., 2020a; Mirdehghan et al., 2007).

12.3 Various Techniques for Characterization and Classification of Bruising or Mechanical Damage

12.3.1 Hyperspectral Imaging Technique

Hyperspectral imaging involves the simultaneous acquisition of spatial images in continuous bands. It collects hundreds and thousands of images of a spatial area at a different wavelength and contains both spectral and spatial details; collectively known as a hypercube. This spectral information when analysed and compared with control sample data gives a clear idea of the presence of tissue damage inside the fruit. Leiva-Valenzuela et al. (2013) evaluated the hyperspectral technique for predicting the extent of inner mechanical damage in blueberries. They have used push broom hyperspectra of three hundred blueberries to develop prediction models that were more precise in their predictions than that developed through usual SSC (soluble solids content) content. Pushbroom spectra are one of the most common ways of image acquisition in food industries as there are flat conveyor belts hence it is also known as line spectroscopy. Huang et al. (2020) also showed its feasibility with the above 99% recognition and differentiation rates for damaged berries. Li and Fu (2017) also evaluated the sorting of blueberries into one or more fractions based on firmness ($R_c = 0.891$, $R_v = 0.871$). He also showed that fruit orientation is insignificant in the determination of firmness. But it is more challenging to identify calyx spores (Fan et al., 2017). Hu et al. (2015) used a random frog algorithm for predicting the mechanical properties of the blueberry using hyperspectral data. They found that models developed through randomly selected spectra give better results than data through normal spectra. The accuracy of hyperspectral imaging to determine the internal bruising or mechanical damage was reported to be more than 75% when checked within 30 min of impact and more than 95% after 12 h; NIR reflectance imaging was considered better for the early detection of bruises and mechanical damage (Fan et al., 2017). Different studies focussing mainly on early and faster detection of bruising were done. Artificial intelligence, neural networking, self-training models, etc. were used to precisely determine the bruising area and extent of tissue damages. Zhang et al. (2020) trained full convolutional models (FCN) based on deep learning and used it for segmentation of multi-channelled input hypercubic data of images. The accuracy of these FCN models was over 81.2% and can be used to determine fruit resistance to impact damages. Results also reflected the use of 9-channel and 3-channel models in real-time online inspection and packaging line detection. Shi et al. (2020) also developed the framework using active learning and deep learning for multi-channel image classification for robust, accurate, real-time online detection of damages in blueberries.

12.3.2 Pulsed Thermographic Imaging

Thermographic imaging is a technique of developing a thermal image with temporal dimension by capturing the variation in temperatures of each pixel. The Fourier transformation is applied to this signal thereby disintegrating it into different sinusoidal components (Kuzy & Li, 2017). These imaging techniques are often carried out in two different ways, i.e., active and passive thermography. In active thermography, external heat stimulus is applied continuously and temperature change is monitored. Whereas in passive thermography no heat stimulus is applied. In pulsed thermography, the heat source is switched on and off for a specific time applying a set number of heat pulses on the object. Redistribution of heat is affected by material properties such as thermal diffusivity, conductivity, heat loss and geometry of the object. Thermal signals of damaged tissue or fruit with bruises differ from the normal ones and hence is the differentiation point for this technique. Kuzy et al. (2018) developed a pulsed thermographic imaging system for blueberries that gives accuracy up to 88%. He compared the relevant features between healthy and bruised fruits and then classified them into different sets with help of various classifiers such as k-nearest, logistic regression, vector, linear discriminant, etc. Feature selection is being performed through RELIEFF algorithm. Bhole and Kumar (2020) used a pre-trained neural network with a squeeze net modelled classification system which delivered an efficiency of more than 95%. In yet another advancement of the same study Bhole et al. (2020) have achieved a better classification with 100% accuracy by fusing the colour and texture data. They have pointed out towards scope of further improvement in accuracy of thermal images dataset for its application in developing a model for damaged and bruised fruit detection.

12.3.3 Electrical Impedance Method

The electrical impedance method determines the passive electrical parameters such as cytoplasm conductivity, cellular membrane resistance and capacitance. The presence of bruises can be determined by noting the variation in the bimodal distribution of impedance modulus (Cox et al., 1993). Values for damaged and bruised tissues are lower compared to undamaged tissue. Apart from assessing mechanical damage in fruits, it is also used for the determination of browning, chilling injury, etc. Vozary et al. (1999) have also mentioned that membrane resistance reduces upon rupturing and hence can be used for the detection of damage. Jackson and Harker (2000) have suggested the use of change in resistance at 50 Hz for the detection of physical damage. Matsumoto et al. (2022) determined the bruise tolerance of strawberries using bioelectrochemical impedance and mechanical properties; and found a correlation between bruise tolerance and elastic modulus. Their study mentions biochemical impedance theory which implies the degree of bruising can be found out through electrical resistance values using cole–cole plots. Bera et al. (2017) mentioned the application of electrical impedance for determining the fruit physiology and physiochemical analysis and indicated the potential in it for bruising detection on a large scale. Sugino et al. (2021) studied the bruising sensitivities of strawberries and their variation with electrical resistance measured from a cole–cole plot (LTO). Ibba (2021) mentions that impedance spectroscopy is still in the developmental stage due to certain technical glitches. They present a design operating on a 10–100 kHz frequency range with good accuracy similar to a high-end impedance analyser. Classification of strawberries when done with KNN and MLP networks gave promising results for real-time non-destructive analysis and detection.

12.4 Quantification of Bruises Damage

Different techniques and studies were carried out to quantify the bruising damage that was broadly classified into three groups (Opara & Pathare, 2014), i.e., manual measurement of area or volume to estimate damages, image analysis to know bruise size, using non-destructive damages techniques. In many studies where controlled bruising was done, bruising severity is a major factor to be calculated and analysed. In a study by Lewis et al. (2007) who considered the bruise to be elliptical, the bruised area was given by Eq. (12.1).

$$A_b = \pi\, l_1 l_2 \tag{12.1}$$

Where A_b is the bruised area, l_1 and l_2 are the major and minor widths of the bruise.

Many other studies relating to the quantification of bruises on a number basis, diameter ratio, volume ratio, weight ratio, etc. have been done (Opara & Pathare, 2014; Ericsson & Tahir, 1996). It was found that quantifying bruise weight was found to be more accurate due to its relation to impact energy (Opara & Pathare, 2014). Impact energy was measured as a difference between the initial and final energy of the fruit once the stimulus is given in form of the force. Lewis et al. (2007) developed an experimental setup where fruits were dropped on corrugated impact surfaces which was recorded by a high-speed video camera that was used to determine the coordinates of the fruit fall. Impact energy was reported to be found from Eq. (12.2)

$$\text{Impact Energy} = \left\{ \left(mgh_{\text{drop}}\right) - \left(mgh_{\text{after}} + \frac{1}{2} mv_{\text{after}}^2\right) \right\} \tag{12.2}$$

Where m is the mass of fruit; h_{drop} is the drop height; h_{after} is the height after impact; v is the apple velocity.

Bruise volume is another way of quantifying mechanical damage. Researchers have studied and compared different bruise volume estimating methods (Opara & Pathare, 2014; Bollen et al., 1999). They found that it is difficult to quote single formulae or model that describes all kinds of impact universally but every method put forth some benefits or advantages under specific conditions which increases their accuracy.

For various models discussed below, V is estimated bruise volume (mm^3), d_b full depth of bruise (mm), d_t is fruit depth from the top (mm), R is fruit radius, w_1 and w_2 are bruise major and minor axes (mm), x is fruit height above contact plain in circular consideration $\left(x = R - \sqrt{R^2 - \frac{w_1^2}{4}}\right)$ and y is fruit height above contact plain in elliptical consideration $\left(y = R - \sqrt{R^2 - \frac{w_1 w_2}{4}}\right)$.

The bruise thickness model (Opara & Pathare, 2014; Mohsenin, 1970) (Fig. 12.3) considers bruises to appear spherical perpendicular to the mid-point of the section. It

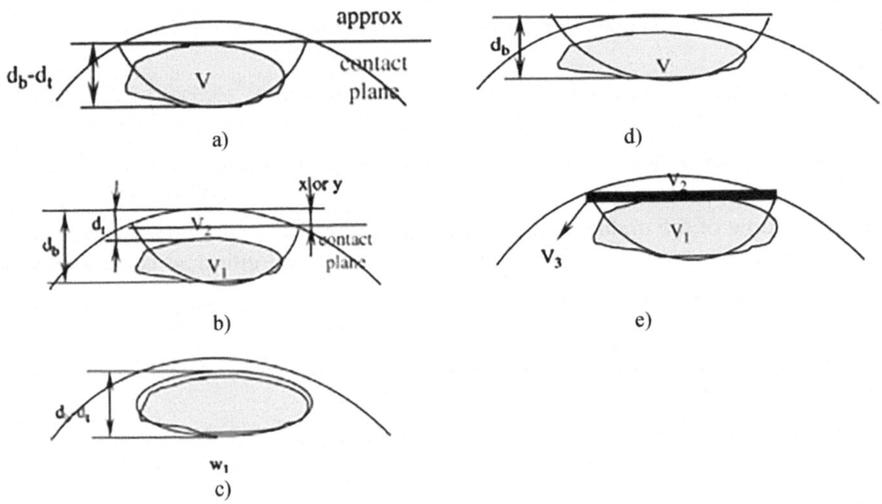

Fig. 12.3 Different models for calculation of bruise volume (**a**) Bruise thickness (**b**) Enclosed volume (**c**) Ellipsoid (**d**) Full volume (**e**) Unbruised volume removed (Opara & Pathare, 2014)

is considered best to be used for low impacts. The modelled equation considering bruise as circular and elliptical is presented in Eqs. (12.3) and (12.4).

$$\text{For circular;}\quad V = \frac{\pi\,(d_b - d_t)}{24}\left(3w_1^2 + 4(d_b - d_t)^2\right) \tag{12.3}$$

$$\text{For elliptical;}\quad V = \frac{\pi\,(d_b - d_t)}{24}\left(3w_1 w_2 + 4(d_b - d_t)^2\right) \tag{12.4}$$

The full depth model (Opara & Pathare, 2014; Saltveit, 1984) (Fig. 12.3) demands depth measurement be made from the contact plane as it considers recovering of fruit shape upon rebounding and extension of bruising up to the surface. The modelled equation considering bruise as circular and elliptical is presented in Eqs. (12.5) and (12.6).

$$\text{For circular;}\quad V = \frac{\pi\,d_b}{24}\left(3w_1^2 + 4d_b^2\right) \tag{12.5}$$

$$\text{For elliptical;}\quad V = \frac{\pi\,d_b}{24}\left(3w_1 w_2 + 4d_b^2\right) \tag{12.6}$$

The enclosed volume model (Opara & Pathare, 2014; Holt & Schoorl, 1977) (Fig. 12.3) divides the entire bruise volume into two parts, i.e., volume above the bruise boundary and volume till the fruit surface. It gives the best and most accurate estimates for high and full-range impacts. The modelled equation considering bruise as circular and elliptical is presented in Eqs. (12.7) and (12.8).

$$\text{For circular;}\quad V = \frac{\pi}{24}\left(3w_1^2 d_b + 4(d_b - x)^3 + 4x^3\right) \tag{12.7}$$

$$\text{For elliptical;}\quad V = \frac{\pi}{24}\left(3w_1 w_2 d_b + 4(d_b - y)^3 + 4y^3\right) \tag{12.8}$$

The unbruised volume remove model (Opara & Pathare, 2014; Bollen et al., 1999; Pang et al., 1992) (Fig. 12.3) is similar to the above-mentioned model except that the volume of the undamaged section is removed by considering it as a sphere. The modelled equation considering bruise as circular and elliptical is presented in Eqs. (12.9) and (12.10).

$$\text{For circular;}\quad V = \frac{\pi}{24}\left(3w_1^2(d_b - d_t) + 4(d_b - x)^3 + 4(d_t - x)^3\right) \tag{12.9}$$

$$\text{For elliptical;}\quad V = \frac{\pi}{24}\left(3\, w_1 w_2(d_b - d_t) + 4(d_b - y)^3 + 4(d_t - y)^3\right) \tag{12.10}$$

The spherical model (Opara & Pathare, 2014; Diener et al., 1979) for bruise volume considers the damaged area to be almost in the shape of the sphere. It can be represented as given in Eq. (12.11).

$$\text{Spherical bruise vol.;}\quad V = \frac{\pi}{6}h\left(\frac{3}{4}D^2 + h^2\right) \tag{12.11}$$

Where V is the estimated bruise volume (mm^3), h is the bruise depth (mm) and D is the bruise diameter (mm).

The semi-oblate spheroid model (Opara & Pathare, 2014; Chen & Sun, 1981) estimates bruise volume using maximum bruise diameter and depth. It can be represented through Eq. (12.12).

$$\text{Semi-oblate bruise volume;}\quad V = \frac{\pi}{6}hD^2 \tag{12.12}$$

Where V is the estimated bruise volume (mm^3), h is the bruise depth (mm) and D is the bruise diameter (mm).

12.5 Indices Used to Quantify Bruises

Different studies have used various indices to quantify the damage in fruits. These indices offer several advantages under specific situations to give a better-estimated way of comparing the extent of bruising in fruits lot.

The bruise index gives the estimated average bruise area or percentage weight distribution of fruits under different bruising categories (Opara & Pathare, 2014; Hadi et al., 2009). It can also be represented as the ratio of the bruised pixel area in a fruit image to the total number of pixels. Such representation is useful in the case when non-destructive technologies are used for the detection of bruises in fruits (Jiménez-Jiménez et al., 2013). Various equations representing the bruise index are given in Eqs. (12.13)–(12.15).

$$\mathrm{BI} = \frac{\text{Bruised Fruit area, weight or number}}{\text{Total fruits area, weight or number}} \times 100 \tag{12.13}$$

$$\mathrm{BI} = \frac{1X_1 + 2.5X_2 + 5.5X_3 + 10X_4}{100} \tag{12.14}$$

Where; X_1, X_2, X_3, X_4 are the weight per cent of fruits with no bruises, minor, moderate and major bruises, respectively.

$$\mathrm{BI}\ (\text{Image analysis}) = \frac{A_1}{A_2} \times 100 \tag{12.15}$$

Where; A_1 and A_2 are no. of pixels corresponding to the bruised area and the total no. of pixels.

Bruise sensitivity and bruise susceptibility (Opara & Pathare, 2014; Van Zeebroeck et al., 2007b) are other indexes useful to analyse the resistance of fruit towards an impact. Bruise sensitivity is defined as a minimum threshold of the fruit which is damage resulting from the impact of unit energy strength ($m^3\ J^{-1}$). Bruise susceptibility on other hand is the ratio of bruise volume and impact energy. The slope of the plot drawn *b/w* bruise volume and impact energy gives us bruise resistance coefficient (BRC) (Holt & Schoorl, 1977). Bruise harmonic index (Blahovec & Paprštein, 2005) is yet another index being used in the studies that are harmonic mean corresponding to bruise volume of 0.5 cm^3. It can be mathematically represented as given in Eq. (12.16).

$$\text{Bruise harmonic Index (BHI)} = \frac{\sqrt{\mathrm{HL}_{0.5}\mathrm{DL}_{0.5}}}{100} \tag{12.16}$$

Where HL and DE are Hysteresis loss and characteristic degree of elasticity corresponding to bruise volume of 0.5 cm^3.

12.6 Quality Changes Due to Bruising or Mechanical Damage in Blueberry/Strawberry

12.6.1 Physical and Textural Changes

Different biochemical reactions are accompanied by softening of the fruits such as neutral sugar loss, depolymerisation and solubilisation of cell wall which is the result of cellular enzymes that gets released during cell rupture upon mechanical damage.

Other metabolic changes such as loss of turgor, degradation of the cellular membrane and cell dynamics take place. Storage at high temperatures leads to flesh softening which increases the bruising susceptibility (Chiabrando et al., 2009; Sanford et al., 1991). Various biochemical components, cell wall composition, mechanical factors and moisture content are contributing factors to fruit texture firmness (Sams, 1999). Many studies have reported that softening of fruit is due to the loss of cellular adhesion. Bruises or damage in softened fruits occurs between cells whereas in firm tissue it occurs through cells (Harker & Hallett, 1992; Johnston et al., 2002; de Smedt et al., 1998). Microscopic examination of tissues of soft fruit shows the rounder cells with larger intercellular space. Firm tissues show fewer intercellular spaces with densely packed cells (Johnston et al., 2002). Pectin and hemicellulose polymers are important in maintaining the rigidity and shape of fruit tissues (Cosgrove, 2001). With bruising intracellular enzymes like proteases and pectinases get activated which affects the rigidness of the cells. The water level present in the cells is also a determinant of their firmness and rigidity. Fruits upon bruising reduces the turgor pressure which leads to loss of firmness and texture (Sams, 1999).

12.6.2 Changes in Colour

Anthocyanin is the pigment responsible for colour in blueberries and strawberries. These are water-soluble phenolic compounds in a glycosylated form that is present in the vacuoles of the plant cell. The rupturing of cells upon bruising or mechanical damage reduces the anthocyanin pigment and thereby impacting their colour. However, some studies have reported no correlation between the anthocyanin content and firmness, bruising per cent of fruits or loss in weight (Connor et al., 2002). However, it was found to be affected due to the pH of the fruit. Scalding was observed when the bruised cherries were kept in a nitrogenous atmosphere. Results were different when they were stored in an absolute oxygen environment where darkening in skin colour was observed due to oxidative browning (Dekazos, 1966).

12.6.3 Microbial Spoilage

Blueberries/Strawberries are rich in nutrients, vitamins, minerals, etc. which altogether makes a good environment for the growth of microbes. In an undamaged fruit, the outer skin serves as the natural barrier that protects against the invasion of microbes (Sardella et al., 2016). But after bruising or mechanical damage this natural barrier no longer preserves the quality of fruit and spoilage microbes such as bacteria and fungi start spoiling the fruit and multiply quickly. Fernández-Cruz et al. (2010) reported the production of toxins in the fruits by microbes which are harmful to the consumers. Microbes might enter through any route after bruising; it might be from improper handling, harvesting equipment, soil, contact with contaminated fruits, processing equipment, etc. (Sardella et al., 2016; Burnett & Beuchat, 2000). Damage to skin releases the nutrients and cell material which triggers the growth of microbes (Zhao et al., 2022). Ripening of fruits brings some major changes in the fruits both physically and biochemically. Starch breakdown, solubilisation of cell wall, protein and pectin breakdown, etc. are some of these changes. Ripening turns fruits richer in nutrients and makes them more susceptible to microbial invasion and bruising (Sardella et al., 2016; Zhao et al., 2022). Rupturing of cells increases the free unbound moisture available to microbes which enables them to divide and grow faster. Studies have reported that microbes causing spoilage are more capable of bruising than external barriers (James, 2003; Thompson, 2016). Mechanically damaged cells or tissues are more susceptible to microbes that produce extracellular enzymes, i.e., proteinases, hemicelluloses, pectinases, etc. (Zhao et al., 2022; Sommer et al., 1992; Udoh et al., 2015; Vico et al., 2014).

12.6.4 Effect of Bruising on Respiration Rate

Respiration continues in the fruit after harvest to sustain the metabolic activities going inside the fruit (Mannapperuma et al., 1991). The strawberries and blueberries have a respiratory coefficient of 1.0 and 1.3, respectively, which indicated the moles of CO_2 to O_2 produced (Fonseca et al., 2002). Reducing the temperature of storage reduces the respiration rate and increases the shelf life of the product (Halachmy & Mannheim, 1991). In a study by Hulme et al. (1963) increase in mitochondrial activity in the peel of fruit was observed before the increase in respiration. Damage in peel and cells, therefore, might result in an immediate decrease in respiration rate which after some time gets increased again due to an increase in microbial load. A decrease in respiration after some time is also attributed to a decrease in the activity of the mitochondrial system and enzymes operating the kerb cycle. Awad and Young (1979) indicated a rapid increase in the activity of cell wall depolymerizing enzymes and respiration rate. The weight loss of fruit is also high during storage after bruising, it might be as high as 40% in pomegranates (Hussein et al., 2020b). The change that occurs in respiration rates is usually proportional to pitting, it got decreased with an increase in pitting (Toivonen et al., 2004). In another study on the comparison of the respiratory quotient of fresh and bruised fruits, it was found that the RQ of bruised or damaged fruits were lower than that of fresh one (Segovia-Bravo et al., 2011).

12.7 Strategies for Effective Post-harvest Handling of Blueberries and Strawberries

Blueberries and strawberries are highly perishable seasonal crops therefore to ensure their availability throughout the year they must be preserved effectively with novel preservation techniques with the use of modified and controlled atmosphere, UV, cold plasma and edible coating. High respiration rates of order 52–245 mg CO_2 kg^{-1} h^{-1} in a low-temperature range are the major cause of the spoilage (Huynh et al., 2019). According to the writings of Mahajan et al. (2014) it was suggested that the application of heat treatments can effectively reduce the microbial load, improves the physiological properties and helps in the decontamination of produce. Treatment with hot water above 40 °C and vapour above 30 °C and greater than 90% Rh was found to be more beneficial (Fan et al., 2008; Zhao et al., 2012). Treatment with hot water does not significantly affect the quality of the blueberries but resulted in a 0.45–0.7 log reduction of total plate count and fungal pathogens. It rendered more than 92% of berries after 4 weeks of storage when treated with water at 60 °C temperature. UV-C is found to reduce microbial load considerably from food residues but penetration of UV-A is better than other UV spectrums. Inactivation of pathogenic microbes like *Listeria*, *Escherichia*, *Bacillus*, *Salmonella*, etc. with UV LEDs is better than with UV lamps. Long exposures of about 40 s can bring 3 log reductions in CFU (Nyhan et al., 2021). Treatment with UV induces increased resistance to senescence, disease and chilling injuries (Andrade Cuvi et al., 2011). It was reported to delay softening and increase the soluble solid content in fruits. Studies have found UV B as an alternative to UV C in reducing the microbial load (Eichholz et al., 2011). Short busts of UV light had led to a greater than 2.5 log reduction in *E. coli* and *Salmonella*. The sanitation process was not recommended to bring a microbial reduction in berries as it is found to promote mechanical damage (Huynh et al., 2019). Sulphur vapours are found to show preservation effects on blueberries and delay their decay. Chlorine dioxide vapours are found to be stronger than chlorine water for bringing microbial reduction (Joshi et al., 2013). It works by reducing the cellular metabolism and destructing the microbial cell walls. It is known to bring greater than 2 log reduction yeasts and moulds count. In many studies, ozone is also reported as a potential sanitizing agent ozone oxidizes the ethylene thereby delaying senescence and ripening (Chiabrando et al., 2004; Huynh et al., 2019). Though the handling of ozone is difficult it is effective for better maintenance of berries. Edible coatings on fruits reduce respiration and transpiration rates. The addition of plasticizer to edible coating gave good results in terms of decreasing fruit and vegetable moisture loss (Salehi, 2020). These coatings reduce the permeability of oxygen, carbon-di-oxide and water vapour which increases its shelf life. Coating fruits with a chitosan layer act as a barrier against bacterial contamination and moisture loss from surface of food products (Shiekh et al., 2013). Metal nanoparticles loaded on bio-polymers matrix reduces the post-harvest losses in fruits and vegetables. The edible coating containing soy protein helps in retaining the firmness in the fruits and reducing the moisture loss from the fruits (Zhang et al., 2018).

12.8 Conclusion

Bruising in blueberries and strawberries reduces their quality and commercial value. Beginning from its harvesting from fields to the overreach of the consumer; its way of handling might make it more susceptible to bruising. As the raw intake of blueberries and strawberries is high, any mechanical damage renders loss of its physical appearance and firm texture which makes it commercially unacceptable. Many harvest and post-havest factors such as harvest methods, harvest time, maturity at ripening, delay in precooling, temperature and humidity at storage and mode of transportation affect the bruising susceptibility. Non-destructive techniques such as hyperspectral and multispectral imaging, pulsed thermographic imaging and electrical impedance method are capable of detecting internal bruises. They have the potential to be utilized in pre-processing industries for sorting bruised fruits. However, the scope for further improvement in the efficiency and efficacy of these techniques exists. Given the increase in mechanization in the field of harvesting and post-harvest handling operations, future research must focus on designing equipment that minimizes the bruising and impact of mechanical damage. The development of automated resilient cold supply chains will help in reducing the post-harvest losses of these fruits and hence must be the focus area of future studies.

References

Ali, S. (2016). *Effect of harvesting time on berry losses during mechanical harvesting of wild blueberries* [PhD thesis].

Andrade Cuvi, M. J., Vicente, A. R., Concellón, A., & Chaves, A. R. (2011). Changes in red pepper antioxidants as affected by UV-C treatments and storage at chilling temperatures. *LWT - Food Science and Technology, 44*, 1666–1671. https://doi.org/10.1016/j.lwt.2011.01.027

Awad, M., & Young, R. E. (1979). Postharvest variation in cellulase, polygalacturonase, and pectinmethylesterase in avocado (*Persea americana* Mill, cv. Fuerte) fruits in relation to respiration and ethylene production. *Plant Physiology, 64*, 306–308. https://doi.org/10.1104/pp.64.2.306

Azam, M. M., Saad, A., & Amer, B. M. A. (2022). Assessment of the quality losses of Cantaloupe fruit during transportation. *Processes, 10*, 1187. https://doi.org/10.3390/pr10061187

Bajema, R. W., Hyde, G. M., & Baritelle, A. L. (1998). Turgor and temperature effects on dynamic failure properties of potato tuber tissue. *Transactions of ASAE, 41*, 741.

Bera, T. K., Bera, S., Chowdhury, A., et al. (2017). Electrical impedance spectroscopy (EIS) based fruit characterization: A technical review. In *Computer, communication and electrical technology* (pp. 279–288). CRC Press.

Bhole, V., & Kumar, A. (2020). Analysis of convolutional neural network using pre-trained squeezenet model for classification of thermal fruit images. In *ICT for competitive strategies* (pp. 759–768). CRC Press.

Bhole, V., Kumar, A., & Bhatnagar, D. (2020). Fusion of color-texture features based classification of fruits using digital and thermal images: A step towards improvement. *Grenze International Journal of Engineering and Technology, 6*, 133–141.

Blahovec, J., & Paprštein, F. (2005). Susceptibility of pear varieties to bruising. *Postharvest Biology and Technology, 38*, 231–238.

Bollen, A. F., Nguyen, H. X., & Rue, B. D. (1999). Comparison of methods for estimating the bruise volume of apples. *Journal of Agricultural Engineering Research, 74*, 325–330.

Brondino, L., Borra, D., Giuggioli, N. R., & Massaglia, S. (2021). Mechanized blueberry harvesting: Preliminary results in the Italian context. *Agriculture, 11*, 1197.

Brown, G. K., Schulte, N. L., Timm, E. J., et al. (1996). Estimates of mechanization effects on fresh blueberry quality. *Applied Engineering in Agriculture, 12*, 21–26.

Brusewitz, G. H., & Bartsch, J. A. (1989). Impact parameters related to post harvest bruising of apples. *Transactions of ASAE, 32*, 953.

Bülent, Ç., & Alayunt, F. (2010). Assessment of the quality losses of fresh fig fruits during transportation. *Journal of Agricultural Sciences, 16*, 180–193. https://doi.org/10.1501/Tarimbil_0000001137

Burnett, S. L., & Beuchat, L. R. (2000). Human pathogens associated with raw produce and unpasteurized juices, and difficulties in decontamination. *Journal of Industrial Microbiology & Biotechnology, 25*, 281–287.

Chen, P., & Sun, Z. (1981). Impact parameters related to bruise injury in apples. Pap-Am Soc Agric Eng Microfiche Collect.

Chiabrando, V., Peano, C., Beccaro, G, et al. (2004). Postharvest quality of highbush blueberry (Vaccinium corymbosum L.) cultivars in relation to storage methods. In: *VIII International Symposium on Vaccinium Culture 715* (pp. 545–552).

Chiabrando, V., Giacalone, G., & Rolle, L. (2009). Mechanical behaviour and quality traits of highbush blueberry during postharvest storage. *Journal of the Science of Food and Agriculture, 89*, 989–992. https://doi.org/10.1002/jsfa.3544

Chun, H. H., Kang, J. H., & Song, K. B. (2013). Effects of aqueous chlorine dioxide treatment and cold storage on microbial growth and quality of blueberries. *Journal of Korean Society for Applied Biological Chemistry, 56*, 309–315.

Clark, H. L., & Walter, M. (1942). Fruit ripeness tester.

Connor, A. M., Luby, J. J., Hancock, J. F., et al. (2002). Changes in fruit antioxidant activity among blueberry cultivars during cold-temperature storage. *Journal of Agricultural and Food Chemistry, 50*, 893–898.

Cosgrove, D. J. (2001). Wall structure and wall loosening. A look backwards and forwards. *Plant Physiology, 125*, 131–134.

Cox, M. A., Zhang, M. I. N., & Willison, J. H. M. (1993). Apple bruise assessment through electrical impedance measurements. *Journal of Horticultural Sciences, 68*, 393–398.

Crisosto, C. H., Garner, D., Doyle, J., & Day, K. R. (1993). Relationship between fruit respiration, bruising susceptibility, and temperature in sweet cherries. *HortScience, 28*, 132–135. https://doi.org/10.21273/HORTSCI.28.2.132

de Smedt, V., Pauwels, E., de Baerdemaeker, J., & Nicolai, B. (1998). Microscopic observation of mealiness in apples: A quantitative approach. *Postharvest Biology and Technology, 14*, 151–158.

Dekazos, E. D. (1966). Anthocyanin in red tart cherries as related to anaerobiosis and scald. *Journal of Food Science, 31*, 226–233.

DeVetter, L. W., Yang, W. Q., Takeda, F., et al. (2019). Modified over-the-row machine harvesters to improve northern highbush blueberry fresh fruit quality. *Agriculture, 9*, 13. https://doi.org/10.3390/agriculture9010013

Diener, R. G., Elliott, K. C., Nesselroad, P. E., et al. (1979). Bruise energy of peaches and apples. *Transactions of ASAE, 22*, 287–0290.

Duan, Y., Wang, G.-B., Fawole, O. A., et al. (2020). Postharvest precooling of fruit and vegetables: A review. *Trends in Food Science and Technology, 100*, 278–291. https://doi.org/10.1016/j.tifs.2020.04.027

Duprat, F., Grotte, M., Pietri, E., et al. (1997). The acoustic impulse response method for measuring the overall firmness of fruit. *Journal of Agricultural Engineering Research, 66*, 251–259. https://doi.org/10.1006/jaer.1996.0143

Eichholz, I., Huyskens-Keil, S., Keller, A., et al. (2011). UV-B-induced changes of volatile metabolites and phenolic compounds in blueberries (Vaccinium corymbosum L.). *Food Chemistry, 126*, 60–64.

Elansari, A. M., Fenton, D. L., & Callahan, C. W. (2019). Precooling. In *Postharvest technology of perishable horticultural commodities* (pp. 161–207). Elsevier.

Ericsson, N.-A., & Tahir, I. I. (1996). Studies on apple bruising: I. Estimation of incidence and susceptibility differences in the bruising of three apple cultivars. *Acta Agriculturae Scandinavica - Section B Soil and Plant Science, 46*, 209–213.

Fadiji, T., Coetzee, C., Chen, L., et al. (2016). Susceptibility of apples to bruising inside ventilated corrugated paperboard packages during simulated transport damage. *Postharvest Biology and Technology, 118*, 111–119. https://doi.org/10.1016/j.postharvbio.2016.04.001

Fan, L., Forney, C. F., Song, J., et al. (2008). Effect of hot water treatments on quality of highbush blueberries. *Journal of Food Science, 73*, M292–M297.

Fan, S., Li, C., Huang, W., & Chen, L. (2017). Detection of blueberry internal bruising over time using NIR hyperspectral reflectance imaging with optimum wavelengths. *Postharvest Biology and Technology, 134*, 55–66. https://doi.org/10.1016/j.postharvbio.2017.08.012

FAOSTAT. (n.d.). Retrieved July 16, 2022, from https://www.fao.org/faostat/en/#data/QCL

Fernández-Cruz, M. L., Mansilla, M. L., & Tadeo, J. L. (2010). Mycotoxins in fruits and their processed products: Analysis, occurrence and health implications. *Journal of Advanced Research, 1*, 113–122.

Ferreira, M. D., Brecht, J. K., Sargent, S. A., & Aracena, J. J. (1994). Physiological responses of strawberry to film wrapping and precooling methods. In *Proceedings of the Florida State Horticultural Society* (pp. 265–268).

Ferreira, M. D., Sargent, S. A., Brecht, J. K., & Chandler, C. K. (2008). Strawberry fruit resistance to simulated handling. *Science in Agriculture, 65*, 490–495.

Ferreira, M. D., Sargent, S. A., Brecht, J. K., & Chandler, C. K. (2009). Strawberry bruising sensitivity depends on the type of force applied, cooling method, and pulp temperature. *HortScience, 44*, 1953–1956. https://doi.org/10.21273/HORTSCI.44.7.1953

Fonseca, S. C., Oliveira, F. A. R., & Brecht, J. K. (2002). Modelling respiration rate of fresh fruits and vegetables for modified atmosphere packages: A review. *Journal of Food Engineering, 52*, 99–119. https://doi.org/10.1016/S0260-8774(01)00106-6

Gallardo, R. K., & Zilberman, D. (2016). The economic feasibility of adopting mechanical harvesters by the highbush blueberry industry. *HortTechnology, 26*, 299–308.

García, J. L., Ruiz-Altisent, M., & Barreiro, P. (1995). Factors influencing mechanical properties and bruise susceptibility of apples and pears. *Journal of Agricultural Engineering Research, 61*, 11–17.

Gonzalez, E. M., de Ancos, B., & Cano, M. P. (2003). Relation between bioactive compounds and free radical-scavenging capacity in berry fruits during frozen storage. *Journal of the Science of Food and Agriculture, 83*, 722–726.

Gunderman, A. L., Collins, J., Myer, A., et al. (2021). Tendon-driven soft robotic gripper for berry harvesting. https://doi.org/10.48550/ARXIV.2103.04270

Hadi, S., Ahmad, D., & Akande, F. B. (2009). Determination of the bruise indexes of oil palm fruits. *Journal of Food Engineering, 95*, 322–326. https://doi.org/10.1016/j.jfoodeng.2009.05.010

Halachmy, I. B., & Mannheim, C. H. (1991). Modified atmosphere packaging of fresh mushrooms. *Packaging Technology and Science, 4*, 279–286.

Harker, F. R., & Hallett, I. C. (1992). Physiological changes associated with development of mealiness of apple fruit during cool storage. *HortScience, 27*, 1291–1294.

Holt, J. E., & Schoorl, D. (1977). Bruising and energy dissipation in apples. *Journal of Texture Studies, 7*, 421–432.

Hu, M.-H., Dong, Q.-L., Liu, B.-L., et al. (2015). Estimating blueberry mechanical properties based on random frog selected hyperspectral data. *Postharvest Biology and Technology, 106*, 1–10. https://doi.org/10.1016/j.postharvbio.2015.03.014

Hu, B., Yang, W. Q., Andrews, H., et al. (2017). Towards a semi-mechanical harvesting platform system for harvesting blueberries with fresh-market quality. *Acta Horticulturae, 1180*, 335–340. https://doi.org/10.17660/ActaHortic.2017.1180.45

Huang, Y., Wang, D., Liu, Y., et al. (2020). Measurement of early disease blueberries based on Vis/NIR Hyperspectral Imaging System. *Sensors, 20*, 5783. https://doi.org/10.3390/s20205783

Hulme, A. C., Jones, J. D., & Wooltorton, L. S. C. (1963). The respiration climacteric in apple fruits. *Proceedings of the Royal Society of London - Series B: Biological Sciences, 158*, 514–535.

Hung, Y.-C., & Prussia, S. E. (1989). Effect of maturity and storage time on the bruise susceptibility of peaches (cv. Red Globe). *Transactions of ASAE, 32*, 1368–1373.

Hussein, Z., Fawole, O. A., & Opara, U. L. (2020a). Harvest and postharvest factors affecting bruise damage of fresh fruits. *Horticultural Plant Journal, 6*, 1–13. https://doi.org/10.1016/j.hpj.2019.07.006

Hussein, Z., Fawole, O. A., & Opara, U. O. (2020b). Effects of bruising and storage duration on physiological response and quality attributes of pomegranate fruit. *Scientia Horticulturae, 267*, 109306. https://doi.org/10.1016/j.scienta.2020.109306

Huynh, N. K., Wilson, M. D., Eyles, A., & Stanley, R. A. (2019). Recent advances in postharvest technologies to extend the shelf life of blueberries (Vaccinium sp.), raspberries (Rubus idaeus L.) and blackberries (Rubus sp.). *Journal of Berry Research, 9*, 687–707. https://doi.org/10.3233/JBR-190421

Ibba, P. (2021) *Fruit quality evaluation using electrical impedance spectroscopy* (PhD thesis). Free University of Bozen-Bolzano.

Ishii, K., & Shinbori, F. (1988). Effects of outer-leaf trimming, precooling methods and delay in precooling on changes in quality of turnips (Brassica campestris L.). *Journal of the Japanese Society for Horticultural Science, 57*, 544–548. https://doi.org/10.2503/jjshs.57.544

Jackson, P. J., & Harker, F. R. (2000). Apple bruise detection by electrical impedance measurement. *HortScience, 35*, 104–107. https://doi.org/10.21273/HORTSCI.35.1.104

James, I. F. (2003). *AD03E preservation of fruit and vegetables*. Agromisa Foundation.

Jarimopas, B., Singh, S. P., Sayasoonthorn, S., & Singh, J. (2007). Comparison of package cushioning materials to protect post-harvest impact damage to apples. *Packaging Technology and Science, 20*, 315–324. https://doi.org/10.1002/pts.760

Jiménez-Jiménez, F., Castro-Garcia, S., Blanco-Roldán, G. L., et al. (2013). Isolation of table olive damage causes and bruise time evolution during fruit detachment with trunk shaker. *Spanish Journal of Agricultural Research, 11*, 65–71.

Jin, P., Wang, S. Y., Wang, C. Y., & Zheng, Y. (2011). Effect of cultural system and storage temperature on antioxidant capacity and phenolic compounds in strawberries. *Food Chemistry, 124*, 262–270. https://doi.org/10.1016/j.foodchem.2010.06.029

Johnston, J. W., Hewett, E. W., & Hertog, M. L. (2002). Postharvest softening of apple (Malus domestica) fruit: A review. *New Zealand Journal of Crop and Horticultural Science, 30*, 145–160.

Joshi, K., Mahendran, R., Alagusundaram, K., et al. (2013). Novel disinfectants for fresh produce. *Trends in Food Science and Technology, 34*, 54–61.

Kenny, T. A. (1979). Studies on precooling of soft fruit: 1. Strawberries. *Iranian Journal of Food Science and Technology, 3*, 19–31.

Kim, E., Freivalds, A., Takeda, F., & Li, C. (2018). Ergonomic evaluation of current advancements in blueberry harvesting. *Agronomy, 8*, 266.

Klein, J. D. (1987). Relationship of harvest date, storage conditions, and fruit characteristics to bruise susceptibility of apple. *Journal of the American Society for Horticultural Science, 112*, 113–118.

Kozos, K., Ochmian, I., & Chełpiński, P. (2014). The effects of rapid chilling and storage conditions on the quality of Brigitta Blue cultivar highbush blueberries (Vaccinium corymbosum L.). *Folia Horticulturae, 26*, 147–153.

Kuzy, J., & Li, C. (2017). A pulsed thermographic imaging system for detection and identification of cotton foreign matter. *Sensors, 17*, 518. https://doi.org/10.3390/s17030518

Kuzy, J., Jiang, Y., & Li, C. (2018). Blueberry bruise detection by pulsed thermographic imaging. *Postharvest Biology and Technology, 136*, 166–177. https://doi.org/10.1016/j.postharvbio.2017.10.011

Lee, E. (2005). *Quality changes induced by mechanical stress on Roma-type tomato and potential alleviation by 1-methylcyclopropene* [PhD thesis]. University of Florida.

Leiva-Valenzuela, G. A., Lu, R., & Aguilera, J. M. (2013). Prediction of firmness and soluble solids content of blueberries using hyperspectral reflectance imaging. *Journal of Food Engineering, 115*, 91–98.

Lewis, R., Yoxall, A., Canty, L. A., & Romo, E. R. (2007). Development of engineering design tools to help reduce apple bruising. *Journal of Food Engineering, 83*, 356–365.

Lewis, R., Yoxall, A., Marshall, M. B., & Canty, L. A. (2008). Characterising pressure and bruising in apple fruit. *Wear, 264*, 37–46. https://doi.org/10.1016/j.wear.2007.01.038

Li, R., & Fu, L. (2017). Nondestructive measurement of firmness and sugar content of blueberries based on hyperspectral imaging. *Transations of the Chinese Society of Agricultural Engineering, 33*, 362–366.

Lippert, F., & Blanke, M. M. (2004). Effect of mechanical harvest and timing of 1-MCP application on respiration and fruit quality of European plums Prunus domestica L. *Postharvest Biology and Technology, 34*, 305–311.

Liu, Q., Sun, K., Peng, J., et al. (2018). Identification of bruise and fungi contamination in strawberries using hyperspectral imaging technology and multivariate analysis. *Food Analytical Methods, 11*, 1518–1527. https://doi.org/10.1007/s12161-017-1136-3

Liu, D.-K., Xu, C.-C., Guo, C.-X., & Zhang, X.-X. (2020). Sub-zero temperature preservation of fruits and vegetables: A review. *Journal of Food Engineering, 275*, 109881. https://doi.org/10.1016/j.jfoodeng.2019.109881

Lobos, G. A., Bravo, C., Valdés, M., et al. (2018). Within-plant variability in blueberry (Vaccinium corymbosum L.): Maturity at harvest and position within the canopy influence fruit firmness at harvest and postharvest. *Postharvest Biology and Technology, 146*, 26–35. https://doi.org/10.1016/j.postharvbio.2018.08.004

Lomeiko, O., Yefimenko, L., & Tarasenko, V. (2019). Vacuum cooling technology for pre-cooling of cherry fruits. In *Modern development paths of agricultural production* (pp. 281–288). Springer.

Lu, L., Huang, X., & Hua, Y. (2009). Effect of packaging methods on vibration bruising of pear fruits by simulated transport tests. *Transactions of the Chinese Society of Agricultural Engineering, 25*, 110–114.

Macrae, E. A., Lallu, N., Searle, A. N., & Bowen, J. H. (1989). Changes in the softening and composition of kiwifruit (Actinidia deliciosa) affected by maturity at harvest and postharvest treatments. *Journal of the Science of Food and Agriculture, 49*, 413–430. https://doi.org/10.1002/jsfa.2740490404

Magazin, N., Keserović, Z., Čabilovski, R., et al. (2015). Modified atmosphere packaging of fully ripe strawberries. *Acta Horticulturae, 1071*, 241–244. https://doi.org/10.17660/ActaHortic.2015.1071.28

Mahajan, P. V., Caleb, O. J., Singh, Z., et al. (2014). Postharvest treatments of fresh produce. *Philosophical Transactions of the Royal Society A: Mathematical, Physical and Engineering Sciences, 372*, 20130309.

Mainland, C. M. (1993). Blueberry production strategies. *Acta Horticulturae, 346*, 111–116. https://doi.org/10.17660/ActaHortic.1993.346.15

Mannapperuma, J. D., Singh, R. P., & Montero, M. E. (1991). Simultaneous gas diffusion and chemical reaction in foods stored in modified atmospheres. *Journal of Food Engineering, 14*, 167–183.

Matsumoto, S., Sugino, N., Watanabe, T., & Kitazawa, H. (2022). Bioelectrochemical impedance analysis and the correlation with mechanical properties for evaluating bruise tolerance differences to drop shock in strawberry cultivars. *European Food Research and Technology, 248*, 807–813. https://doi.org/10.1007/s00217-021-03928-2

Mercier, S., Brecht, J. K., & Uysal, I. (2019). Commercial forced-air precooling of strawberries: A temperature distribution and correlation study. *Journal of Food Engineering, 242*, 47–54. https://doi.org/10.1016/j.jfoodeng.2018.07.028

Mirdehghan, S. H., Rahemi, M., Martínez-Romero, D., et al. (2007). Reduction of pomegranate chilling injury during storage after heat treatment: Role of polyamines. *Postharvest Biology and Technology, 44*, 19–25.

Mohsenin, N. N. (1970). *Physical properties of plant and animal materials: Structure*. Gordon and Breach.

Mukama, M., Ambaw, A., & Opara, U. L. (2020). Advances in design and performance evaluation of fresh fruit ventilated distribution packaging: A review. *Food Packaging and Shelf Life, 24*, 100472. https://doi.org/10.1016/j.fpsl.2020.100472

Nalbandi, H., Seiiedlou, S., Ghasemzadeh, H. R., & Rangbar, F. (2016). Innovative Parallel Airflow System for forced-air cooling of strawberries. *Food and Bioproducts Processing, 100*, 440–449. https://doi.org/10.1016/j.fbp.2016.09.002

Nasrin, T. A. A., Rahman, M. A., Hossain, M. A., et al. (2017). Postharvest quality response of strawberries with aloe vera coating during refrigerated storage. *The Journal of Horticultural Science and Biotechnology, 92*, 598–605. https://doi.org/10.1080/14620316.2017.1324326

Nunes, M. C. N., Brecht, J. K., Morais, A., & Sargent, S. A. (1995a). Physical and chemical quality characteristics of strawberries after storage are reduced by a short delay to cooling. *Postharvest Biology and Technology, 6*, 17–28.

Nunes, M. C. N., Brecht, J. K., Sargent, S. A., & Morais, A. M. M. B. (1995b). Effects of delays to cooling and wrapping on strawberry quality (cv. Sweet Charlie). *Food Control, 6*, 323–328. https://doi.org/10.1016/0956-7135(95)00024-0

Nyhan, L., Przyjalgowski, M., Lewis, L., et al. (2021). Investigating the use of ultraviolet light emitting diodes (UV-LEDs) for the inactivation of bacteria in powdered food ingredients. *Foods, 10*, 797.

Opara, U. L., & Fadiji, T. (2018). Compression damage susceptibility of apple fruit packed inside ventilated corrugated paperboard package. *Scientia Horticulturae, 227*, 154–161.

Opara, U. L., & Pathare, P. B. (2014). Bruise damage measurement and analysis of fresh horticultural produce—A review. *Postharvest Biology and Technology, 91*, 9–24. https://doi.org/10.1016/j.postharvbio.2013.12.009

Pang, W., Studman, C. J., & Ward, G. T. (1992). Bruising damage in apple-to-apple impact. *Journal of Agricultural Engineering Research, 52*, 229–240.

Park, J.-E., & Hwang, S.-J. (2010). Effect of precooling and storage temperatures on the postharvest management of the fruits in 'Maehyang' and 'Soogyeong' strawberries for export. *Journal of Bio-Environment Control, 19*, 366–371.

Pelletier, W., Brecht, J. K., do Nascimento Nunes, M. C., & Emond, J.-P. (2011). Quality of strawberries shipped by truck from California to Florida as influenced by postharvest temperature management practices. *HortTechnology, 21*, 482–493.

Peterson, D. L., & Brown, G. K. (1996). Mechanical harvester for fresh market quality blueberries. *Transactions of ASAE, 39*, 823–827.

Peterson, D. L., Wolford, S. D., Timm, E. J., & Takeda, F. (1997). Fresh market quality blueberry harvester. *Transactions of ASAE, 40*, 535–540.

Pyrotis, S., Abayomi, L., Rees, D., & Orchard, J. (2012). Effect of temperature and humidity on strawberry firmness at two different sites in the Huelva region of Spain. *Acta Horticulturae, 926*, 567–570. https://doi.org/10.17660/ActaHortic.2012.926.81

Robbins, J., & Moore, P. P. (1992). Fruit quality of stored, fresh red raspberries after a delay in precooling. *HortTechnology, 2*, 468–470. https://doi.org/10.21273/HORTTECH.2.4.468

Salehi, F. (2020). Edible coating of fruits and vegetables using natural gums: A review. *International Journal of Fruit Science, 20*, S570–S589. https://doi.org/10.1080/15538362.2020.1746730

Saltveit, M. E., Jr. (1984). Effects of temperature on firmness and bruising of «Starkrimson Delicious» and «Golden Delicious» apples. *HortScience, 19*, 550–551.

Sams, C. E. (1999). Preharvest factors affecting postharvest texture. *Postharvest Biology and Technology, 15*, 249–254. https://doi.org/10.1016/S0925-5214(98)00098-2

Samuel, D. V. K., & Beera, V. (2013). Evaporative cooling system for storage of fruits and vegetables—A review. *Journal of Food Science and Technology, 50*, 429–442.

Sanford, K. A., Lidster, P. D., McRae, K. B., et al. (1991). Lowbush blueberry quality changes in response to mechanical damage and storage temperature. *Journal of the American Society for Horticultural Science, 116*, 47–51.

Sardella, D., Muscat, A., Brincat, J.-P., et al. (2016). A comprehensive review of the pear fungal diseases. *International Journal of Fruit Science, 16*, 351–377.

Sargent, S. A., Takeda, F., Williamson, J. G., & Berry, A. D. (2021). Harvest of southern highbush blueberry with a modified, over-the-row mechanical harvester: Use of soft-catch surfaces to minimize impact bruising. *Agronomy, 11*, 1412. https://doi.org/10.3390/agronomy11071412

Ścibisz, I., & Mitek, M. (2007). The changes of antioxidant properties in highbush blueberries (Vaccinium corymbosum L.) during freezing and long-term frozen storage. *Acta Scientiarum Polonorum Technologia Alimentaria, 6*, 75–81.

Segovia-Bravo, K. A., García-García, P., López-López, A., & Garrido-Fernández, A. (2011). Effect of bruising on respiration, superficial color, and phenolic changes in fresh manzanilla olives (Olea europaea pomiformis): Development of treatments to mitigate browning. *Journal of Agricultural and Food Chemistry, 59*, 5456–5464. https://doi.org/10.1021/jf200219u

Shehata, S. A., Abdeldaym, E. A., Ali, M. R., et al. (2020). Effect of some citrus essential oils on post-harvest shelf life and physicochemical quality of strawberries during cold storage. *Agronomy, 10*, 1466.

Shi, F., Wang, Z., Hu, M., & Zhai, G. (2020). Active learning plus deep learning can establish cost-effective and robust model for multichannel image: A case on hyperspectral image classification. *Sensors, 20*, 4975. https://doi.org/10.3390/s20174975

Shiekh, R. A., Malik, M. A., Al-Thabaiti, S. A., & Shiekh, M. A. (2013). Chitosan as a novel edible coating for fresh fruits. *Food Science and Technology Research, 19*, 139–155.

Singh, S. P., & Xu, M. (1993). Bruising in apples as a function of truck vibration and packaging. *Applied Engineering in Agriculture, 9*, 455–460.

Sommer, N. E., Fortlage, R. J., & Edwards, D. C. (1992). Postharvest diseases of selected commodities. *Journal of Plant Diseases and Protection, 3*, 3–11.

Sugino, N., Watanabe, T., Nakamura, N., & Kitazawa, H. (2021). Electrical and mechanical analysis to evaluate the cultivar difference in strawberries with respect to their bruising sensitivities and mass loss acceleration. *Postharvest Biology and Technology, 175*, 111489. https://doi.org/10.1016/j.postharvbio.2021.111489

Sun, S., & Li, C. (2016). Height estimation for blueberry bushes using LiDAR based on a field robotic platform. In *2016 ASABE Annual International Meeting* (p. 1). American Society of Agricultural and Biological Engineers.

Tabatabaekoloor, R., Hashemi, S. J., & Taghizade, G. (2013). Vibration damage to kiwifruits during road transportation. *International Journal of Agriculture Food Science & Technology, 4*, 467–474.

Takeda, F., Krewer, G., Andrews, E. L., et al. (2008). Assessment of the V45 blueberry harvester on Rabbiteye blueberry and southern highbush blueberry pruned to V-Shaped Canopy. *HortTechnology, 18*, 130–138. https://doi.org/10.21273/HORTTECH.18.1.130

Takeda, F., Yang, W., Li, C., et al. (2017). Applying new technologies to transform blueberry harvesting. *Agronomy, 7*, 33. https://doi.org/10.3390/agronomy7020033

Thompson, A. K. (2016). *Fruit and vegetable storage*. Springer International Publishing.

Toivonen, P. M., Kappel, F., Stan, S., et al. (2004). Firmness, respiration, and weight loss of 'bing', 'lapins' and 'sweetheart' cherries in relation to fruit maturity and susceptibility to surface pitting. *HortScience, 39*, 1066–1069.

Tsukamoto, M. (1979). Studies on the mechanical injury of fruit. *Journal of the Japanese Society for Horticultural Science, 48*, 374–380. https://doi.org/10.2503/jjshs.48.374

Udoh, I. P., Eleazar, C. I., Ogeneh, B. O., & Ohanu, M. E. (2015). Studies on fungi responsible for the spoilage/deterioration of some edible fruits and vegetables. *Advances in Microbiology, 5*, 285–290. https://doi.org/10.4236/aim.2015.54027

Valero, C., Crisosto, C. H., & Slaughter, D. (2007). Relationship between nondestructive firmness measurements and commercially important ripening fruit stages for peaches, nectarines and plums. *Postharvest Biology and Technology, 44*, 248–253. https://doi.org/10.1016/j.postharvbio.2006.12.014

van Dalfsen, K. B., & Gaye, M. M. (1999). Yield from hand and mechanical harvesting of highbush blueberries in British Columbia. *Applied Engineering in Agriculture, 15*, 393–398. https://doi.org/10.13031/2013.5794

Van Linden, V., Scheerlinck, N., Desmet, M., & De Baerdemaeker, J. (2006). Factors that affect tomato bruise development as a result of mechanical impact. *Postharvest Biology and Technology, 42*, 260–270. https://doi.org/10.1016/j.postharvbio.2006.07.001

Van Zeebroeck, M., Van Linden, V., Ramon, H., et al. (2007a). Impact damage of apples during transport and handling. *Postharvest Biology and Technology, 45*, 157–167. https://doi.org/10.1016/j.postharvbio.2007.01.015

Van Zeebroeck, M., Darius, P., De Ketelaere, B., et al. (2007b). The effect of fruit properties on the bruise susceptibility of tomatoes. *Postharvest Biology and Technology, 45*, 168–175.

Vico, I., Duduk, N., Vasić, M., & Nikolić, M. (2014). Identification of Penicillium expansum causing postharvest blue mold decay of apple fruit. *Pestic Fitomedicina, 29*, 257–266.

Vigneault, C., Gariepy, Y., Roussel, D., et al. (2001). Effect of precooling delay on broccoli quality and shelf life. Cah Etudes Rech Francoph Agric Fr.

Vozary, E., Laslo, P., & Zsivanovits, G. (1999). Impedance parameter characterizing apple bruise. *Annals of the New York Academy of Sciences, 873*, 421–429. https://doi.org/10.1111/j.1749-6632.1999.tb09491.x

Xia, M., Zhao, X., Wei, X., et al. (2020). Impact of packaging materials on bruise damage in kiwifruit during free drop test. *Acta Physiologiae Plantarum, 42*, 119. https://doi.org/10.1007/s11738-020-03081-5

Yu, P., Li, C., Takeda, F., & Krewer, G. (2014a). Visual bruise assessment and analysis of mechanical impact measurement in southern highbush blueberries. *Applied Engineering in Agriculture, 30*, 29–37.

Yu, P., Li, C., Takeda, F., et al. (2014b). Measurement of mechanical impacts created by rotary, slapper, and sway blueberry mechanical harvesters. *Computers and Electronics in Agriculture, 101*, 84–92.

Zhang, L., Chen, F., Lai, S., et al. (2018). Impact of soybean protein isolate-chitosan edible coating on the softening of apricot fruit during storage. *LWT - Food Science and Technology, 96*, 604–611. https://doi.org/10.1016/j.lwt.2018.06.011

Zhang, M., Jiang, Y., Li, C., & Yang, F. (2020). Fully convolutional networks for blueberry bruising and calyx segmentation using hyperspectral transmittance imaging. *Biosystems Engineering, 192*, 159–175. https://doi.org/10.1016/j.biosystemseng.2020.01.018

Zhao, J., Impaprasert, R., Yu, L., et al. (2012). Evaluation of postharvest quality of three southern highbush blueberry (Vaccinium corymbosum hybrid) cultivars subjected to heat

pre-treatment. In: *II Asia Pacific Symposium on Postharvest Research Education and Extension: APS2012* 1011 (pp. 187–195).
Zhao, J., Xie, X., Dai, W., et al. (2018). Effects of precooling time and 1-MCP treatment on 'Bartlett' fruit quality during the cold storage. *Scientia Horticulturae, 240*, 387–396. https://doi.org/10.1016/j.scienta.2018.06.049
Zhao, P., Ndayambaje, J. P., Liu, X., & Xia, X. (2022). Microbial spoilage of fruits: A review on causes and prevention methods. *Food Review International, 38*, 225. https://doi.org/10.1080/87559129.2020.1858859

Chapter 13
Bruise Damage Susceptibility of Table Olive

Mahdi Rashvand

Abstract Olive is one of the strategic products of the Mediterranean region. Unacceptable harvesting methods of olives which include hand, wood and harvesting machines is a concern of customers. Also, the transfer process of fruit from orchards to processing food industries leads to mechanical damage. Mechanical damage of olives can be observed as a bruise phenomenon. Bruise susceptibility is an indicator of mechanical damage to agricultural products. Damaged Olive not only reduces customer attraction but also decreases the quality of the extracted oil. More mechanical damage of olive lead to more value of acidity and peroxide of olive oil. For this reason, engineers and researchers investigate the reasons for mechanical damage of olives (from harvest time to post-harvest processing). Then, acceptable methods are suggested by researchers in order to reduce the bruise volume and area and increase the absorbed energy. This chapter is divided into three categories: (1) bruise susceptibility caused by impactor and vibrator harvester machines; (2) bruise damage that occurs due to vibration during transportation; (3) bruise susceptibility, which is caused in the factory by impact and drop. Furthermore, the advantages and disadvantages of offered solutions by researchers and engineers to reduce the mechanical damage of table olives are described.

Keywords Table olive · Bruise susceptibility · Mechanical damage · Mechanical properties · Viscoelastic properties

13.1 Introduction

Olive fruit is an agricultural product of the Mediterranean region. Although olive is a tropical crop, it can sustain cold weather conditions. In other words, olives can endure up to -12 (Friedman, 2000). The consumption of olive canned fruit has long been common in all parts of the world and its importance and consumption are

M. Rashvand (✉)
School of Agriculture, Forestry, Food and Environmental Science, University of Basilicata, Potenza, Italy
e-mail: Mahdi.rashvand@unibas.it

P. B. Pathare, U. L. Opara (eds.), *Mechanical Damage in Fresh Horticultural Produce*, https://doi.org/10.1007/978-981-99-7096-4_13

increasing day by day. Also, its oil is used due to its special medicinal and nutritional properties. Due to the high competition between vegetable oils from different seeds and plants, traditional olive cultivation is gradually decreasing and changing to modern gardens.

Due to the strategy of the olive crop, the cultivation area of this crop is increasing. The cultivation area of olive and its amount are 56,000 ha and 84,000 million tons in 2018, respectively (FAO). More than half of the produced products are from Spain, Greece and Italy, which are ranked in the production of olives, respectively. Unfortunately, due to mechanical damage that occurs during the harvesting and post-harvest processes of olive, the mentioned crop is not consumed and part of it is wasted as agricultural waste.

Due to the growing population, farmers not only looking for a way to increase agricultural products but also try to reduce the mechanical damage of agricultural products to provide better quality products. Olives are one of the agricultural products that are always damaged due to improper harvest by farmers. Furthermore, during transportation of olive from the garden to processing industries, the bruise phenomenon is occurred.

The bruise phenomenon caused by mechanical damage affects the appearance of the olive and reduces the quality of extracted olive oil. Researchers and engineers are trying to reduce the area, volume and bruise susceptibility by designing proper harvesting machines and food industry machines. Researchers have used methods such as callipers (Van Zeebroeck et al., 2007), doping tissue analysis and analysis by image processing (Du et al., 2020) and non-destructive methods (Jiménez-Jiménez et al., 2012) to obtain the area and volume of bruised sections.

In order to optimize the harvesting and post-harvest machines, the physical and mechanical properties of the olives must be considered. For this reason, researchers used pressure tests to obtain the mechanical properties of olives. Also, in order to obtain the volume, surface and bruise susceptibility of agricultural products, pendulum and drop tests are used. In the following, the mentioned items are explained.

13.2 Mechanical Damage Caused by Improper Harvesting

The primary objective of investing in planting is to achieve a successful harvest. The most desirable situation happens when the harvest is as large as possible and the losses are minimal. One of the major problems in olive oil production is its traditional harvesting costs, which account for about half of its production revenue. In general, the cost of olive harvesting is about 30% of the net income of its production. The increasing need of farmers in order to increase the coefficient of agricultural mechanization, reduce the cost of olive production and increase the quality and efficiency of production, has necessitated the study of mechanical methods.

The determination of the appropriate harvesting method depends on the planting method, garden topography, labour force and economic status (Herrera-Cáceres et al., 2017). Mechanized harvesting methods not only speed up the work, reduce the required human force and production costs but also reduce waste and increase product quality. Olives are harvested traditionally in some Third World countries and mechanized in developed countries. Unfortunately, due to the lack of attention to agricultural mechanization, farmers must use the traditional method (picking by hand or wood). Although using agricultural machinery is common in industrialized countries, in many cases harvesting machines lead to mechanical damage to the olives. In the following, the methods of harvesting olives that cause mechanical damage are described.

13.2.1 Bruise Phenomenon Caused by Manual Methods

The manual method involves picking the fruit by hand or hitting the fruit with wood. Picking fruit by hand does not have a detrimental effect on fruit quality. However, after picking the fruit by hand, due to the impact of the fruit from a certain height (about 2–3 m), the olive fruit damage. This method of picking is not only time consuming but also because hiring workers is expensive. Although the manual method has some disadvantages, the quality of extracted olive oil is appropriate. The acidity of this method is 1–1.8 (Saglam et al., 2014). However, sometimes farmers prefer to use wood, which reduces the harvesting time. Harvesting with wood due to direct impact on the olive leads to bruising, which reduces the quality of olives and olive oil.

13.2.2 Generated Bruise Susceptibility by the Mechanized Method

Nowadays, many farmers in developed countries prefer to use agricultural machinery to harvesting olives. Over the past decade, engineers have attempted to reduce the amount of damage by optimizing prototype machines. Researchers performed a lot of research about the harvesting method. The two methods of harvesting by impact and vibration are considered for olives, which are explained.

13.2.2.1 Impactor Harvester

Impact harvesters are made of two types of reciprocating (butterfly) and rotary. A butterfly machine work with an engine that produces compressed air and forces two plastic combs made in different shapes to oscillate (Fig. 13.1). By hitting the fruits between the two plates, they cause them to separate from the tree. The rotary type

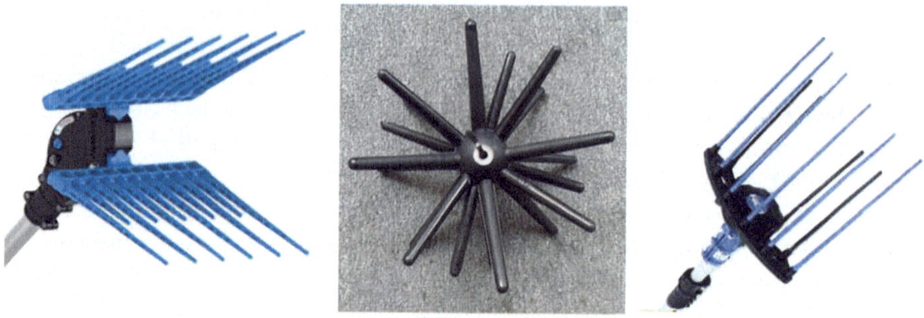

Fig. 13.1 View of the impact heads of the olive fruit harvester

causes more bruising with a direct impact on the product than the butterfly. In small-scale orchards, some farmers use this type of machine to harvest olives.

The brown colour of the olive surface occurs mainly by the impact of fruit-fruit and fruit-branch during mechanical harvesting. In this case, cell tissue is disrupted and leads to the release of polyphenol oxidase and its phenolic substrate and as a result of physical contact and in the presence of oxygen and oleuropein, which is the main phenol in olive paste, oxidizes and forms brown polymers. Processing the effect of a combination of edible and nitrogen coatings on the quality of mechanically harvested olives shows that the brown spots of olives stored in the open air have developed due to the oxidation of oleuropein. Therefore, post-harvest storage of nitrogen-covered olives can be a good way to prevent bruising damage to mechanically harvested olives.

13.2.2.2 Vibration Harvester

One of the most common types of olive harvesting machines are vibrating systems that produce vibrating force and apply it to the trunk or branch of the olive tree, causing various tree components to vibrate and as a result, the fruits separate from the tree (Fig. 13.2). This device is commonly used in large-scale orchards. However, the small vibrating device (which is connected to only one branch) is also applicable for small-scale olive orchards. The advantages of this device include increasing efficiency and high operating speed, increasing productivity, maintaining quality and reducing waste, saving cost and time, reducing labour and producing high-quality oil. Since the device does not hit the olive, the bruise phenomenon has not occurred.

Although the head of the vibrating harvesters does not impact the olives, the bruises caused by the drop of the olives on the ground can damage the olives. The agricultural machinery designers have been using plastic conveyors with good elastic properties to prevent olives from colliding with the ground. Impact damage occurs when an olive fruit impacts the surface with a force that lad to breaking or even separating cells and its external sign is bruising.

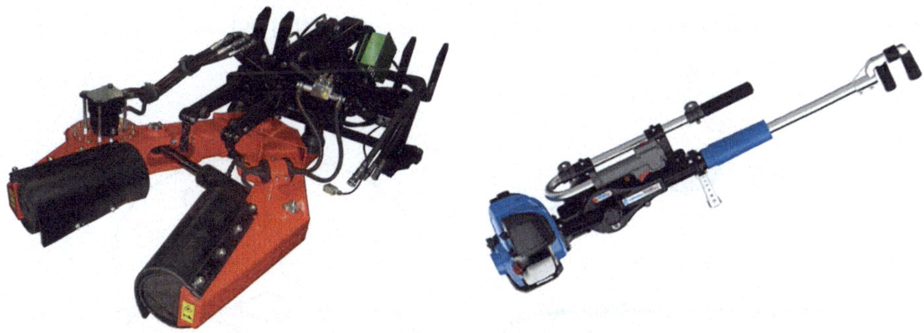

Fig. 13.2 Vibration harvester of olive

Impact mechanics theories were founded by St. Venant, who proposed wave theory and Hertz (contact of elastic objects). Many researchers carried out some experiments to create a relation of bruise area to various mechanical factors such as force (Brusewitz et al., 1991), fall height (Bollen et al., 1999), acceleration and impact velocity (Zapp et al., 1990) and absorbed energy (Jiménez-Jiménez et al., 2013). The researchers found increasing the energy levels and impact force significantly increased the bruise phenomenon. In addition, report a linear correlation between absorbed energy and bruise volume.

Hoshyarmanesh et al. (2017) evaluate the olive vibration harvester machine and simulate the olive tree harvesting model by measuring the physical properties of the tree. In the experiment, the upper part of the olive tree responded positively to a frequency higher than 22 Hz and a displacement of 97 cm. The lower part of the tree did not respond positively to the frequency of 92 Hz and the displacement of 96 cm and had the least effect on fruit drop; Therefore, it can be concluded that the picker should be designed to pick by applying different frequencies with the ability to change the amplitude of the shake (Hoshyarmanesh et al., 2017).

13.3 Post-harvest Mechanical Damage of Olive

Post-harvest mechanical damage usually occurs in two stages: transport from the garden to the factory and on the processing line of food factories. Improper transfer of fruit during harvesting, packaging, separation and transportation causes damage to the tissue, cell and its structure, which is caused by impact, pressure, scratches, holes or many other factors. Fracture in the structure of the fruit increases the risk and preparation of the fruit to rot and the growth of microorganisms. This damage usually occurs in dynamic shocks between fruits and between the fruit and the box.

The accumulation of olives on top of each other during transportation causes pressure damage to the olives, which are placed at the bottom of the box. Pressure causes bruises and cracks. Increased olive ripening or increased vibration level in packaged olives increases sensitivity to compressive loads. These two factors

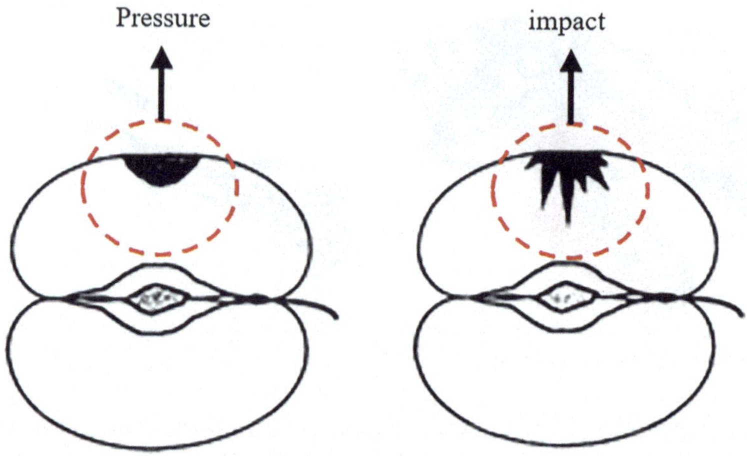

Fig. 13.3 Comparison of two types of mechanical damage due to pressure and impact

determine the severity of compressive damage to fruit packaging. The type of bruising caused by gentle pressure is completely different from the bruising caused by impact. The researchers reported that the cross-section of the bruising in the pressed fruit is elliptical. While the cross-section of a bruise due to impact often has a long nail-like shape that is radially drawn from the impact surface into the body of the fruit (Fig. 13.3) and the asymmetric pattern of impact bruise is more difficult to determine the extent of bruised section.

13.3.1 Bruise Susceptibility of Table Olive During Transportation

Inappropriate transportation of olives causes mechanical damage. The shape of the damage depends on the physical and biological structure of the product and the type of load applied (static, dynamic and oscillating load). If the method of transportation is not proper, it will cause the olive to shake violently and impact the surfaces of the carrier or other olives, thus deforming its tissues. If this deformation exceeds the biological yield point, the tissue will decay in a short time.

Damage of olives during transportation depends on some factors such as vibration characteristics (vibration acceleration and frequency, vibration duration), cultivar type (fruit size and vibration characteristics), type of vehicle, the height of the fruit in the box and road conditions. To measure the damage of olives, direct measurement methods, reduction of modulus of elasticity, measurement of absorbed energy and acoustic hardness can be assessed.

The differential equation of a vibrational system of one degree of freedom with a viscous attenuation stimulated by a harmonic force is expressed as Eq. (13.1).

$$m\ddot{x} + c\dot{x} + kx = F_0 \sin \omega t \tag{13.1}$$

where m is the mass (kg), c is the damping constant (Ns/m), k is the spring stiffness coefficient (N/mm), F_0 is the motive force (N), ω is the angular frequency (rad/s), x displacement (m) and t time (s).

By solving this differential equation, the maximum displacement x is obtained from Eq. (13.2):

$$X = \frac{F_0}{\sqrt{\left(k - M\omega^2\right)^2 + (c\omega)^2}} \tag{13.2}$$

Unbalance is considered by the eccentric mass m and the eccentric distance e, which rotate at an angular velocity ω. Thus, F_0 is replaced by $me\omega^2$ in the above equation. If the damper is not used in the above vibration system ($c = 0$), the equation is simplified as Eq. (13.3).

$$X = \frac{me\omega^2}{k - M\omega^2} \tag{13.3}$$

The conclusion of Eq. (13.3) is that the displacement is a function of the mass of the eccentric weights, their eccentric weight, the mass of the table, the stiffness of the spring and the angular frequency. On the other hand, there is a relationship between acceleration, amplitude and frequency (Eq. 13.4).

$$\ddot{X} = X\omega^2 \tag{13.4}$$

Where $\ddot{X}$ vibration acceleration (m/s^2), X displacement (m) and ω angular frequency (rad/s). Therefore, by changing the mentioned parameters, vibration with the desired amplitude, acceleration and frequency can be achieved using the eccentric system. By placing Eq. (13.3) in Eq. (13.4), the acceleration of the system in terms of its unbalanced and off-centre mass and angular velocity was obtained (Eq. 13.5).

$$\ddot{X} = \frac{me}{M}\left(\frac{\omega_n \omega^3}{\omega_n^2 - \omega^2}\right) \tag{13.5}$$

Where $\ddot{X}$ the acceleration of the device (m/s^2), me is the product of mass in the unbalance distance, M is the mass of the whole seismic part of the device (kg), ω_n is the normal angular velocity of the device (rad/s) and ω is the angular velocity (rad/s).

13.3.2 Table Olive Damage During Processing

Olives in processing plants are divided into two separate forms, canned and oil. Some manufacturers believe that because the customer uses canned olives, one should pay attention to their appearance. For this reason, heed to the mechanical damage caused by the processing part. While impact bruising occurred in the olive processing line leads to a decrease in the quality of the extracted olive oil.

Damage to the olive skin remains even after the fermentation process. Phenolic compounds and the enzymes polyphenol oxide (PPO) and peroxide (POD) play an important role in tissue colour. In damaged olive cells, the browning reaction begins due to the combination of enzymes with the precursor. Mechanical damage to olives during oil processing increases the oxidation, thus increasing the acidity and peroxide value. In these oils, the value of volatile acids (acetic and butyric acid) increases, which eventually causes an unpleasant odour. To increase the quality of olive oil and canned olives, engineers and researchers have studied when olives are hit and come up with solutions. According to the recommendations of all researchers, the first step to reduce the mechanical damage of olives in the processing line, before building and optimizing the production line, is to get acquainted with the rheological properties of olives. In the following, the properties of olive rheology are described.

13.4 Rheological Characteristic of Table Olive

Rheological properties describe the behaviour or reaction of an object to the forces acting on it that lead to deformation or continuous flow. This deformation and flow depending on the nature of the body. Due to its viscoelastic nature, time-dependent behaviour is observed in agricultural products such as table olive. Therefore, the stress–strain relationship in them depends on the loading rate. For this purpose, the time factor should be included in the strain–stress equation in agricultural products. Various experiments are used to determine the properties of rheology. These experiments are performed on a complete sample or with a regular geometric shape. These experiments are mainly compressive because mechanical damage to fruits often occurs under compressive forces. In the following, the physical and mechanical properties of table olive are briefly described.

13.4.1 Physical Properties

Agricultural products have unique properties that separate them from engineering materials. The irregular shape of many agricultural products makes it difficult to analyse their behaviour. It is necessary to know its physical properties to design equipment related to planting, harvesting, transporting, storing and processing olives. The physical properties of materials such as shape, volume and area are important in any issue related to the design and development of special machines (Mohsenin, 1986). Size, shape and mechanical properties are important for the design of machines for separation, harvesting and sorting. Density and porosity affect the amount of load on the structures and is one of the main parameters in predicting the loads on the transfer structures.

Physical properties of olive fruit include volume, surface area, density and moisture. To obtain moisture, 100 g of olive (W1) can be placed in an oven at 105 °C and the weight of the sample was measured at different times to reach a constant weight (W2) (Boukhiar et al., 2017).

Because the two medium and small dimensions of olive are almost the same, the two dimensions can be determined with a good approximation of the fruit size, which can be used with a digital calliper with an accuracy of 0.01 mm. The mass of the olive fruit and its kernel can measure using an accurate scale of 0.001 g. The density can be obtained by dividing the mass of the fruit by its volume. A curvature meter is used to measure the radius of curvature, and the radius of curvature is determined in two directions for each of the samples.

Another factor that is considered in olive fruit is the ratio of meat to the kernel. The studies have shown that the smaller size of the olive leads to more spherical coefficient. Lavasani et al. (2008) evaluate the physical properties of some table olive cultivars. The dimensions of the Mary cultivar were the largest and the Yellow cultivar was the smallest. By contrast, the sphericity coefficient was the lowest in the Mary cultivar and the highest in the Yellow cultivar. In another study, the effect of the cultivar (oil and snake) on geometric mean and arithmetic diameter and sphericity coefficient at the level of 1% was significant (Lavasani et al., 2008).

In recent years, researchers have preferred to use reverse engineering to achieve greater accuracy. In this method, a 3D scanning camera is used to obtain super fruit points. Then they use software to turn the dot design into a surface. Using this method, the dimensions of the fruit can be measured with the highest accuracy.

13.4.2 Mechanical Properties

The mechanical properties of olives are very important for engineers and researchers. Important mechanical properties include fracture energy, fracture force, modulus of elasticity and yield stress. Stress–strain tests use to obtain the mentioned factors. In these experiments, a sample of fruit with a regular geometric shape (usually a cylinder) is taken out using a sampler and loaded by a comprehensive traction and pressure device until failure. Then, from the stress–strain curve, the modulus of elasticity, stress and strain at the moment of failure, fracture energy and toughness are extracted.

Compression tests on the complete sample determine important mechanical properties in quality assessment and control and determine the maximum allowable static load to minimize mechanical damage due to static and quasi-static loading on convex foods such as fruits, vegetables and grains and legumes is used. The equations used in these experiments are adapted from Hertz's theory of contact stresses for convex elastic bodies. The information provided by these simple experiments can be used to quantify differences due to parameters such as variety, temperature, storage method, degree of processing and processing methods.

The mechanical properties of two Spanish (Hojiblanca) and orange (Dovoro) olive cultivars were determined and measured. The results showed that the olive skin was stronger and harder than the flesh in the tensile test. Also, the study of fracture strain in skin and flesh showed that the fruit's skin is less deformable. During fruit ripening, skin and flesh strength and resistance decreased in the Hojiblanca cultivar, while in the Dovoro cultivar, skin resistance and flesh decreased. In general, the texture of Hojiblanca was harder. The meat in the Hojiblanca cultivar was leaner than the Dovoro cultivar and with increasing maturity, the fat content in both cultivars decreased (Georget et al., 2001).

Ghamary et al. (2010) studied some of the physical properties of olives in both yellow and oil cultivars. The studied properties included dimensions, mass, volume, average geometric diameter and sphericity. Regression analysis showed that the relationship between volume and mass of each cultivar is significant. Also, the volume of olives was calculated using the fruit assumption as an elliptical volume and compared with the measured volume, which had a significant relationship between the two volumes. The relationship between meat weight and total fruit weight was also calculated. No significant difference was observed between the characteristics of the two studied cultivars (Ghamary et al., 2010).

The physical, mechanical and aerodynamic properties of four cultivars, Manzanilla, Kalamata, Fishmi and Oily were determined. The measured physical properties included dimensions, weight, volume, geometric mean diameter, arithmetic mean diameter, sphericity coefficient, porosity and surface area. To determine the mechanical properties, a compression test with a comprehensive traction device was used. The variables obtained from this experiment were fracture strength and energy, fracture stress and toughness. As the fracture force increases, the fracture energy also increases as the area below the graph increases. The highest failure energies for oil, Kalamata, Fishmi and Manzanilla samples were 109.29, 124.39, 169.85 and 181.36, respectively (Akbarnia & Rashvand, 2019).

To determine the elastic properties in compression tests, it is necessary to draw a force-deformation curve. The parameters of stiffness, toughness, force and deformation at the turning point, bi-yield point and breaking point can be obtained from the force-deformation curve. To investigate the possibility of using the theoretical equations of Hertz contact stresses to determine the elastic properties (modulus of elasticity) and the maximum contact stress at the loading level in fruits and vegetables, the force-deformation curve must be plotted on a logarithmic scale. These parameters can be used as a benchmark for evaluating food quality control (ASAE Standards, 2009).

The force-deformation curve in olives can be modelled using the Taylor expansion polynomial of the third degree (Henry et al., 2000) (Eq. 13.6), each of which also expresses three types of behaviour (elastic, viscous and fracture) that happens at the same time in the body.

$$F = aD + bD^2 + cD^3 \tag{13.6}$$

Where F is the amount of force (N), D is the deformation (mm) and a, b and c are the model coefficients. The Henry model assesses the behaviour of the olive according to the mentioned coefficients. The first sentence of the equation represents the linear or elastic behaviour of the olive. The second sentence controls the viscous behaviour and the third sentence is the olive failure behaviour. Using this model, a complete description of the deformation force curve can be obtained as an indicator of the mechanical properties of the olive product.

The slope at any point on the curve is the value of the tangent modulus. By deriving from the original equation, the tangent modulus equation for each curve point can be obtained (Eq. 13.7). The maximum slope occurs at the turning point. At this point, there is the greatest resistance to deformation.

$$T = a + 2bx + 3cx^2 \tag{13.7}$$

The slope value from the origin to each point is on the curve of the sequential modulus (Eq. 13.8). The acoustic modulus reaches a maximum at a point where the line drawn from the origin to that point is tangent to the curve.

$$q = \frac{F}{x} = a + bx + cx^2 \tag{13.8}$$

Using the fitting of the third-degree polynomial function on the unrestricted force-deformation curve, the coefficients of the Henry model are determined for olive. In extracting mechanical parameters through various tests such as compressive loading on olives, one of the important issues is that the curve extracted from such an experiment is so uniform that it can be used to evaluate and compare the texture of products.

The modulus of apparent elasticity for olives was obtained assuming that the Hertz theory conditions were met in the flat plate loading experiment and using the amount of force in the deformation of half the turning point using the Eq. (13.9).

$$E_a = \frac{0.338\, F\, (1 - \mu^2)}{D^{3/2}} \left(2K \left(\frac{1}{R_1} + \frac{1}{R_2} \right)^{1/3} \right)^{3/2} \tag{13.9}$$

Where E_a is the modulus of elasticity, D is the degree of deformation, μ is the Poisson's ratio, F is the applied force, R_1 and R_2 are the minimum and maximum curvature radii at the points of contact at the top or bottom of the fruit with the load plates (ASABE Standards, 2008). Also, The Kojima (1983) equation can be used to

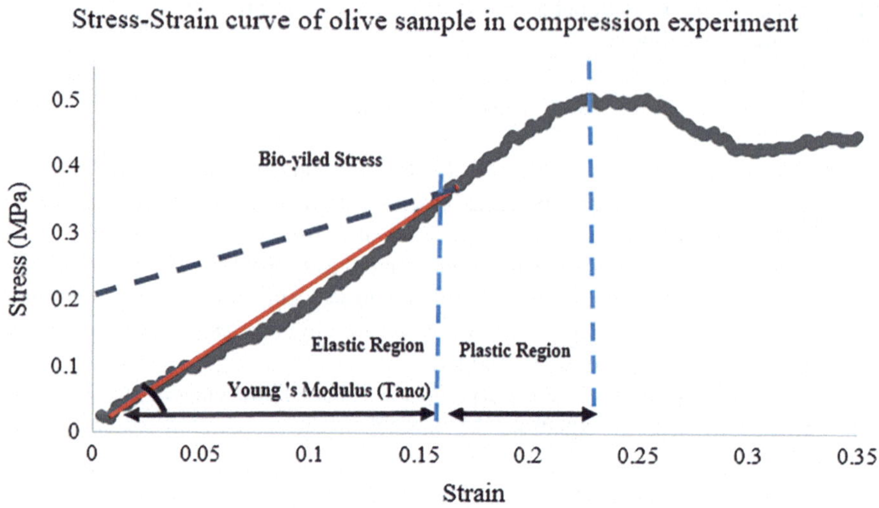

Fig. 13.4 Results of pressure test to obtain the mechanical properties of oily cultivar (Rashvand et al., 2021)

obtain the Poisson coefficient of olives. Most researchers report that the Poisson's ratio of olives is 0.5.

In order to obtain the value of bruising of the olive, researchers must determine the stress yield point. The yield point is the point at which the bruise phenomenon occurs. Figure 13.4 shows the elastic and plastic phases and the formation of mechanical damage to the Oily olive fruit. According to the figure, if the impact on the olive fruit exceeds 0.2 MPa, the olive tissue enters the plastic state from the elastic state and the olive tissue damage.

13.5 Mechanical Modelling of Table Olive Impact

The collision between two olives at one point occurs. Vertical stress is dominant in the collision plane and is called contact stress. Suppose the contact stress increases beyond the biological yield of the olive, the probability of damage and bruising increases. Hertz contact stress theory has been developed for agricultural products including olives. However, a general solution for viscoelastic bodies has not yet been proposed, but various approximate solutions have been used numerically. In stress analysis in the collision of olives with each other, relations related to elastic objects are used (taking into account the properties dependent on the time of material movement). Because in olives under constant load, the displacement changes as a function of time, the stress distribution also changes at the collision level. If the loading time is less than the stress relief time (impact mode), the olive behaviour is very close to that of an elastic material.

Many models have been used to describe the process by which olives impact a surface. The first model is the model that assumes olives to be spherical elastic bodies and uses Hertz theory (Horsfield et al., 1972). This method is very approximate and not practical, but it provides information about the elastic behaviour of olives for engineers and researchers. Internal stresses and strains can be explained in the modelling of collisions with elastic objects, and this model states that bruising occurs at a certain depth inside the fruit and under the shell, which is where most Shear stress and strain occur. In fact, Hertz theory is not proper for soft materials and high-velocity shocks and should be replaced by an analytical calculation for such plastic deformations. The following describes the viscoelastic properties of agriculture products and the proper model for this material.

13.5.1 Viscoelastic Properties of Olive

The flesh of the olive fruit is viscoelastic and its behaviour depends on time. Therefore, its properties change with the loading rate. To determine the stress distribution in olives, its characteristic equation must be obtained. This equation can obtain by determining the characteristic curves (using experiments). There are three factors (stress–strain–time) to evaluate the viscoelastic properties of olives. For this reason, one of the stress or strain factors must be kept constant and other changes must be studied over time.

The three viscoelastic properties of olives are modulus of elasticity (E), modulus of density (K) and Poisson's ratio (μ), which are related to each other. Similarly, the shear modulus (G) is related to the modulus of elasticity (E) and the Poisson ratio (μ). The three modulus of elastic (E), shear (G) and volumetric (K) are also related together (Eqs. 13.11 and 13.12).

$$E = 3(1 - 2\mu)K \tag{13.10}$$

$$E = 2(1 + \mu)G \tag{13.11}$$

$$\frac{1}{E} = \frac{1}{3G} + \frac{1}{9K} \tag{13.12}$$

On the other hand, some researchers have shown the usefulness of regression models compared to theoretical models, which can be said that regression models are preferred for estimating bruise areas. In comparison, theoretical models are more useful to understand the impact of factors Various related to damages. In addition, it is difficult to detect the presence of mechanical damage in olives. Because it most often occurs under the skin of the olive, its accuracy is influenced by factors such as the time of bruising, type of bruising, severity of bruising, type and variety of product, pre-and post-harvest conditions. If tissue damage is visible, visual aids are sufficient to diagnose bruise volume.

With advances in computers and electronics, new methods have been developed to detect bruising in recent years. Several biochemical and physical methods have been used to assess bruising if detection is impossible with the human eye. Also, tissue hardness analysis methods, sound response method, resonance frequency and optical methods such as ultrasonic, X-ray, gamma-ray, magnetic resonance, fluorescence, optical reflection, dielectric properties, heat absorption can be used to obtain the amount of olive bruising. However, the economics of using these methods to detect olive bruise volume are not acceptable. Jiménez-Jiménez et al. (2013) identified, evaluated and estimated the sources of impact on olives. For this purpose, digital image analysis was used to identify the damage in olives. Harvesting olives with a vibrator increased bruising in olives 12 times larger than manual harvesting. The frequency of the seismic harvesting machine was between 28 and 30 Hz and the average effective acceleration of the machine on the tree was about 250 m s^{-2}. Fruit-to-fruit and fruit-to-branch blows during vibration with the harvesting machine were the most important causes of injury. These factors account for about 60% of the damage to the fruit. In addition, most damage occurs in the first hour after harvest and the continuation of the process is exponential (Jiménez-Jiménez et al., 2013).

13.5.2 Table Olive Impact Tests

Impact tests are performed to determine the relationship between the amount of bruise and the energy absorbed. In general, there are two types of pendulum and drop tests for agricultural products. The drop test device usually consists of a steel sheet as a base on which the force meter is placed. The drop position can be adjusted using a calibrated steel wire mounted on a steel cylinder. Also, the vacuum valve and the shut-off valve were screwed on the steel cylinder. The flow shut-off valve was connected to an air pump. The load gauge and the fall plate were tightened under the impact site. Using the camera (installed near the impact site), the moment of the fall of all the samples is taken, placed on the background of the film and behind the crash site is a graduated white screen. The returned height after the sample drop is estimated using a calibrated whiteboard behind the falling site (Fig. 13.5).

In another study, Saracoglu et al. (2011) examined some characteristics and susceptibility to bruising damage for two olive cultivars. The bruising was caused by drop test, and the area and volume of the bruising were calculated. The results showed that the height of the drop and the impact plate were two influential factors in

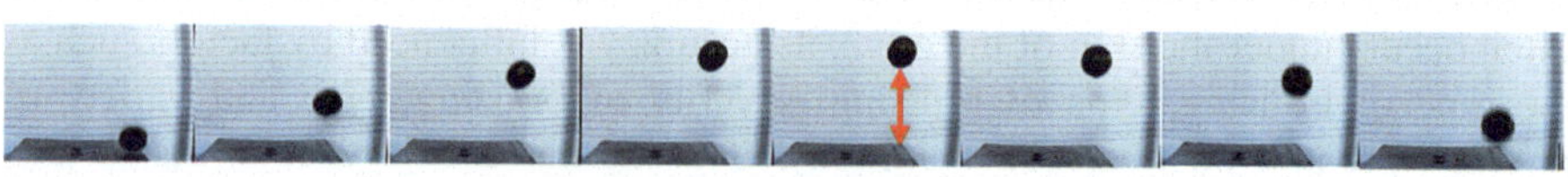

Fig. 13.5 Captured images of the drop test (from left to right) and the height of the return sample (Kargarpour et al., 2018)

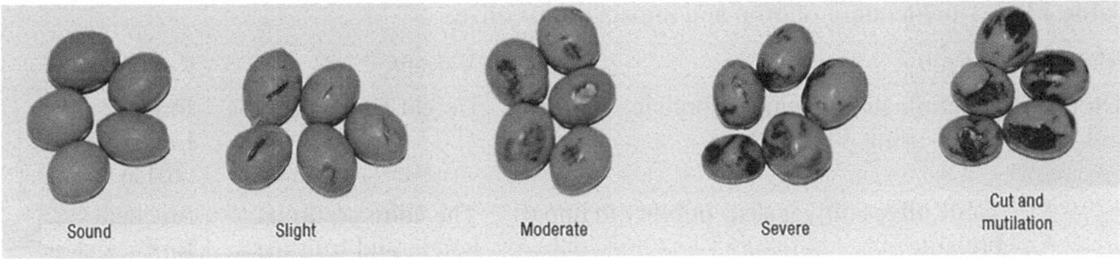

Fig. 13.6 Damage and observed changing the colour of olive fruit 24 h after harvest (Jiménez-Jiménez et al., 2013)

the amount of olive bruise, which increased significantly with increasing drop height and the area of the bruise (Saracoglu et al., 2011).

A study by Jimenez et al. (2013) was carried out to examine the physical changes of olive fruit damage for the cultivar "Manzanilla DeSilva". Damaged fruits were soaked in FAE solution (formalin, acetic acid, 95% ethanol and distilled water (10: 5:50:35 v/v/v/v)) 30 min and 24 h after damage. Tissue changes (cell fracture, loss of cell wall thickness and colour intensity) were observed at the site of damage. Tissue findings revealed differences between the two measurement times. Cell fracture and loss of cell wall at the site of damage were more appear 24 h after damage (Fig. 13.6) (Jiménez Gonzélez et al., 2013).

A new system was developed to measure the impact damage on olive fruit from harvest to packaging. The system consisted of a sphere equipped with a micro-electromechanical system. The system had a microcontroller for special use, a converter and a mini USB port to connect to a computer. The preliminary studies on this system showed that butter was an efficient tool for identifying the stresses on the olive fruit during and after harvest (Catania et al., 2015).

Investigated the sensitivity of cultivars and descriptive evaluation of olive fruit bruising. In this study, tissue colour changing due to the bruise phenomenon and its level in 3 olive cultivars and 14 new genotypes were studied. Significant differences were observed between different genotypes in susceptibility to bruising. Tissue sections of damaged and healthy fruit tissue showed a discoloured area beneath the shell, but there was a much larger area of tissue in which the cells deformed but were not damaged. These observations support the hypothesis that the development of the bruise phenomenon involves physical and chemical processes that differ in the level of effect and progression (Jiménez et al., 2011). Also, Table 13.1 is related to the research conducted by researchers to find the amount of contusion and energy absorbed by olives.

Olive fruit is one of the fruits that are susceptible to bruising and the increasing demand for this useful nutrient has increased the area under cultivation and mechanized its harvesting stages. Mechanical damages are occurred during the harvest and post-harvest process and the transportation of olive fruit. To reduce the damage of bruising of olive fruit, first, the sources of bruising should be identified and evaluated, and then the design, fabrication and use of machines and processing lines of the product should be optimized.

Table 13.1 The literature of drop and impact test of olive

Method	Objective	Variables	References
Drop test	Determination of impact bruising on table olives using Vis-NIR	Height and fruit mass	Jiménez-Jiménez et al. (2012)
	Table olive cultivar susceptibility to impact bruising	The cultivar, drop height and fruit mass	Jiménez-Jiménez et al. (2013)
	Susceptibility to bruising damage of table olive fruit	The cultivar, drop height and impact surface	Saracoglu and Ozarslan (2011)
	Measurement bruise volume of olive during impact test using FEM	Cultivar, impact surface, ripening stage	Rashvand et al. (2021)
Impact test	Measuring damaging impacts on table olives	–	Catania et al. (2015)
	Determination of the mechanical properties of olive fruits required in DEM simulation	–	González-Montellano et al. (2011)
	Measure liquid viscosity depends on the pendulum model	–	Ali et al. (2018)

13.6 Suggested Methods to Reduce Mechanical Damage of Table Olive Fruit

The first step to reduce olive damage is heed during the harvesting process. According to the researcher's recommendation, using chemicals materials (to decrease the connection force of the fruit stem-branch or the fruit stem-fruit) before the harvest with a shock and vibrating machine is applicable. In the stage of impact harvesting, the farmer should pay attention to the speed of the machine and the material of the impactor surface, according to the mechanical properties of the olive and its ripening stage. The speed of the machine should be determined in such a way that not only the amount of olive fracture is reduced but also the harvesting efficiency is increased. If the speed is lower than the standard value, it will increase the harvest time as well as get stuck in the olive branches. In addition, the material of the head has a great effect on reducing the sensitivity of olive bruises.

Currently, factories that produce harvester heads use neoprene material. Although the use of neoprene is resistant to foliage and increases efficiency, it increases olive damage. Especially the use of this material for the ripped stage is not recommended. According to the results of researchers who have worked in the field of the olive harvest, using this material in some cases causes an olive fracture. Another suggestion is to use replaceable heads in different handling conditions. The modulus of elasticity and refractive energy of all olive varieties at three times immature (for canning), semi-ripe (for direct use by the customer) and ripe (for oiling) are very different and their differences are significant. For this reason, farmers should use heads with different materials.

Another situation in which olives are damaged is when they are thrown to the ground by a direct impact and drop from a tree by a vibrating machine. Regarding the use of impact machines, most farmers spread a cover on the ground and collect it after the olives drop. However, this method does not affect the amount of bruising. They also use plastic conveyors when vibrating machines are used. One of the suggested solutions to reduce the sensitivity of olive bruises during the drop is foam and nylon materials. According to researchers, the use of these two materials can reduce the bruise susceptibility of olive.

Next, the transfer of olives from the orchard to the processing plants should be investigated. The transport of olives has always been affected by impact stresses. For this reason, engineers and researchers have studied the stresses of transportation. It is recommended to use acceleration, displacement and distance sensors for the vibration table system for road simulation to control the vibration parameters in the vibration test better and more accurately. It also examined the extent of olive bruise damage at different frequencies.

Mechanical damage of olives in processing plants is often unavoidable. When the olives are transported on conveyors in food industries (during process line), the fruit is damaged. Especially in canning and oiling (due to the optimization of the factory from the point of view of industrial engineering and reducing the space occupied by the machines), the olives are bruised due to drop from the conveyors. At this stage, the only factor that can be changed to minimize olive bruising is height. Engineers should consider the drop heights so that the olives have the least damaged. As the height of the fall increases, the tension inside the olive tissue and the impact force increase. In conclusion, all the proposed solutions were mentioned according to the economic dimension of olives and olive oil. Reducing the mechanical damage of olives leads to an increase in the quality of canned olives and extracted olive oil, which can ultimately help customers' health.

References

Akbarnia, A., & Rashvand, M. (2019). An evaluation of the physical, dynamic and aerodynamic properties of olives. *Research in Agricultural Engineering, 65*(2), 48–55.

Ali, S. H., Al-Zuky, A. A. D., Al-Saleh, A. H., & Mohamad, H. J. (2018). Measure liquid viscosity depends on the pendulum model using image processing algorithms. *Journal of College of Education*, (3), 165–178.

ASABE. (2008). Method of determining and expressing fineness of feed materials by sieving.

Bollen, A. F., Nguyen, H. X., & Dela Rue, B. T. (1999). Comparison of methods for estimating the bruise volume of apples. *Journal of Agricultural Engineering Research, 74*(4), 325–330.

Boukhiar, A., Kechadi, K., Abdellaoui, R., Iguergaziz, N., Guemmane, M., & Benamara, S. (2017). Drying ability of whole black olive (Olea europaea L.) fruits in Kabylian region. *Indian Journal of Traditional Knowledge, 16*(1), 89–94.

Brusewitz, G. H., McCollum, T. G., & Zhang, X. (1991). Impact bruise resistance of peaches. *Transactions of the ASAE, 34*(3), 962–965.

Çakmak, B., Saraçoğlu, T., Alayunt, F. N., & Özarslan, C. (2011). Vibration and noise characteristics of flap type olive harvesters. *Applied Ergonomics, 42*(3), 397–402.

Catania, P., Febo, P., Alleri, M., & Vallone, M. (2015). A novel system for measuring damaging impacts on table olives. *Chemical Engineering Transactions, 44*, 1–6.

Du, Z., Zeng, X., Li, X., Ding, X., Cao, J., & Jiang, W. (2020). Recent advances in imaging techniques for bruise detection in fruits and vegetables. *Trends in Food Science & Technology, 99*, 133–141.

Friedman, T. L. (2000). *The Lexus and the olive tree: Understanding globalization*. Farrar, Straus and Giroux.

Georget, D. M. R., Smith, A. C., & Waldron, K. W. (2001). Effect of ripening on the mechanical properties of Portuguese and Spanish varieties of olive (Olea europaea L). *Journal of the Science of Food and Agriculture, 81*(4), 448–454.

Ghamary, B., Rajabipour, A., Borghei, A. M., & Sadeghi, H. (2010). Some physical properties of olive. *Agricultural Engineering International, 12*(1), 104–110.

González-Montellano, C., Llana, D. F., Fuentes, J. M., & Ayuga, F. (2011). Determination of the mechanical properties of corn grains and olive fruits required in DEM simulations. In *2011 Louisville, Kentucky, August 7–10*, 2011 (p. 1). American Society of Agricultural and Biological Engineers.

Henry, Z. A., Zhang, H., & Onks, D. O. (2000). New model for elastic behaviour of cellular material. *Journal of Agricultural Engineering Research, 76*, 399–408.

Herrera-Cáceres, C., Pérez-Galarce, F., Álvarez-Miranda, E., & Candia-Véjar, A. (2017). Optimization of the harvest planning in the olive oil production: A case study in Chile. *Computers and Electronics in Agriculture, 141*, 147–159.

Horsfield, B. C., Fridley, R. B., & Claypool, L. L. (1972). Application of theory of elasticity to the design of fruit harvesting and handling equipment for minimum bruising. *Transactions of the ASAE, 15*(4), 0746–0750.

Hoshyarmanesh, H., Dastgerdi, H. R., Ghodsi, M., Khandan, R., & Zareinia, K. (2017). Numerical and experimental vibration analysis of olive tree for optimal mechanized harvesting efficiency and productivity. *Computers and Electronics in Agriculture, 132*, 34–48.

Jiménez González, M. R., Suárez García, M. P., Rallo Morillo, P., Rapoport, H. F., Galindo, M., Morales Sillero, A. M., & Casanova Lerma, L. (2013). Histological characterization of table olive fruit bruising cv. Manzanilla de Sevilla. In VII Congreso Ibérico de Agroingeniería y Ciencias Hortícolas. Sociedad Española de Ciencias Hortícolas.

Jiménez, R., Rallo, P., Suárez, M. P., Morales-Sillero, A. M., Casanova, L., & Rapoport, H. F. (2011). Cultivar susceptibility and anatomical evaluation of table olive fruit bruising. *Acta Horticulturae, 924*, 419–424.

Jiménez-Jiménez, F., Castro-García, S., Blanco-Roldán, G. L., Ferguson, L., Rosa, U. A., & Gil-Ribes, J. A. (2013). Table olive cultivar susceptibility to impact bruising. *Postharvest biology and technology, 86*, 100–106.

Jiménez-Jiménez, F., Castro-García, S., Blanco-Roldán, G. L., Agüera-Vega, J., & Gil-Ribes, J. A. (2012). Non-destructive determination of impact bruising on table olives using Vis–NIR spectroscopy. *Biosystems Engineering, 113*(4), 371–378.

Jiménez-Jiménez, F., Castro-García, S., Blanco-Roldán, G. L., Ferguson, L., Rosa, U. A., & Gil-Ribes, J. A. (2013). Table olive cultivar susceptibility to impact bruising. *Postharvest Biology and Technology, 86*, 100–106.

Kargarpour, H., Tavakoli Hashjin, T., Hemmat, A., & Ghobadian, B. (2018). Assessment of mechanical damage on olive fruit under impact loading. *Journal of Agricultural Machinery, 8*(2), 365–376.

Kojima, T. (1983). Mechanical properties and resonant vibration characteristics of fruits Bulletin of Faculty of Agriculture, Saga University.

Lavasani, S., Afkari, A. H., & Golmahammadi A. (2008). The study of effect of the size and duration time of after harvest on the physical and mechanical properties of olive fruit in the common varieties (pp. 1–11). In *Proceedings of the 5th National Agricultural Machinery Engineering Congress*, Mashhad, Iran.

Mohsenin, N. N. (1986). *Physical properties of plant and animal materials*. Gordon and Breach Science Publishers.

Rashvand, M., Akbarnia, A., Abbaszadeh, R., Karimi, D., & Jafari, A. (2021). Measurement bruise volume of olive during impact test using FEM and quality evaluation of extracted olive oil. *Food Science & Nutrition, 9*(10), 5493–5508. https://doi.org/10.1002/fsn3.2508

Saglam, C., Tuna, Y. T., Gecgel, U., & Atar, E. S. (2014). Effects of olive harvesting methods on oil quality. *APCBEE Procedia, 8*, 334–342.

Saracoglu, T., Ucer, N., & Ozarslan, C. (2011). Engineering properties and susceptibility to bruising damage of table olive (Olea europaea) fruit. *International Journal of Agriculture and Biology, 13*(5), 801–805.

Van Zeebroeck, M., Darius, P., De Ketelaere, B., Ramon, H., & Tijskens, E. (2007). The effect of fruit factors on the bruise susceptibility of apples. *Postharvest Biology and Technology, 46*(1), 10–19.

Zapp, H. R., Ehlert, S. H., Brown, G. K., Armstrong, P. R., & Sober, S. S. (1990). Advanced instrumented sphere (IS) for impact measurements. *Transactions of the ASAE, 33*(3), 955–960.

Chapter 14
Bruise Damage Susceptibility of Banana

Umezuruike Linus Opara and Pankaj B. Pathare

Abstract Banana is the most widely consumed and traded fruit globally. The presence of mechanical damage, especially bruise damage, is a major quality defect affecting the handling and marketing of fresh bananas. Affected fruit show browning discoloration, may be downgraded, or rejected, which may result in considerable postharvest losses. Bruising is particularly a difficult problem to control in the banana industry because fruit is traded and transported both locally and internationally in the green mature stage when the presence of mechanical damage is difficult to detect. Furthermore, when the fruit is ripe, the susceptibility to damage is considerably higher due to the softening of both the peel and edible pulp. Given that banana peel and flesh are differentially susceptible to mechanical damage resulting in bruising, the presence of early bruise is difficult to determine with the naked eye during fruit inspection, grading, and sorting. Various preharvest and postharvest factors contribute to the potential for banana fruit to damage, including cultivar differences, plantation management practices, harvesting and handling methods, and packaging systems used. Therefore, understanding the potential for damage to occur along the value chain becomes paramount. The application of novel non-invasive and non-destructive technologies such as hyperspectral/multispectral imaging, near-infrared spectroscopy, X-ray imaging, and computed tomography represent potential solutions for banana bruise damage detection and prediction. To reduce the incidence of postharvest losses and waste of bananas due to bruise damage, it is still best to handle fruit at the green mature stage so that ripening occurs closer to the point of utilization such as the retail stage. The design of ventilated packaging using an

U. L. Opara (✉)
SARChI Postharvest Technology Research Laboratory, Africa Institute for Postharvest Technology, Faculty of AgriSciences, Stellenbosch University, Stellenbosch, South Africa

UNESCO International Centre for Biotechnology, Nsukka, Enugu State, Nigeria
e-mail: opara@sun.ac.za

P. B. Pathare
Department of Soils, Water & Agricultural Engineering, College of Agricultural and Marine Sciences, Sultan Qaboos University, Al-Khod, Oman

P. B. Pathare, U. L. Opara (eds.), *Mechanical Damage in Fresh Horticultural Produce*, https://doi.org/10.1007/978-981-99-7096-4_14

integrated multi-criteria approach of maintaining the cold chain and providing mechanical protection of the fresh produce offers huge benefits.

Keywords Banana · Mechanical damage · Non-destructive techniques · Bruise susceptibility · Preharvest

14.1 Introduction

Banana is one of the most widely consumed and traded fruit globally. It is the second largest fruit produced following citrus, and it contributes over 16% of the total global fruit production and represents the fourth most important global staple food crop after rice, wheat, and maize (Kuyu & Tola, 2018). Recent estimates indicate that during the past decade, there was nearly a 10% increase in global production, which reached over 119 million tonnes, with the majority on smallholder farms across the major producing regions and only 15% produced on commercial plantations (Panigrahi et al., 2021). Bananas produced on smallholder farms often do not enter the global food supply chain because these farmers rely on their own production for subsistence and their little surplus is sold in local markets for financial income.

Five major exporting countries (Ecuador, the Philippines, Costa Rica, Guatemala, Colombia), and the Netherlands account for over three-quarters of global trade in bananas (FAO, 2018). With about 25% of total global production traded internationally, this represents considerable postharvest handling and logistics challenges, from the plantation to the end-user, thereby exposing the fruit to mechanical forces which could lead to damage and associated losses and waste. Given that the large majority of globally traded bananas are produced in the Southern Hemisphere and exports reach all continents, with Europe as the main importer, the science of banana production and postharvest handling has increased rapidly during the past century to address the technological challenges.

The postharvest handling and marketing of bananas follow two distinct value chains, which reflect the end use (Opara et al., 2013). As ripe fruit, the majority of banana is destined for direct consumption and utilization in the processing of other food products (such as cakes, and bread). As green mature fruit, banana is also consumed in different forms—boiling, frying into chips, and processing into powder, etc. A climacteric fruit, the banana is also highly sensitive to both endogenous production and exogenous application of ethylene (a gaseous hormone that accelerates fruit ripening). Therefore, banana fruit is commonly handled at the green mature stage in order to ensure optimal delivery of fruit volumes to meet market demand. This way, tonnes of fresh bananas produced in Guatemala and the Philippines, for instance, can be shipped to markets across the globe in shipping containers, where fruit are then held in the cold stage until required for ripening. Although the fruit has a thick rind/skin/peel (Arendse et al., 2018a, b, c, d, 2022a, b) which envelopes and protects the edible pulp (flesh), mechanical damage can occur to the skin leading to a downgrading of appearance quality. In extreme cases, affected fruit (or a whole consignment) may be rejected when the bruised area exceeds the applicable market standards.

The presence of a bruise on a banana peel may not result in bruising of the edible pulp/flesh due to the cushioning effect of the thick peel. However, the presence of bruising on banana flesh is considered an extreme case of mechanical damage, which can lead to a loss of consumer confidence in the batch of fruit or the vendor. Mature unripe banana fruit is more resistant to mechanical damage than ripe fruit due to the softening and loss of tissue structural integrity in ripe fruit. The presence of small bruise damage on mature unripe banana fruit is difficult to determine during the manual inspection, especially on the colorful varieties that may be green, purple, or black even when mature and ripe. Using color charts developed for specific banana varieties and cultivars will aid manual inspection, but for large-scale operations requiring strict quality standards to be maintained to assure that banana with mechanical injury is segregated and removed, the use of automated non-invasive sensor technologies, such as color sorters, provide viable options.

Given that banana peel and flesh are differentially susceptible to mechanical damage resulting in bruising, the presence of early bruise is difficult to determine with the naked eye during fruit inspection, grading, and sorting, understanding the potential for damage to occur becomes paramount. The application of novel non-invasive and non-destructive technologies such as hyperspectral/multispectral imaging (Okere et al., 2021, 2022a, b, 2023), near-infrared spectroscopy (Nturambirwe et al., 2020, 2021a, b), X-ray (Arendse et al., 2018a, b, c, d, 2021a, b, 2022a, b) and computed tomography (Magwaza & Opara, 2014) offer potential solutions for bruise damage detection and prediction. To reduce the incidence of postharvest losses and waste due to bruise damage, it is also important to understand the factors contributing to bruising. These factors include inherent (genetic) fruit physical, structural, and biochemical properties, orchard management practices (preharvest), harvesting practices, and postharvest handling.

The occurrence of mechanical damage on bananas, especially bruising, is a major economic problem. The fruit quality could be lowered by mechanical damage, which would reduce its market value. As a result, mechanical damage is a major problem for the fresh produce sector (Al-Dairi et al., 2023; Opara & Pathare, 2014; Fadiji et al., 2016a, b, c). Both internal and external quality attributes of bananas—including physical defects such as bruising are influenced by the postharvest management practices and engineering technologies applied (Sugianti et al., 2022; Al-Dairi et al., 2023).

Fresh produce chemical and physical characteristics can be affected by mechanical injury (Rashvand et al., 2022). In pears, bruises lessen the affected fruit's lightness and cause a greater loss of firmness (Pathare & Al-Dairi, 2021b). Tomatoes that have been bruised lose more weight and have a red hue (Pathare & Al-Dairi, 2021a). Additionally, it has been demonstrated that mechanical damage raises the fruit's respiration rate, sugar content, and acidity (Hussein, 2019a, b). In their study on fresh tomatoes, Sablani et al. (2006) showed that when bruised fruit was stored at refrigeration temperature (4 °C) and room temperature (25 °C), vitamin C content declined but the rate of loss was significantly higher when bruised fruit were stored at high temperature but not between bruised and control fruit. It is expected that irrespective of fruit type and postharvest condition, the effect of mechanical damage on fruit quality will largely depend on the severity of the damage.

Based on their evaluation of banana handling practices in retail shops and associated loss-causing fungal pathogens in southwest Ethiopia, Kuyu and Tola (2018) reported that fruit damage was high (56.2%), as well as the associated disease incidence (54.2%) and severity (34%), respectively. High incidences of mechanical damage due to improper handling and poor sanitary conditions in the market were identified as potential causes for the observed fungal pathogens. The authors concluded that to reduce mechanical damage and the associated pathogenic deterioration, stakeholder coordination was necessary.

Mechanical damage to bananas can lessen their shelf life and esthetic appeal. Banana fruit that has been bruised or mechanically damaged exhibits a variety of physical and physiological changes, including an accelerated ripening process that is attributable to an increase in respiration, ethylene generation, mass loss, and enzyme activity (Maia et al., 2014). Furthermore, studies by Al-Dairi et al. (2022) on bruising of stored banana (cv. Malindi) and the impacts on fruit quality and physiology showed that weight loss reduction and color lightness changes over time, especially at 22 °C, were affected by the presence of bruise damage. During the experiment, as storage temperature and impact level increased, the firmness of bruised banana fruit decreased. The appearance of severe bruise damage in banana fruit was diminished during storage at 13 °C, while increased transpiration rate was connected to both bruising and storage temperature. The authors concluded that banana quality losses due to bruise damage can be decreased by raising awareness of bruising mechanisms and improving storage management during the postharvest supply chain.

To avoid bruise damage and assure excellent sensory and nutritional quality bananas at the end-user level, the ideal practice is to allow the fruit to ripen on the tree, harvest only the ripe, and immediately consume it. The next level would be to harvest only optimally mature fruit, carry each bunch carefully from the orchard to the ripening room, and deliver the ripe fruit to market with sufficient cushioning to avoid fruit damage. With considerable effort, these can be applied to very short supply chains under subsistence food systems where banana production, trade, and utilization all happen within a close community.

However, none of the above scenarios are amenable to the modern integrated supply chains that guarantee the daily delivery of highly sought-after fresh produce (such as bananas) to supermarkets and other purchase outlets for consumers around the world. These long supply chains require multiple handling of produce, from harvesting to the packhouse, packaging, transport, distribution, repacking, retail display, and delivery to the consumer or other end-users. The incidence of mechanical damage, and hence bruising of banana fruit is, therefore, inevitable. Hence, the goal of postharvest research and improved postharvest handling systems is to prevent the occurrence of damage where possible and to reduce the severity of damage to acceptable market standards. The objective of this chapter is to discuss the recent developments in mechanical damage of banana fruit, including the contributing factors, and to highlight future prospects for new advancements in our understanding of the susceptibility to damage.

14.2 Mechanical Damage of Banana: An Overview

Bananas are highly sensitive to physical stress, therefore mechanical damage like browning manifests itself fast during distribution to customers. The fruit undergoes considerable mechanical handling from farm to table. Each of these handling steps could result in one or more of the different forms of mechanical damage. Excessive mechanical load on banana fruit may result in bruising, puncture, abrasion, cuts, cracks, splitting, and in extreme cases of compression loading the squashing of whole banana finger or bunch may occur. Bruises are the most common type of mechanical damage. Fruit can be bruised due to fruit-to-fruit impact/compression during postharvest handling, through rough handling on packing lines, loading fruit, and during storage. Although mostly considered a cosmetic downgrade, bruises can make fruit susceptible to rot and spoilage.

Punctures occur due to the presence of sharp pointed objects such as a nail, stick, small stone, knife, or parts of the packaging line. Fruit might fall and hit on the object or the other way round. Similarly, a puncture might occur when fruit hits or presses on a pointed object or the other way round. During manual cleaning, sorting, grading, and packing of bananas, the uncut fingernail of personnel is a common source of cuts on bananas, especially when the fruit is ripe. Training and inspection of quality control personnel are critical as part of the efforts to reduce the presence of cuts and other forms of mechanical damage on fresh produce.

Cuts on bananas and other types of fresh produce also occur when sharp elongate objects such as a knife or thin metal edge encounter the fruit during harvesting or postharvest handling operations. Although thin punctures and cuts may be difficult to see immediately after the incidence, they often enlarge as fruit softens due to ripening or in the presence of mechanical (compression) load. Such openings provide the avenue for microbial infection leading to contamination and spoilage such as rotting. In extreme cases, such microbial contamination may result in food safety hazards.

Bananas are also susceptible to abrasion damage which occurs when fruit rolls over a rough surface or when a moving rough surface brushes over a stationary fruit. This can happen when improper brushes are used to clean fruit surfaces directly or in a water dump. Abrasion damage can also occur along the packing line during sorting and grading of banana hands, and due to improper (or lack of) packaging during transportation and distribution when fruit bunches or hands roll freely on a rough surface.

Bruising is by far the most common type of mechanical damage affecting banana fruit during harvesting and postharvest handling. Fruit can be bruised due to impact or compression with other surfaces, and due to fruit-to-fruit impact or compression. Banana bruising can occur during harvesting when a bunch falls from a height and when an object such as a harvesting tool falls on the bunch. During loading in the orchard, bruising can also occur when a bunch is dropped from an excessive height into the bin or when the bin is excessively loaded, causing the fruit at the bottom and middle layer on the bin to experience bruise-damaging loads.

During packhouse operations (cleaning, sorting, grading, and packing), banana fruit may undergo rough handling on packing lines, which results in bruise damage. These include hands of fruit falling from a height onto a hard surface, causing impact damage. After placing the hands inside packaging such as paperboard ventilated packaging and plastic or wooden crates (Opara & Fadiji, 2018), bruising may also occur when the container/box is accidentally dropped from a height causing impact damage or when loads of boxes are excessively stacked on top of each other, resulting in excessive compressive load and compression bruising of fruit below the stack (Fadiji et al., 2016a, b, c).

During loading, the most common forms of mechanical damage on banana fruit are impact bruising and shatter bruising (which results in the breakage of fruit). These may result from improper handling of the container by personnel, improper operation of handling equipment such as a forklift which may result in knocking down the stack of cartons/boxes, or the collapse of the stack of containers due to improper stacking. Using packaging that is not properly designed for mechanical strength may lead to structural failure of the box, which in turn may lead to direct excessive load on fruit—resulting in mechanical damage.

During storage, boxes of banana fruit may be stacked directly on top of each other, or several boxes may be stacked together and placed into storage. This handling system results in considerable compression (static) load on the boxes, and when the boxes are improperly designed to withstand the loads, these forces are transferred to fruit, resulting in compression bruising. Under excessive forces, the boxes may collapse and fruit inside them squashed beyond compression bruising.

During transportation, various forms of bruise damage may occur including impact, compression, and abrasion. The mechanical condition of the rucks and similar vehicles used to transport bananas can result in mechanical damage. For example, brakes and engine suspension systems must be properly maintained to ensure a smooth ride. Furthermore, bad road conditions due to rough/uneven surfaces and the presence of excessive bumps that result in movement on the load (banana boxes/stacks) can lead to impact, compression, and abrasion forces which result in mechanical damage when transmitted to banana fruit. Therefore, to reduce fruit mechanical damage during the transport of bananas, both the vehicle and road surface needs to be maintained. Most importantly, drivers require adequate and continuous training, including a basic understanding of the perishability and susceptibility of fresh produce to different forms of mechanical damage.

Although the presence of bruise damage is considered a cosmetic defect that generally results in the downgrading of affected produce, the outright rejection of affected fruit, box, stack, or whole consignment can occur when acceptable market standards are exceeded. Bruising and other forms of mechanical damage can make fruit susceptible to rot, spoilage, and decay. These conditions contribute to a high incidence of postharvest losses and waste, which in turn exacerbate the prevalence of food and nutrition insecurity. Furthermore, postharvest losses and waste also contribute to unsustainable food systems, the over-exploitation of natural resources used in food production and handling, and the emission of greenhouse gases which have been demonstrated to harm the environment and planet Earth.

14.3 Factors Contributing to Mechanical Damage of Bananas

A wide range of preharvest, harvest, and postharvest factors contribute to the occurrence of mechanical damage to fresh produce (Hussein et al., 2018a, b, 2019a, b, c, d). Preharvest factors influence both the genetic predisposition to mechanical damage as well as the magnitude of damage that occurs under specific loading conditions. Preharvest factors include cultivar, orchard management practices (such as crop load, irrigation, fertilization), and maturity status at harvest. Harvesting factors contributing to the incidence and severity of damage include harvest time during the day and harvest technique. Postharvest factors contributing to the incidence and magnitude of mechanical damage include time-lapse after harvest, degree of fruit ripening, storage, packaging, and handling system used to move the fruit from my point to another, including transport and distribution.

14.3.1 Preharvest Factors

The susceptibility of banana fruit to mechanical damage varies among varieties, cultivars, and genotypes (Banks et al., 1991), which in turn is determined by the genetic composition. Fruit with thick peel/rind/skin will be more resistant to mechanical damage due to the greater resistance to mechanical forces which might lead to damage. Similarly, fruit shape at the point of impact or compression has been demonstrated by researchers to affect the severity of bruise damage. For a given amount of mechanical force and associated impacting or compressing objects, the amount of damage on fresh produce would be higher if the force is applied on a smaller impacted area of the produce—a phenomenon often referred to as the "stiletto effect."

Orchard management practices such as crop load, application of fertilizer, and irrigation amount and frequency affect fruit susceptibility to mechanical damage including bruising (Opara, 2007). Low crop load due to thinning, fertilizer application, and irrigation contribute to big fruit size, which has been shown to be more prone to mechanical damage. Within the same crop species and maturity, larger fruit size has larger cell size with thin cellular membranes, which are more susceptible to rupture/breakage under a given mechanical load.

Similarly, more mature, and ripe fruit also have higher susceptibility to mechanical damage. With advancing maturity and ripening, the breakdown of starch into simple sugars results in an increase in the contents of the vacuoles and concomitant weakening of the cell walls which stretch to accommodate the higher vacuole contents. This explains, in part, why leaving fruit on the tree for late harvested fruit often results in higher losses due to cracking and splitting in the orchard (Opara, 1996; Opara et al., 1997a, b, c, Opara et al., 2000; Opara & Tadesse, 2000a, b, c) and bruise damage during harvesting and postharvest handling (Opara, 2007; Hussein et al., 2017, 2018a, b, 2019a, b, c, d, 2020).

14.3.2 Harvesting Factors

Harvest management practices affect fruit susceptibility to mechanical damage. Ensuring that only commercially mature produce is harvested reduces the prospects of handling over-ripe fruit early in the supply chain, which has been demonstrated to be more prone to damage. The time of harvest during the day is also important. Fresh produce that is mature and ready for harvest, including banana, is firmer during early morning hours, and it is recommended that harvest should be carried out before the air temperature and sunlight increases to unacceptable levels. Apart from increasing the potential of produce to damage under these conditions, harvesting personnel is also more likely to make mistakes and apply improper harvesting and handling practices under harvest weather conditions which could result in mechanical injury to produce. Bananas undergo significant changes as they mature, making the pre-climacteric period the ideal time to harvest them for commercial handling and supply chain management (Opara et al., 2013).

The harvesting technique applied, and tools used can have major effects on the incidence of mechanical damage. For banana harvesting, where plants can grow up to 7 m high, and individual bunches can be up to 2 m above ground, weighing up to 25 kg and over 1.5 m in length (Fig. 14.1), the prospects for considerable mechanical damage to occur are very high if the bunch falls on to any surface. Therefore, proper harvesting technique includes the use of a sharp knife to cut the bunch from the stalk, ensuring that the bunch is lowered gently into the handling bin or safely transferred to a conveyor belt or rope.

14.3.3 Postharvest Factors

Bananas pass through a range of postharvest handling steps after harvest before they get to the end user. The handling stages expose fruit to a wide range of mechanical forces such as impact, compression, and abrasion, which may result in damage to the fruit. Delaying the handling of fruit after harvest has been demonstrated to result in a higher incidence of bruising during subsequent handling steps due to loss of

Fig. 14.1 Freshly harvested banana bunch in Akwa Ibom State, South-South Nigeria, modeled by Ms. Udo Eno-Obong Joseph

turgidity and rapid weakening of the fruit tissue. Hence, fruit should be precooled immediately after harvest (Duan et al., 2020).

The degree or extent of fruit ripening also affects the potential of bananas to damage during handling. Fruit resistance to damage decreases as ripening advances. Therefore, given that bananas are either allowed to ripen naturally after harvest or ripened through controlled application of ripening agents such as ethylene, it is important to manage this process carefully to reduce the handling associated with an oversupply of ripe fruit. Additional care should be taken to cushion ripe bananas against physical injury, and staff must be trained to apply extra care under such circumstances.

Controlling both temperature and relative humidity during storage is well known to impact the quality and shelf life stability of fresh horticultural produce, including bananas. However, studies have examined the effect of relative humidity during cold storage of bananas and found that fruit susceptibility to mechanical damage was not affected (Akkaravessapong et al., 1992). This suggests the overriding effect of temperature on the storability and ripening of bananas and the potential for physical injury arising from the postharvest environment.

14.3.4 The Case for Mechanical Damage During Banana Transport

The delivery of bananas to the market in a supply chain requires transportation, which cannot be avoided (Fernando et al., 2019a). During transportation, bananas are vulnerable to mechanical damage, including bruising, cutting, breaking, impact wounding, and other types of injury (Fadiji et al., 2016a, b, c). When fruits are stored, loaded, and unloaded from packages, mechanical stressors including compression and impact pressures might harm them (Zeebroeck et al., 2007). Bananas commonly sustain mechanical damage in the form of brown-black skin discoloration from fruit contact with the sides of each banana finger or the corrugated carton walls. Different degrees of damage can result from where a banana box is placed on a stacked pallet as well as where the pallet is placed on the truck floor. In comparison to cartons piled in the rear of the trailer, there were noticeably more damaged cartons in the trailer's front position (Fernando et al., 2019b). Additionally, the degrees of input root-mean-squared vibration received by the pallets stacked in various positions on the truck were connected to the levels of damage in the banana mechanical damage index (Fernando et al., 2019b).

In their empirical study on banana quality degradation along postharvest supply chains in Australia, Fernando et al. (2019a) found that package height and pallet location in a stack during transit had a substantial impact on the level of mechanical damage, which rose from the packhouses (1.3%) to the distribution centers (9.0%), and then to retail (13.3%). The distance between a distribution center and a retail outlet had an impact on the amount of abrasion damage in ripe bananas. The study also identified a number of risk factors, including weakened paperboard cartons from significant moisture absorption throughout the ripening process, as contributing to the reported incidence of fruit damage.

Due to the ease with which producers and customers can handle them, corrugated paperboard cartons are frequently utilized in food packing. Researchers have looked into the efficacy of various banana packaging options, such as one-piece corrugated cartons, two-piece corrugated cartons, and reusable plastic crates (Fernando et al., 2020; Fadiji et al., 2018a, b, c). This emphasizes the need for nonrigid packaging materials in which soft fruits, such as bananas, can be packed. The rigid sidewalls of reusable plastic crates caused a multi-fold increase in the bruising of bananas clustered in the top tiers relative to the damage suffered in two-piece corrugated cartons. The one-piece corrugated cartons offered the maximum amount of decrease in banana damage levels brought on by in-transit vibrations based on their overall protective performance.

The lack of cushioning around fresh produce during postharvest may lead to the mechanical injury of bananas. This damage often occurs due to vibrations and impact, especially during transportation from the orchard to the packing house in large plantations because of typically poor road conditions. During transportation, the loading and unloading operations may also contribute to banana damage and losses in the supply chain.

During loading and unloading operations, bananas may experience excessive mechanical impacts and/or compression which cause physical injury. Bananas may sustain mechanical damage as a result of postharvest packaging lacking padding. Due to the frequently bad road conditions, this damage frequently results from vibrations and impacts, particularly during transit from the orchard to the packing house in large plantations.

14.4 Measuring Banana Fruit Damage

Opara and Pathare (2014) provide a detailed state-of-the-art review of the techniques and procedures for measurement and analysis of bruise damage, applicable to a wide range of fresh horticultural produce, including bananas. Opara et al. (2007a, b) reported the design and development of a simple mechanical device for assessing impact bruise susceptibility of fresh produce based on a swing arm that is mini-automated to estimate the rebound height after impact. A simple but effective technique for assessing the bruise susceptibility of bananas and other types of fresh produce is shown in Fig. 14.2. The setup comprises a graduated hollow pipe mounted on a stand. This allows a round steel ball of known size (mass and diameter) to be dropped at a specified height onto the fruit surface by rapidly pulling away the nail positioned across the pipe diameter on which the steel ball is held. If the fruit is placed on a sufficiently cushioning material such as the human palm, it is reasonable to assume, for relatively low impact energy levels, that the total energy applied is equal to the energy absorbed by the produce (Opara, 2007; Opara & Pathare, 2014).

Measuring the damage to banana fruit is important for several reasons. Firstly, measuring bruises enables the determination of the susceptibility of produce to damage, which is also often referred to as bruise resistance by some researchers (Opara, 2007; Opara & Pathare, 2014; Banks & Joseph, 1991; Banks et al., 1991). Secondly, it allows the size of damage to be quantified and known especially given market standards do specify the minimum amount of bruise damage that can be tolerated. Bruise resistance is the inverse of bruise susceptibility.

Bruise susceptibility is determined as:

Fig. 14.2 Demonstrating a simple drop test device for measuring bruise damage susceptibility of fresh produce

$$\frac{\text{Amount of bruise damage area or volume}}{\text{Amount of energy absorbed by produce}} \quad (14.1)$$

Where:

Bruise area is in mm^2 or cm^2 and bruise volume is in mm^3 or cm^3; and
Energy absorbed = energy applied minus energy dissipated during rebound, Joules.

One of the seminal studies on factors contributing to banana fruit resistance to mechanical damage was reported by Banks and Joseph (1991) based on research conducted in St. Lucia in the West Indies—a major banana-producing region. By calculating the minimum (threshold) compression pressures and impact energy (both applied using an 8 mm diameter ball) necessary to generate bruises on fruit, factors impacting banana fruit susceptibility to bruising were investigated. High thresholds for compression bruising were found to be associated with fruits that released significant amounts of latex during harvest, a sign of their high turgidity, according to an assessment of fruit from 11 sites in St. Lucia at altitudes ranging from 4 to 290 m. In one experiment, fruit that was picked at 07.45 in the morning, when its turgidity would have been at its peak, took 5–49 N of compression to damage it, compared to 4–55 N for fruit that was picked at 10.45 and 13.45. The authors reported that the thresholds for compression bruising decreased from 5–35 to 4–32 and 4–29 N, respectively, due to water loss caused by leaving fruit at ambient environmental conditions for 1 and 2 days between harvest and the application of bruising. In one experiment where the initial rate of decline in thresholds for compression bruising was reduced by maintaining fruit at high relative humidity, the significance of water loss in this impact was established.

In comparison to fruit kept at 13–5, 19, and 24–5 °C (4–52, 4–72, and 4–46 N, respectively), fruit at 30 °C showed a lower compression bruising threshold (3–31 N). On the contrary, impact bruising thresholds were affected by delay after harvest and temperature in the opposite direction to compression bruising thresholds, rising from 74 to 104 μJ as a result of raising the temperature from 19 to 30 °C and from 93 to 120 μJ following a 2-day delay after harvest. Based on these findings, Banks and Joseph (1991) concluded that before suggestions to modify banana handling practices can be given, more postharvest research on mechanical damage will need to be done to identify which of the two forms of influence is most common in the postharvest handling chain.

In a separate study, Banks et al. (1991) reported an optimized simple, and inexpensive test for assessing the resistance of bananas to bruise damage by compression suitable for use in remote locations. The study determined the compression bruising thresholds based on the lowest range of compression forces that were applied using an 8 mm steel ball to the fruit which produced a visible bruise at the contact point which showed browning coloration after fruit ripening. Banks and co-workers reported that more bruises became visible to the naked eye as fruit ripening progressed, and by increasing the duration of compression time up to 10 s, the minimum force required to produce visible bruise damage declined. The researchers concluded that applying a 10 s force duration and allowing fruit to ripen to yellowness before bruise assessment was the most effective test method.

Bugaud et al. (2014) investigated the bruise susceptibility of bananas as affected by genotype and post-climacteric storage condition using five cultivars, each with a different potential for impact bruise damage. Peel electrolyte leakage (PEL), total polyphenolic content, firmness, water content, and peel thickness were all examined as potential physicochemical markers of bruise susceptibility. The smallest amount of impact energy—converted into impact energy (20–200 mJ with a 20 mJ increment)—necessary for an object placed on post-climacteric banana fruit from a certain height to cause visible bruising was evaluated as measure of bruise susceptibility. The bananas were either kept at 18 °C for the duration of ripening or at 13 °C from the second to the sixth day following ethylene induction.

The results showed that irrespective of the storage settings, neither the hybrid Flhorban925 nor the Grande Naine bruised at the maximum impact energy (200 mJ) during ripening. Among the other cultivars, a gradient in bruise susceptibility was observed: French Corne > Fougamou > hybrid Flhorban916. Furthermore, bananas stored at 18 °C were more prone to bruising and this susceptibility increased as the fruit ripened. The 2-day delay in fruit maturity and increased susceptibility to bruises were both caused by the lower ripening temperature. Bagaud and co-authors concluded that bruise susceptibility did not correspond with polyphenol content but was positively connected with PEL ($R = 0.78$) and very slightly negatively correlated with fruit hardness ($R = -0.45$), indicating that the first hint to understanding banana bruise susceptibility comes from membrane permeability.

14.5 Potential for Application of Sensors and Machine Learning in Banana Bruise Assessment

Recent advances in optical sensors combined with data science and machine learning algorithms provide novel opportunities for more accurate detection, measurement, and prediction of defects on horticultural fresh produce (Nturambirwe & Opara, 2020), including bruise damage. These technological advances in measurement and sensor science are particularly amenable to addressing some of the unique challenges facing bruise measurement and analysis such as the problem of early or latent bruises which are not readily detectable by human vision as well as established technologies such as X-ray.

Recently, Nturambirwe et al. (2021b) reported the successful use of hyperspectral imaging to develop a classification learning of latent bruises on apples. Earlier, the researchers demonstrated the use of a contactless near-infrared spectrometer to detect both the presence and severity of bruising on apples (Nturambirwe et al., 2020). For fruit that has thick pee/skin/rind such as bananas and pomegranates, Arendse et al. (2018a, b, c, d, 2021a, b, 2022a, b) have reported various successes in non-destructive measurement of both external and internal quality attributes. More recently, Okere et al. (2023) developed a Vis-NIR and SWIR-based hyperspectral imaging techniques to identify and quantify bruises on pomegranate fruit. These advances in optical sensor technology combined with imaging and chemometrics offer considerable potential for non-invasive and rapid assessment of mechanical damage on bananas, especially in early bruising and minor cuts and abrasion which often manifest as defects as fruit ages during storage and handling.

14.6 Conclusions and Future Prospects

A climacteric fruit, the banana exhibits peak ethylene and respiration production as it ripens. The physiological, biochemical, and physical aspects of the fruit should be considered when managing bananas after harvest because these factors are directly related to quality decline. Bananas undergo significant changes as they mature, making the pre-climacteric period the ideal time to harvest them for commercial handling and supply chain management (Opara et al., 2013).

Banana peel is highly sensitive to physical stress, therefore mechanical damage like browning manifests itself fast during distribution to customers. One alternate strategy for minimizing mechanical damage during transportation is to regulate the time at which bananas are harvested. Bananas should be harvested at 75% of their full maturity and stored at 12 °C for distant markets. Fruits can be picked at 90–100% maturity and kept in storage at 14–16 °C for local markets. Overall, the sooner the bananas are harvested at the mature green stage, the longer the distribution distance to the banana market that can be handled before ripening can occur during which fruit are more susceptible to mechanical damage.

Banana presents unique challenges in postharvest handling and efforts to reduce mechanical damage. The fruit grows in a bunch, containing several hands (clusters) which in turn contain several fingers (individual fruit). While bananas may be handled in bunches and clusters, they are consumed and utilized at the finger level. Little is known to date about the severity and magnitude of mechanical damage of bananas at these three levels during postharvest handling. Previous research on the mechanical damage of bananas has focused on the individual fruit (finger) level.

The macro-structure of banana fruit, with a thick rind/peel covering the edible portion (pulp/flesh) also presents a challenge in assessing fruit susceptibility to damage. Often, researchers do not distinguish between peel versus pulp bruising, and hence, the thresholds for peel bruising to result in pulp bruising remains largely unknown. This information is important because when consumer tolerance for peel bruise damage may be high because not all such damage may result in pulp damage, the presence of pulp bruising has much higher consequences on overall fruit quality and consumer dissatisfaction. Research on banana and other types of fruit have shown that bruised fruit have higher respiration and transpiration rates and also exhibit a decline in both sensory and nutritional quality attributes.

The fact that bananas are primarily handled and traded in the mature green stage but consumed mostly in the ripe stage presents another challenge in assessing bruise damage. Bruising is manifested as brown discoloration which shows better for detection in ripe fruit (yellow or red) than mature green fruit. In this regard, efforts need to be intensified in the early detection of bruises (Nturambirwe et al., 2021a, b, 2023; Okere et al., 2023) at the mature green stage before they manifest in ripe fruit, resulting in quality downgrade and possible losses due to rejection in the market. Different types and configurations of paperboard packaging are used to handle bananas and through a combination of overloading and poor design of these boxes, mechanical damage to both the container and fruit can occur, leading to losses. The design of ventilated packaging using an integrated multi-criteria approach to maintain the cold chain, provide mechanical protection, and maintain the quality of the fresh produce (Fadiji et al., 2022) offers huge benefits. This approach has been successfully applied to design and/or test packaging design systems for the handling of other fruit types such as pomegranates (Lufu et al., 2021; Mukama et al., 2020a, b, 2021) and apples (Fadiji et al., 2016a, b, c; Getahun et al., 2017a, b). More generalized and integrated modeling and experimental systems combining computational fluid dynamics and solid mechanics have also been applied to improve the design and performance evaluation of ventilated packaging (Opara et al., 2019a, b, c, d; Ambaw et al., 2022a, b; Lin et al., 2022; Berry et al., 2016, 2018, 2022; Mukama et al., 2020a, b; Fadiji et al., 2018a, b, c, 2019a, b).

The problem of early detection of bruises on mature green bananas remains in industry. Coupled with recent and rapid advances in artificial intelligence, including machine learning, these developments in non-destructive and non-invasive measurement and analysis offer considerable prospects for the detection, quantification, and segregation of latent damage in bananas as well as other fruit with thick rind/peel as well as other types of fresh produce (Nturambirwe & Opara, 2020).

Finally, advances in modern biotechnology offer considerable promise in efforts to reduce the susceptibility of fresh produce to mechanical damage, ab initio, and to reduce the severity and magnitude of damage that may occur during postharvest handling. Breeding for bruise resistance and controlling the action of ripening genes through their upward and downward regulation are some of the ways that biotechnology can be applied to control the mechanical damage of bananas and other fresh produce. While the use of edible coating has been demonstrated by researchers to prolong the shelf life of bananas and other fresh produce, the effect of applying edible coating on bruising incidence is yet to be characterized and quantified. In the meantime, providing adequate training to both preharvest, harvest, and postharvest handling personnel offers the first step in reducing and controlling bruise damage of bananas and other fresh produce.

Acknowledgement This work is based on the research supported in part by the National Research Foundation of South Africa (Grant Numbers: 64813). The opinions, findings and conclusions or recommendations expressed are those of the author(s) alone, and the NRF accepts no liability whatsoever in this regard.

References

Akkaravessapong, P., Joyce, D. C., & Turner, D. W. (1992). The relative humidity at which bananas are stored or ripened does not influence their susceptibility to mechanical damage. *Scientia Horticulturae, 52*(3), 265–268.

Al-Dairi, M., Pathare, P. B., & Al-Mahdouri, A. (2022). The contribution of impact damage to the quality changes of stored banana fruit. *Biology and Life Sciences Forum, 16*, 31. https://doi.org/10.3390/IECHo2022-12483

Al-Dairi, M., Pathare, P. B., Al-Yahyai, R., Jayasuriya, H., & Al-Attabi, Z. (2023). Postharvest quality, technologies, and strategies to reduce losses along the supply chain of banana: A review. *Trends in Food Science & Technology, 134*, 177–191.

Ambaw, A., Mukama, M., Fadiji, T., & Opara, U. L. (2022a). Fresh fruit packaging design verification through virtual prototyping technique. *Food Packaging and Shelf Life, 32*, 100858. https://doi.org/10.1016/j.fpsl.2022.100858

Ambaw, A., Mukama, M., Fadiji, T., & Opara, U. L. (2022b). Packaging of the future: Innovations in packaging design and cold chain logistical management. *Acta Horticulturae, 1349*, 389–398. https://doi.org/10.17660/ActaHortic.2022.1349.53

Arendse, E., Fawole, O. A., Magwaza, L. S., & Opara, U. L. (2018a). Comparing the analytical performance of near and mid infrared spectrometers for evaluating pomegranate juice quality. *LWT - Food Science and Technology, 91*, 180–190. https://doi.org/10.1016/j.lwt.2018.01.035

Arendse, E., Fawole, O. A., Magwaza, L. S., & Opara, U. L. (2018b). Evaluation of biochemical markers associated with the development of husk scald and the use of diffuse reflectance NIR spectroscopy to predict husk scald in pomegranate fruit. *Scientia Horticulturae, 232*, 240–249. https://doi.org/10.1016/j.scienta.2018.01.022

Arendse, E., Fawole, O. A., Magwaza, L. S., & Opara, U. L. (2018c). Non-destructive prediction of internal and external quality attributes of fruit with thick rind: A review. *Journal of Food Engineering, 217*, 11–23. https://doi.org/10.1016/j.jfoodeng.2017.08.009

Arendse, E., Fawole, O. A., Magwaza, L. S., Nieuwoudt, H. H., & Opara, U. L. (2018d). Fourier transform near infrared diffuse reflectance spectroscopy and two spectral acquisition modes for evaluation of external and internal quality of intact pomegranate fruit. *Postharvest Biology and Technology, 138*, 91–98. https://doi.org/10.1016/j.postharvbio.2018.01.001

Arendse, E., Nieuwoudt, H., Fawole, O. A., & Opara, U. L. (2021a). Effect of different extraction methods on the quality and biochemical attributes of pomegranate juice and the application of Fourier transformed infrared spectroscopy in discriminating between different extraction methods. *Frontiers in Plant Science, 12*, 702575. https://doi.org/10.3389/fpls.2021.702575

Arendse, E., Nieuwoudt, H., Magwaza, L. S., Nturambirwe, J. F. I., Fawole, O. A., & Opara, U. L. (2021b). Recent advancements on vibrational spectroscopic techniques for the detection of authenticity and adulteration in horticultural products with a specific focus on oils, juices and powders. *Food and Bioprocess Technology, 14*(1), 1–22. https://doi.org/10.1007/s11947-020-02505-x

Arendse, E., Nieuwoudt, H., Fawole, O. A., & Opara, U. L. (2022a). Non-destructive prediction of pomegranate juice quality: Near-infrared vs mid-infrared spectroscopy. *Acta Horticulturae, 1349*, 341–348. https://doi.org/10.17660/ActaHortic.2022.1349.47

Arendse, E., Nieuwoudt, H., Fawole, O. A., & Opara, U. L. (2022b). The application of Fourier transformed infrared spectroscopy in discriminating different pomegranate juice extraction methods. *Acta Horticulturae, 1349*, 349–356. https://doi.org/10.17660/ActaHortic.2022.1349.48

Banks, N. H., & Joseph, J. (1991). Factors affecting resistance of banana fruit to compression and impact bruising. *Journal of the Science of Food and Agriculture, 56*(3), 315–323. https://doi.org/10.1002/jsfa.2740560307

Banks, N. H., Borton, C. A., & Joseph, M. (1991). Compression bruising test for bananas. *Journal of the Science of Food and Agriculture, 56*(2), 223–226.

Berry, T., Defraeye, T., Nicolai, B. M., & Opara, U. L. (2016). Multiparameter analysis of cooling efficiency of ventilated fruit cartons using CFD: Impact of vent hole design and internal packaging. *Food and Bioprocess Technology, 9*(9), 1481–1493. https://doi.org/10.1007/s11947-016-1733-y

Berry, T. M., Fadiji, T. S., Defraeye, T., Coetzee, C., & Opara, U. L. (2018). A multi-parameter approach to vent hole design for cartons packed with internal packaging. *Acta Horticulturae, 1194*, 1307–1314. https://doi.org/10.17660/ActaHortic.2018.1194.184

Berry, T. M., Defraeye, T., Shrivastava, C., Ambaw, A., Coetzee, C., & Opara, U. L. (2022). Designing ventilated packaging for the fresh produce cold chain. *Food and Bioproducts Processing, 134*, 121. https://doi.org/10.1016/j.fbp.2022.04.005

Bugaud, C., Ocrisse, G., Salmon, F., & Rinaldo, D. (2014). Bruise susceptibility of banana peel in relation to genotype and post-climacteric storage conditions. *Postharvest Biology and Technology, 87*, 113–119.

Duan, Y., Wang, G. B., Fawole, O. A., Verboven, P., Zhang, X. R., Wu, D., Opara, U. L., Nicolai, B., & Chen, K. (2020). Postharvest precooling of fruit and vegetables: A review. *Trends in Food Science and Technology, 100*, 278–291.

Fadiji, T., Coetzee, C., Chen, L., Chukwu, O., & Opara, U. L. (2016a). Susceptibility of apples to bruising inside ventilated corrugated paperboard packages during simulated transport damage. *Postharvest Biology and Technology, 118*, 111–119.

Fadiji, T., Coetzee, C., & Opara, U. L. (2016b). Compression strength of ventilated corrugated paperboard packages: Numerical modelling, experimental validation and effects of vent geometric design. *Biosystems Engineering, 151*, 231–247. https://doi.org/10.1016/j.biosystemseng.2016.09.010

Fadiji, T., Coetzee, C., Pathare, P., & Opara, U. L. (2016c). Susceptibility to impact damage of apples inside ventilated corrugated paperboard packages: Effects of package design. *Postharvest Biology and Technology, 111*, 286–296. https://doi.org/10.1016/j.postharvbio.2015.09.023

Fadiji, T., Berry, T. M., Coetzee, C. J., & Opara, U. L. (2018a). Mechanical design and performance testing of corrugated paperboard packaging for the postharvest handling of horticultural produce. *Biosystems Engineering, 171*, 220–244.

Fadiji, T., Ambaw, A., Coetzee, C. J., Berry, T. M., & Opara, U. L. (2018b). Application of finite element analysis to predict the mechanical strength of ventilated corrugated paperboard packaging for handling fresh produce. *Biosystems Engineering, 174*, 260–281. https://doi.org/10.1016/j.biosystemseng.2018.07.014

Fadiji, T., Coetzee, C. J., Berry, T. M., Ambaw, A., & Opara, U. L. (2018c). The efficacy of finite element analysis (FEA) as a design tool for food packaging: A review. *Biosystems Engineering, 174*, 20–40. https://doi.org/10.1016/j.biosystemseng.2018.06.015

Fadiji, T., Coetzee, C. J., & Opara, U. L. (2019a). Analysis of the creep behaviour of ventilated corrugated paperboard packaging for handling fresh produce—An experimental study. *Food and Bioproducts Processing, 117*, 126–137. https://doi.org/10.1016/j.fbp.2019.07.001

Fadiji, T., Coetzee, C. J., Berry, T. M., & Opara, U. L. (2019b). Investigating the role of geometrical configurations of ventilated fresh produce packaging to improve the mechanical strength–experimental and numerical approaches. *Food Packaging and Shelf Life, 20*, 100312. https://doi.org/10.1016/j.fpsl.2019.100312

Fadiji, T., Ambaw, A., & Opara, U. L. (2022). Impact and packaging: Do they affect fruit quality? *Acta Horticulturae, 1349*, 399–406. https://doi.org/10.17660/ActaHortic.2022.1349.54

FAO. (2018). *FAOSTAT*. Food and Agricultural Organization (FAO).

Fernando, I., Fei, J., Stanley, R., Enshaei, H., & Eyles, A. (2019a). Quality deterioration of bananas in the post-harvest supply chain—An empirical study. *Modern Supply Chain Research and Applications, 1*(2), 135–154.

Fernando, I., Fei, J., & Stanley, R. (2019b). Measurement and analysis of vibration and mechanical damage to bananas during long-distance interstate transport by multi-trailer road trains. *Postharvest Biology and Technology, 158*, 110977.

Fernando, I., Fei, J., Stanley, R., & Rouillard, V. (2020). Evaluating packaging performance for bananas under simulated vibration. *Food Packaging Shelf Life, 23*, 100428.

Getahun, S., Ambaw, A., Delele, M., Meyer, C. J., & Opara, U. L. (2017a). Analysis of airflow and heat transfer inside fruit packed refrigerated shipping container: Part I—Model development and validation. *Journal of Food Engineering, 203*, 58–68. https://doi.org/10.1016/j.jfoodeng.2017.02.010

Getahun, S., Ambaw, A., Delele, M., Meyer, C. J., & Opara, U. L. (2017b). Analysis of airflow and heat transfer inside fruit packed refrigerated shipping container: Part II—Evaluation of apple packaging design and vertical flow resistance. *Journal of Food Engineering, 203*, 83–94. https://doi.org/10.1016/j.jfoodeng.2017.02.011

Hussein, Z., Fawole, O. A., & Opara, U. L. (2017). Investigating bruise susceptibility of pomegranate (Punica granatum L.) cultivars during postharvest handling. *African Journal of Rural Development, 2*(1), 33–39.

Hussein, Z., Fawole, O. A., & Opara, U. L. (2018a). Bruise damage susceptibility of pomegranates (Punica granatum L.) and its impact on fruit physiological response. *Acta Horticulturae, 1201*, 55–64. https://doi.org/10.17660/ActaHortic.2018.1201.9

Hussein, Z., Fawole, O. A., & Opara, U. L. (2018b). Preharvest factors influencing bruise damage of fresh fruits—A review. *Scientia Horticulturae, 229*, 45–58. https://doi.org/10.1016/j.scienta.2017.10.028

Hussein, Z., Fawole, O. A., & Opara, U. L. (2019a). Bruise damage susceptibility of pomegranates (Punica granatum, L.) and impact on fruit physiological response during short term storage. *Scientia Horticulturae, 246*, 664–674. https://doi.org/10.1016/j.scienta.2018.11.026

Hussein, Z., Fawole, O. A., & Opara, U. L. (2019b). Determination of physical, biochemical and microstructural changes in impact-bruise damaged pomegranate fruit. *Journal of Food Measurement and Characterization, 13*, 2177–2189. https://doi.org/10.1007/s11694-019-00138-z

Hussein, Z., Fawole, O. A., & Opara, U. L. (2020). Harvest and postharvest factors affecting bruise damage of fresh fruits. *Horticultural Plant Journal., 6*(1), 1–13. https://doi.org/10.1016/j.hpj.2019.07.006

Kuyu, C. G., & Tola, Y. B. (2018). Assessment of banana fruit handling practices and associated fungal pathogens in Jimma town market, Southwest Ethiopia. *Food Science & Nutrition, 6*(3), 609–616. https://doi.org/10.1002/fsn3.591

Lin, M., Fawole, O. A., Saeys, W., Wu, D., Wang, J., Opara, U. L., Nicolai, B., & Chen, K. (2022). Mechanical damages and packaging methods along the fresh fruit supply chain: A review. *Critical Reviews in Food Science and Nutrition*, 1. https://doi.org/10.1080/10408398.2022.2078783

Lufu, R., Ambaw, A., & Opara, U. L. (2021). The influence of internal packaging (liners) on moisture dynamics, physical and physiological quality of pomegranate fruit during cold storage. *Food, 10*, 1388. https://doi.org/10.3390/foods10061388

Magwaza, L. S., & Opara, U. L. (2014). Investigating non-destructive quantification and characterisation of pomegranate fruit internal structure using X-ray computed tomography. *Postharvest Biology and Technology, 95*, 1–6.

Maia, V. M., Salomão, L. C. C., Siqueira, D. L., Aspiazú, I. L., & Maia, C. B. (2014). Physical and metabolic changes induced by mechanical damage in 'Dwarf-Prata' banana fruits kept under cold storage. *Australian Journal of Crop Science, 8*, 81029–81037.

Mukama, M., Ambaw, A., & Opara, U. L. (2020a). A virtual prototyping approach for redesigning the vent-holes of packaging for handling pomegranate fruit—A short communication. *Journal of Food Engineering, 270*, 109762. https://doi.org/10.1016/j.jfoodeng.2019.109762

Mukama, M., Ambaw, A., & Opara, U. L. (2020b). Advances in design and performance evaluation of fresh fruit ventilated distribution packaging: A review. *Food Packaging and Shelf Life, 24*, 100472. https://doi.org/10.1016/j.fpsl.2020.100472

Mukama, M., Ambaw, A., & Opara, U. L. (2021). Characterisation of ventilated multi-scale packaging designs used for postharvest handling of pomegranate in South Africa. *AMA-Agricultural Mechanization in Asia, Africa and Latin America, 52*(4), 31–38.

Nturambirwe, J. F. I., & Opara, U. L. (2020). Machine learning applications to non-destructive defect detection in horticultural products. *Biosystems Engineering, 189*, 60–83.

Nturambirwe, J. F. I., Nieuwoudt, H. H., Perold, W. J., & Opara, U. L. (2020). Detecting bruise damage and level of severity in apples using A contactless NIR spectrometer. *Applied Engineering in Agriculture, 36*(3), 257–270. https://doi.org/10.13031/aea.13218

Nturambirwe, J. F. I., Perold, W. J., & Opara, U. L. (2021a). A squid-detected NMR relaxation study of banana fruit ripening. *Applied Engineering in Agriculture, 37*(2), 219–231. https://doi.org/10.13031/aea.13991

Nturambirwe, J. F. I., Perold, W. J., & Opara, U. L. (2021b). Classification learning of latent bruise damage to apples using shortwave infrared hyperspectral imaging. *Sensors, 21*, 4990. https://doi.org/10.3390/s21154990

Nturambirwe, J. F. I., Hussein, E. A., Vaccari, M., Thron, C., Perold, W. J., & Opara, U. L. (2023). Feature reduction for the classification of bruise damage to apple fruit using a contactless FT-NIR spectroscopy with machine learning. *Food, 12*(1), 210. https://doi.org/10.3390/foods12010210

Okere, E. E., Arendse, E., Nieuwoudt, H., Fawole, O. A., Perold, W. J., & Opara, U. L. (2021). Non-invasive methods for predicting the quality of processed horticultural food products, with emphasis on dried powders, juices and oils: A review. *Foods, 10*(12), 3061. https://doi.org/10.3390/foods10123061

Okere, E. E., Ambaw, A., Perold, W. J., & Opara, U. L. (2022a). Non-destructive detection of bruises in pomegranate fruit by Vis-NIR and SWIR hyperspectral imaging. *Acta Horticulturae, 1349*, 381–388. https://doi.org/10.17660/ActaHortic.2022.1349.52

Okere, E. E., Arendse, E., Ambaw Tsige, A., Perold, W. J., & Opara, U. L. (2022b). Pomegranate quality evaluation using non-destructive approaches: A review. *Agriculture, 12*(12), 2034. https://doi.org/10.3390/agriculture12122034

Okere, E. E., Tsige, A. A., Perold, W., & Opara, U. L. (2023). Vis-NIR and SWIR hyperspectral imaging method to detect bruises in pomegranate fruit. *Frontiers in Plant Science, 14*, 1151697. https://doi.org/10.3389/fpls.2023.1151697

Opara, L. U. (1996). Some characteristics of internal ring-cracking in apples. *Fruit Varieties Journal, 50*(4), 260–262.

Opara, L. U. (2007). Bruise susceptibilities of 'Gala' apple as affected by orchard management practices and harvest date. *Postharvest Biology and Technology, 43*, 47–54.

Opara, U. L., & Fadiji, T. S. (2018). Compression damage susceptibility of apple fruit packed inside ventilated corrugated paperboard package. *Scientia Horticulturae, 227*, 154–161. https://doi.org/10.1016/j.scienta.2017.09.043

Opara, U. L., & Pathare, P. B. (2014). Bruise damage measurement and analysis of fresh horticultural produce—A review. *Postharvest Biology and Technology, 91*, 9–24.

Opara, L. U., & Tadesse, T. (2000a). Calyx-end splitting and physico-chemical properties of Pacific Rose apple as affected by orchard management factors. *Journal of Horticultural Science & Biotechnology, 75*(5), 581–585.

Opara, L. U., & Tadesse, T. (2000b). Fruit growth and mineral element accumulation in Pacific Rose apple in relation to orchard management factors and calyx-end splitting. *Journal of Plant Nutrition, 23*(8), 1079–1093.

Opara, L. U., & Tadesse, T. (2000c). Fruit growth characteristics and chronological development of calyx-end splitting in Pacific Rose apple. *Journal of the American Pomological Society, 54*(3), 153–159.

Opara, L. U., Studman, C. J., & Banks, N. H. (1997a). Physico-mechanical properties of 'Gala' apples and stem-end splitting as influenced by orchard management practices and harvest date. *Journal of Agricultural Engineering Research, 68*(2), 139–146.

Opara, L. U., Studman, C. J., & Banks, N. H. (1997b). Sunlight affects the incidence of internal ring cracking in Gala apples. *Journal of Tree Fruit Production, 2*(1), 45–52.

Opara, L. U., Studman, C. J., & Banks, N. H. (1997c). Fruit skin splitting and cracking. *Horticultural Reviews, 19*, 217–262.

Opara, L. U., Studman, C. J., & Hodson, A. (2000). Stem-end splitting and internal ring-cracking in gala apples as influenced by orchard management practices. *Journal of Horticultural Science & Biotechnology, 75*(4), 465–469.

Opara, L. U., Al-Ghafri, A., Agzoun, H., Al-Issai, J., & Al-Jabri, F. (2007a). Design and development of a new device for measuring susceptibility to impact damage of fresh produce. *New Zealand Journal of Crop and Horticultural Science, 35*(2), 245–251. https://doi.org/10.1080/01140670709510191

Opara, L. U., Al-Said, F. A., & Al-Abri, A. (2007b). An assessment of what the consumer values in fresh fruit quality: A case study of Oman. *New Zealand Journal of Crop and Horticultural Science, 35*, 235–243.

Opara, U. L., Yahyai, R., Al-Waili, A.-N., Said, F. A., Al-Ani, M., Manickavasagan, A., & Al-Mahdouri, A. (2013). Postharvest responses of 'Malindi' Cavendish banana to various storage conditions. *International Journal of Fruit Science, 13*, 373–388.

Opara, U. L., Ambaw, A., & Berry, T. (2019a). Chapter 6: CFD investigation of fresh produce cooling processes and effects of package stacking. In D.-W. Sun (Ed.), *Computational fluid dynamics in food processing: Contemporary food engineering series* (2nd ed., pp. 121–148). CRC Press. https://doi.org/10.1201/9781351263481. ISBN 9781351263481.

Opara, U. L., Ambaw, A., & Fadiji, T. (2019b). Chapter 10: Future ventilated packaging design: Novel modelling approaches and integrated multi-criteria performance. In R. M. S. Cruz (Ed.), *Food packaging: Innovation and shelf-life*. CRC Press. https://doi.org/10.1201/9780429023101. ISBN 9780429023101.

Opara, U. L., Berry, T., & Ambaw, A. (2019c). Chapter 8: Optimization of horticultural carton vent hole design by CFD. In D.-W. Sun (Ed.), *Computational fluid dynamics in food processing: Contemporary food engineering series* (2nd ed., pp. 169–191). CRC Press. https://doi.org/10.1201/9781351263481. ISBN 9781351263481.

Opara, U. L., Caleb, O. J., & Belay, Z. A. (2019d). Chapter 7: Modified atmosphere packaging for food preservation. In C. M. Galanakis (Ed.), *Food quality and shelf life* (pp. 235–259). Academic. https://doi.org/10.1016/B978-0-12-817190-5.00007-0. ISBN: 9780128171905.

Panigrahi, N., Thompson, A. J., & A.J., Zubelzu, S., Knox, J.W. (2021). Identifying opportunities to improve management of water stress in banana production. *Scientia Horticulturae, 276*, 109735.

Pathare, P. B., & Al-Dairi, M. (2021a). Bruise damage and quality changes in impact-bruised, stored tomatoes. *Horticulturae, 7*, 113.

Pathare, P. B., & Al-Dairi, M. (2021b). Bruise susceptibility and impact on quality parameters of pears during storage. *Frontiers in Sustainable Food Systems, 5*, 658132.

Rashvand, M., Altieri, G., Genovese, F., Li, Z., & Renzo, C. D. (2022). Numerical simulation as a tool for predicting mechanical damage in fresh fruit. *Postharvest Biology and Technology, 187*, 11187.

Sablani, S. S., Opara, L. U., & Al-Balushi, K. (2006). Influence of bruising and storage temperature on vitamin C content of tomato fruit. *Journal of Food, Agriculture & Environment, 4*(1), 54–56.

Sugianti, C., Imaizumi, T., Thammawong, M., & Nakano, K. (2022). Recent postharvest technologies in the banana supply chain. *Reviews in Agricultural Science, 10*, 123–137. https://doi.org/10.7831/ras.10.0_123

Zeebroeck, M. V., Linden, V. V., Darius, P., Ketelaere, B. D., Ramon, H., & Tijskens, E. (2007). The effect of fruit properties on the bruise susceptibility of tomatoes. *Postharvest Biology and Technology, 45*(2), 168–175.

Chapter 15
Impact of Packaging on Bruise Damage of Fresh Produce

Tobi Fadiji, Tafadzwa Kaseke, Robert Lufu, Zhiguo Li, Umezuruike Linus Opara, and Olaniyi Amos Fawole

Abstract Consumers demand high-quality fresh produce; nonetheless, mechanical damage to fresh produce, such as fruit and vegetables, that occurs along the postharvest supply chain affects fresh produce quality and impedes consumer acceptance, resulting in considerable economic losses. Improper packaging and packaging materials are the leading causes of bruise damage in the postharvest handling of fresh produce. Bruising, a form of mechanical damage, is caused when packages are subjected to compression, impact, or vibration loadings and is a vital component that limits mechanized harvesting, transportation, postharvest handling, and fresh produce processing. This chapter presents the impact of packaging on the bruising of fresh produce, including fruit and vegetables. The chapter provides an overview of fresh produce packaging and summarizes the mechanisms of bruising in fresh produce. The effects of bruise damage on fresh produce quality

T. Fadiji
Postharvest Research Laboratory, Department of Botany and Plant Biotechnology, University of Johannesburg, Johannesburg, South Africa

Centre for Applied Data Science, University of Johannesburg, Johannesburg, South Africa

T. Kaseke · O. A. Fawole
Postharvest Research Laboratory, Department of Botany and Plant Biotechnology, University of Johannesburg, Johannesburg, South Africa

R. Lufu
SARChI Postharvest Technology Research Laboratory, Africa Institute for Postharvest Technology, Faculty of AgriSciences, Stellenbosch University, Stellenbosch, South Africa

Z. Li
College of Mechanical and Electronic Engineering, Northwest A&F University, Yangling, Shaanxi, China

U. L. Opara (✉)
SARChI Postharvest Technology Research Laboratory, Africa Institute for Postharvest Technology, Faculty of AgriSciences, Stellenbosch University, Stellenbosch, South Africa

UNESCO International Centre for Biotechnology, Nsukka, Enugu State, Nigeria
e-mail: opara@sun.ac.za

P. B. Pathare, U. L. Opara (eds.), *Mechanical Damage in Fresh Horticultural Produce*, https://doi.org/10.1007/978-981-99-7096-4_15

were also discussed, as were various methods of assessing packaging and fresh produce damage.

Keywords Bruising · Fresh produce · Mechanical damage · Bruise susceptibility mechanical damage · Multi-scale packaging · Ventilated packaging · MAP

15.1 Introduction

The benefits and importance of fresh produce, mainly fruit and vegetables, cannot be overemphasized. fruits and vegetables have been associated with a protective role in health maintenance (Hervert-Hernández et al., 2011). Regular consumption of fruits is linked with reduced risks of several illnesses, including stroke, cardiovascular diseases, and cancer (Alissa & Ferns, 2017). Significant amounts of key minerals, vitamins, phytochemicals, and dietary fibres are found in fruits (Chang et al., 2016; Onwude et al., 2021). Regardless, they are highly susceptible to mechanical damage and quality deterioration during handling, resulting in postharvest losses (Yousefi et al., 2016; Opara & Fadiji, 2018; Zhang et al., 2018).

During postharvest handling operations such as transportation, storage, and packing processes, fruit and vegetables are exposed to a range of static or dynamic forces (Fadiji et al., 2018a; Dagdelen & Aday, 2021). When combined with inappropriate postharvest handling, such as improper packaging, mechanical damage and quality degradation occur (Pathare & Opara, 2014; Fadiji et al., 2018a), and consequent economic losses (Stropek & Gołacki, 2015; Wohner et al., 2019).

The most common symptom of mechanical damage in fresh produce and a major cause of postharvest loss is bruising (Opara & Pathare, 2014; Al-Dairi et al., 2022). Bruising often occurs when the forces of fruits and vegetables under dynamic and static loading exceed the failure stress of the tissue (Hussein et al., 2019a; Du et al., 2020). Figure 15.1 depicts a typical bruising on apple fruit. According to Fadiji et al. (2016a), disposed produce due to mechanical damage is estimated to be about 40%. Fruit losses from mechanical damage have been proven to be minimized by careful handling and appropriate packaging (Singh et al., 1992; Barchi et al., 2002;

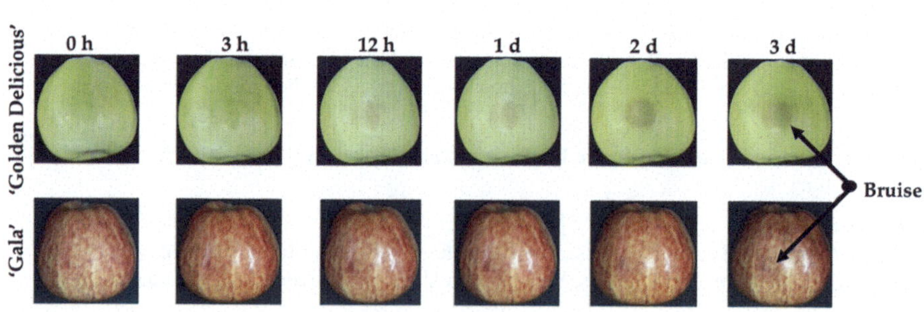

Fig. 15.1 Typical bruising evolution on two varieties of apple fruit over 3 days when subjected to impact force. (Adapted from Lu & Lu, 2017)

Chonhenchob & Singh, 2003; Vursavuş & Özgüven, 2004; Chonhenchob et al., 2009; Fadiji et al., 2016a, b). Evidence of severe mechanical damage is increasing and impacting the trade in fruits and vegetables, indicating a clear need to improve handling methods, notably optimizing the packaging to provide further protection (Okhuoya, 1995). The ability of packaging to protect the packaged contents from handling, storage, and transportation hazards is an essential indicator of its performance. The resistance of the produce to damage, the hazards encountered in the distribution system, the properties of the packaging materials, and the package arrangement are all factors in the protection provided by packages (Pathare & Opara, 2014; Fadiji et al., 2016a, b). Therefore, the present chapter discusses the impact of packaging on the bruising of fresh produce, particularly on fruit and vegetables.

15.2 Fresh Produce Packaging: Overview

Since harvested fresh produce is metabolically active and continues to undergo ripening and senescence processes, its nutritional and sensory properties start to deteriorate. In addition, mechanical injuries and postharvest physiological disorders are implicated in the loss of quality in fresh produce after harvest. Therefore, management of these processes is warranted to minimize postharvest and economic losses. Packaging is one of the main tools used to preserve the quality of fresh produce. Packaging assures the delivery of produce from grower to consumer safely, and conveniently, at a minimum overall cost (Ambaw et al., 2021). It serves a multifunctional purpose (Defraeye et al., 2015). The main functions of packaging are handling, containment, protection, and communication (Fig. 15.2). Fresh produce packaging can be classified into primary, secondary, and tertiary packaging, and each has a specific purpose. The primary packaging directly contains and protects the produce against contamination, while secondary packaging houses the primary packaging, which could be single or multiple primary packages. On the hand, tertiary packaging offers a certain level of cushioning to the primary product, in addition to acting as an implement for bulk transportation (Pascall, 2020). Typically, packaging should protect the produce from mechanical damage during handling, storage, and transportation; promote uniform and rapid cooling of fresh produce; maintain the quality of the produce and extend the produce shelf life (Pathare & Opara, 2014; Defraeye et al., 2015; Fadiji et al., 2018a).

Various packaging materials and technologies are available for applications in the fresh produce industry (Koutsimanis et al., 2012). Common packaging used when handling, transporting, and marketing fresh produce includes wood, bags, crates, hampers, baskets, corrugated paperboard cartons, plastics, bulk bins, and palletized containers (Koutsimanis et al., 2012; Fadiji et al., 2016a, b). The main packaging for fresh produce is corrugated paperboard; nonetheless, additional internal packaging is required, which includes tissue paper wraps, trays, plastic bags, liners, clamshells, cups, or pads to prevent damage from abrasion (Ščetar & Kurek, 2010; Fadiji et al.,

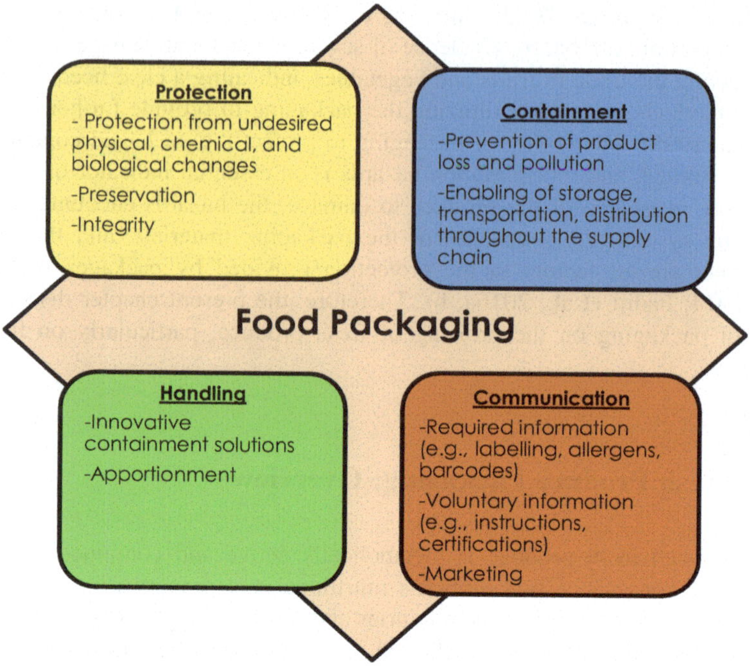

Fig. 15.2 Functions of food packaging. (Adapted from Al-Dairi et al., 2022)

2018a; Mukama et al., 2020). This is often referred to as multiscale packaging. Depending on the type of fresh produce, layers may be added within the packaging to maximize space utilization. In fresh produce such as fruits, mould trays may be used. These separate the fruits, which may be separately wrapped in tissue or waxy paper to enhance physical protection or reduce the spread of bacterial or fungal diseases. Corrugated paperboard cartons have been used extensively and have remained an essential part of the growing food packaging industry (Fadiji et al., 2016a, 2018a). Figure 15.3 shows examples of typical corrugated paperboard cartons used in the fresh produce industry. The success of the prevalent use and dominance of corrugated paperboard, particularly in the fresh fruit industry, is due to numerous advantages, including low cost and versatility characteristics, efficient and economic material characteristics, good protection of the produce, capability to withstand various shipping and transportation hazards (impact, compression, and vibration damages, moisture, etc.), recyclability, and biodegradability (Ambaw et al., 2021; Fadiji et al., 2018a; Subedi et al., 2017). However, one fundamental difficulty that affects the use of corrugated paperboard cartons is the capacity to preserve the mechanical strength of the cartons under cold chain conditions for an extended period (Ambaw et al., 2021). Consequently, this causes a collapse of the cartons and damage to the packed produce, which could be expressed as bruising.

The type of packaging for fresh produce may be determined by the physical, anatomical, and physiology (mainly transpiration, respiration, and ethylene

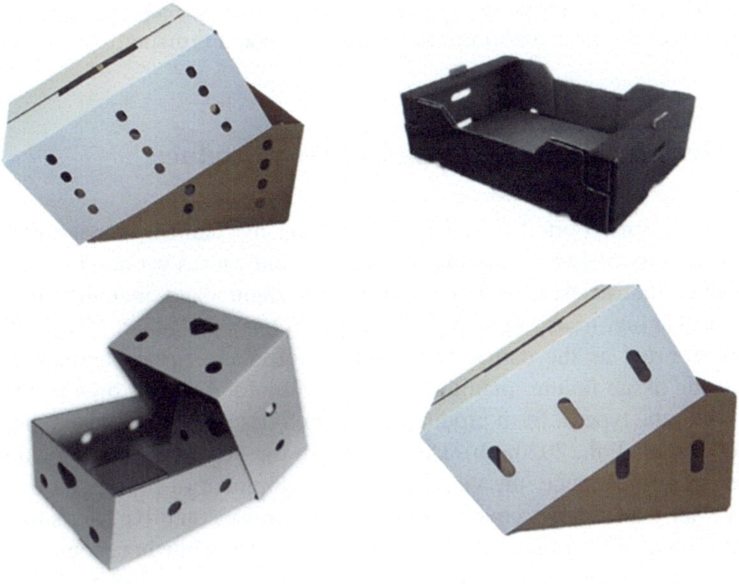

Fig. 15.3 Examples of corrugated paperboard cartons used for handling fresh produce

production rate) and susceptibility of the fresh produce to microbial decay. Other factors implicated in the packaging of fresh produce and the postharvest life of fresh produce are temperature, relative humidity, and ventilation. To maintain quality, most fresh produce are stored at low temperature; however, low temperature alone is not adequate in controlling the quality of fresh produce. Thus, the primary packaging has been modified to further protect and enhance fresh produce's shelf life. Technologies such as controlled atmosphere (CA) packaging, modified atmosphere packaging (MAP), and edible coating have been implemented to augment primary packaging and cold storage management in fresh produce quality preservation.

CA and MAP are primarily applied to alter the atmosphere composition (O_2, CO_2, and N_2) surrounding and within the fresh produce to influence cellular metabolism. Various fruits and vegetables have varied optimal CA and MAP conditions which can be manipulated to control the physiological and biochemical changes during storage. Literature has reported significant improvement in the quality, storability, and shelf life of fresh produce such as papaya, pomegranate, tomato, grape, guava, pear, orange, and strawberry stored under MAP (Belay et al., 2019). Edible coatings may be applied as a hurdle technology to reduce quality losses in fresh produce. They provide a durable protective barrier around the fruit's surface, which may slow the ripening process and extend the shelf life of fresh produce when used together with other postharvest technologies (Riva et al., 2020). Despite the availability of all these technologies, the provision of proper and adequate packaging for fresh produce requires understanding the characteristics of each type of packaging, and the physical and chemical properties of the produce, as these factors contribute to the susceptibility of the produce to mechanical damage and bruising.

15.3 Mechanisms of Bruising in Fresh Produce

Bruising has been reported as the most common mechanical damage that downgrades the quality of fresh horticultural produce and causes economic losses to the fresh produce industries. It occurs at any supply chain stage, including harvesting, grading, packaging, transportation, and storage (Opara & Pathare, 2014). Bruising has been reported as the damage to the fresh produce surface implicated in the subcutaneous tissue failure, manifested by the discolouration (brown to black spot depending on the type of fresh produce) of injured tissue, which impairs their visual quality (Hussein et al., 2020a, b). Depending on the degree of mechanical damage and type of fresh produce, the visualization development of bruises on the injured area may vary from a few to many hours. Brown bruises on olive fruit surface were observed soon after the impact and intensified after 24 h, while in coconut and peach, bruises development at room temperature takes between 24 and 48 h (Hussein et al., 2019b). This means that bruising in some produce may not be detected until it reaches the retail store or consumer. The severity of the bruise may be recorded as the diameter, area, and volume of the bruise. According to Pathare and Al-Dairi (2022), bruising of fresh produce along the supply chain may increase postharvest losses by almost 51%. Bruising in fresh produce accelerates physiological and metabolic activities, which catalyses browning, quicker ripening, and quality losses (Pathare & Al-Dairi, 2022). Given the economic importance of bruising, understanding the mechanisms of bruising in fresh produce is essential in formulating strategies to minimize postharvest losses.

15.3.1 Types of Mechanical Damage Associated with Bruising in Fresh Produce/Drivers of Bruise Damage in Fresh Produce

Bruise damage is one of the mechanical injury problems that could appear in fresh produce during the postharvest supply chain (Pathare & Al-Dairi, 2021a, b). Bruising results from excessive impact, compression, and vibration or abrasion of the fruit or vegetable against each other or against a rigid surface, which could be packaging, grading, treatment, or harvesting equipment (Kitthawee et al., 2011; Opara & Fadiji, 2018; Mukama et al., 2020). Alternatively, this could be due to the combined influence of all these actions, depending on the type of postharvest handling. The extent of damage on the fresh produce varies with the type of produce, harvest maturity, and source of mechanical damage.

15.3.1.1 Impact Damage

Impact damage, the most severe mechanical damage, occurs when the fresh produce hits a surface with sufficient force to rupture or even separate cells. Postharvest impact damage occurs between the fresh produce or between produce and the ground during harvesting (free drop), packaging, or insufficiently cushioned surfaces. It is regarded as the most severe mechanical damage mechanism (Li & Thomas, 2014; Opara & Pathare, 2014). It occurs at every stage of the technological process of handling fresh produce and is challenging to eliminate in large-scale horticulture fresh produce production and distribution (Gołacki et al., 2009). For instance, packages holding fresh produce can be dropped from a given height during loading, unloading, and processing. The fruits may be bruised in this situation due to the impact (Jarimopas et al., 2007; Yeşiloğlu Cevher, 2022). Consequently, the produce quality diminishes, leading to economic losses due to fungal contamination of bruised areas (Opara & Pathare, 2014).

Furthermore, bruise pathogens can penetrate the fruit quickly through dead or bruised tissues, contaminating the remaining produce in the package (Pholpho et al., 2011; Yeşiloğlu Cevher, 2022). In addition, experiments have shown that the extent of impact damage varies with height, storage temperature, type of surface on which the fresh produce is dropped, and strength of the packaging material. Celik et al. (2021) reported permanent damage progression on apples dropped on steel, high-density polyethylene, and wood surfaces. Expanded polystyrene provided better shock protection to apples and reduced bruising susceptibility by 40% compared to the modified fibre tray method (Dunno et al., 2021). An increase in drop height increased the damage of apples packed in single-wall and double-wall corrugated boxes, with apples in the lower layers of the boxes experiencing greater damage (Lu et al., 2010). Storage of tomatoes at a lower temperature reduced bruising impact (Pathare & Al-Dairi, 2021b). Additionally, one of the most fundamental problems preventing successful mechanization and automation of harvesting and postharvest handling activities is impact bruise damage to fresh produce (Opara, 2007). As a result, several researchers have studied the susceptibility or resistance of fresh produce to bruising caused by impact. Table 15.1 shows some studies on the susceptibility, incidence, or resistance to impact bruising damage of packed fresh produce.

15.3.1.2 Compression Damage

Meanwhile, compression damage occurs when too much force is applied to the produce during loading, transportation, storage, packing, or mechanical harvesting when the force applied to the fruit causes tissue failure. Compression damage has been investigated on fruits such as pineapples, apples, and strawberries (Sugino et al., 2021; Stropek & Gołacki, 2022; Liu et al., 2023). Static or quasi-static forces cause compression damage when exerted over an extended period, generally

Table 15.1 Effect of packaging types on impact bruising damage of fresh produce

Produce	Type of packaging	Key findings	References
Persimmon	Foam net material	Bruise volume of fruit with foam net was more significant than unpacked fruit	Yeşiloğlu Cevher (2022)
		Linear relationship between impact energy and bruise volume	
Apple	Corrugated paperboard cartons with two internal packagings:	EPS trays decreased the bruise susceptibility compared to the MF trays	Dunno et al. (2021)
	Moulded fibre (MF)	EPS trays experienced lower impact acceleration	
	Expanded polystyrene (EPS) trays	EPS trays better reduced the severity and level of bruising	
		More fruit damage was observed in the middle layers of the package	
Sapota	Gunny bag	Impact damage was not observed on fruit inside foldable plastic container	Antala et al. (2021)
	Gunny bag lined with bubble sheet	Fruit packed inside perforated polypropylene bags had the highest impact damage, followed by those inside gunny bags	
	Perforated polypropylene (PP) bag	Fruits on the bottom layer had more impact damage compared to those on the top layer	
	Foldable plastic container		
	Egg tray in corrugated paperboard carton		
	Plastic crate		
	Corrugated paperboard carton		
Tomato	Plastic crate	Fruits packed inside carton crates had the lowest bruising due to impact compared to the other packaging types	Onu et al. (2021)
	Wooden crate		
	Carton crate		
Kiwi	Wooden box	Fruits packed in EPS box had lower bruise index and bruise area	Xia et al. (2020)
	High-density polyethylene (HDPE) box		
	Expanded polystyrene (EPS) box		
Apple	Corrugated paperboard cartons with two internal packagings:	Package type affected the susceptibility and incidence of impact bruising damage	Fadiji et al. (2016a)
	Fruit on tray layers	Fruit in retail polyethylene plastic bags had more damage than this on trays	
	Fruit in retail polyethylene plastic bags		

occurring in packing lines or during storage when the package is not strong enough to support the stacking loads (Lu & Lu, 2017). Nonetheless, the probability of exposure of the fruits to impact and vibration forces are more than those of compression force in packaging lines (Aliasgarian et al., 2013). Fruit response to compression force can be influenced by factors such as size, cultivar, and maturity stage. In another study by Pathare and Opara (2018), the combination of package dimension and packaging materials has been demonstrated to alter the package's resistance to compression force. Compression damage to fresh produce, mainly fruit, usually materializes as bruising, resulting in both quantitative and qualitative (physicochemical modifications of fruit attributes) losses (Spricigo et al., 2021). Several researchers have studied the mechanical damage of fresh produce due to compression force (Ashtiani et al., 2019; Pascua et al., 2020; Li et al., 2021; Spricigo et al., 2021; Hou et al., 2021; An et al., 2022; Polychroniadou et al., 2022; Lang et al., 2022; Liu et al., 2023).

When packages (cartons) are stacked on top of each other, the bottom package must have sufficient and adequate compression strength to prevent the stacked packages from collapsing, causing damage to the packed produce (Fadiji et al., 2018a). Hence, it is proposed that cartons not be stacked over their design specifications (Kader, 2002). Appropriate package design techniques and packaging protect packages and produce against compression forces, and using robust packages capable of withstanding multiple stacking can assist in reducing this danger, minimizing the occurrence and extent of compression damage on fresh produce. A comprehensive understanding of the interaction between the package and produce under static (compression) loading circumstances is essential in minimizing mechanical damage and assuring the delivery of high-quality fresh produce to consumers (Opara & Fadiji, 2018). Babarinsa and Ige (2012) assessed the energy absorption capacity of tomatoes packed inside two packaging types (plastic crate and raffia woven basket) subjected to compression force. Plastic crates had higher energy than the basket at bioyield, break, and peak points. It has been revealed that fruit damage is related to the amount of energy available for bruising and the properties of the specific fruit (Hussein et al., 2019a, 2020c). Absorbed or dissipated energy in the fruit tissue can cause tissue failure, often classified as cracking, slip, or bruising.

In another report by Opara and Fadiji (2018), the authors studied the susceptibility and resistance of apple fruit inside ventilated corrugated paperboard cartons to compression damage. Two types of regular slotted cartons were utilized, MK6 and MK4, mainly distinguished by their length–to–height ratio and internal packaging. The MK4 package has a higher length-to-height ratio of 1.86 and longer trays than the MK6 package, which has a length-to-height ratio of 1.45 and shorter trays. Furthermore, the authors evaluated the spatial variability and severity of compression bruise incidence inside the cartons. The study showed that packaging types significantly affected the susceptibility to bruising damage. Fruit bruise damage inside the package with the lower length-to-height was more than the fruit bruise damage inside the package with the higher length-to-height. Additionally, fruit positioned near the bottom of the packaging suffered less damage, regardless of the type of packaging.

Similarly, Onu et al. (2021) evaluated plastic, wooden, and carton crates for possible adoption in safely handling and transporting fresh tomato fruits. The packages were subjected to a static condition for 4 weeks (28 days), and the extent of damage/deterioration on the tomato fruits in each packaging type leading to losses was determined every week (7 days). The study showed that the plastic crate had the minimum loss of 78.33%, followed by the wooden crates (83.33%), while the carton crate had the maximum loss of 90.48% at the end of week 4. In addition, the tomatoes packed inside wooden crates lost moisture at a higher rate than those in cartons and plastic crates.

15.3.1.3 Vibration/Abrasion Damage

Vibration/abrasion has been reported to typically occur during movement when the fresh produce is conveyed at an excessive speed or when the produce surface is rubbed away by friction against the surface of the equipment or packaging (Li & Thomas, 2014). When packages and packed produce are transported by truck, rail, or aircraft, they are often exposed to vibration, which is affected by several factors, including road roughness, travelling speed, number and load of axles, and truck suspension (Idah et al., 2012; Fadiji et al., 2016b; Fernando et al., 2018, 2019; Al-Dairi et al., 2022). Consequently, vibration/abrasion damage to the produce occurs, primarily when the acceleration is greater than the acceleration of gravity (9.8 m s^{-2}) because packed produce does not move in relation to the packages or the nearest produce when the acceleration is less than this level. Although some studies focused on measuring the vibration without correlation to produce damage, most studies assessed the damage incurred by the produce due to the vibration experienced. At the same time, a few studies investigated the influence of vibration on produce quality. For instance, Fernando et al. (2020) assessed the protective performance of two types of corrugated paperboard packaging, reusable plastic crates (RPC) and vacuum tightening, in minimizing banana damage during simulated transport vibration. The findings demonstrate that the vibration transmissibility of the column, the fruits' freedom of movement, the packaging's dampening qualities, and the packaging's construction material all impacted the mechanical damage to bananas. Additionally, the corrugated cartons performed the best in reducing the damage levels in bananas caused by in-transit vibration.

A study on cantaloupe showed that using paper pulp tray packing reduced damage during vibration testing (Azam et al., 2022). Damage due to cut forces occurs when fresh produce is rotated or pushed into the sharp border of the backing units. Bananas "dwarf-prata" subjected to cutting forces exhibited increased weight loss, electrolyte leakage, and PPO average activity, accelerated peel colour discolouration when to the undamaged fruits (Maia et al., 2014). Pathare and Al-Dairi (2021c) studied the influence of a 2 h simulated transport vibration and storage temperature conditions (10 and 22 °C) on tomato's quality attributes such as weight loss, colour parameters, firmness, total soluble solids (TSS), and headspace gases. The study highlighted that vibration occurrence and increasing storage temperature cause various changes in the quality attributes of tomatoes. At 22 °C storage conditions, a reduction in TSS, yellowness, weight, and lightness in the tomatoes was reported compared to when stored at 10 °C after vibration. Hence, suitable transportation and storage technologies are essential to reduce quality variations and degradation of tomatoes in the harvest-consumption system. Table 15.2 provides further examples of vibration studies and their effects on produce damage.

15.3.2 Occurrence of Enzymatic Browning in Bruised Fresh Produce

It has been hypothesized that the process of browning in bruised fresh produce involves three steps, which include cell damage, enzymatic oxidation, and browning.

15.3.2.1 Cell Damage

The external forces applied to fresh produce damage cells by breaking the cells' structures and causing membrane failure. This is characterized by the loss of cell wall integrity and decreased peel resistance to mechanical damage (Fig. 15.4). One crucial role of the cell membranes is keeping the liquid constituents separate; however, they cannot perform this function once damaged. Due to the varied chemical composition among fresh produce and cultivars, the extent of cell damage also differs (Li & Thomas, 2014; Hussein et al., 2019b). The size, shape, maturity level, and drop height have been implicated in cell damage severity. Using scanning electron microscopy, Mitsuhashi-Gonzalez et al. (2010) were able to observe burst, crushed dead and live cells from bruised apple tissues. The authors also observed that the cell damage severity increased with an increase in the maturity of the apples. Using pomegranate fruit, Hussein et al. (2019b) reported that the impact of cell damage varies with drop height.

Comparing the size of tomatoes on the impact of cell damage revealed that the cell structure of larger tomatoes (17–19 cm, circumference) showed severe shrinkage, deformation, and rupture, and the original structure of the cells was destroyed as compared to the smaller ones (13–15 cm, circumference) (Sun et al., 2021).

15.3.2.2 Enzymatic Oxidation and Other Biochemical Changes

The damage of cells facilitates the direct contact of phenolic compounds with polyphenol oxidase (PPO) and peroxidase (POD), resulting in oxidative damage in the presence of oxygen. During the process, the PPO converts the phenolic compounds to *o*-quinones that polymerize to form browning pigments on damaged

Table 15.2 Some examples of vibration studies and their effect on produce damage

Produce	Research objective	Package types	Key findings	References
Banana	To determine the vibration characteristics and mechanical damage levels in packed bananas as a function of the stacked position and stack height during long-distance transit	Paper cartons placed in pallets	Damage levels in packed bananas were influenced by the package height and the stacked position of the pallet	Fernando et al. (2019)
			Top-tier banana packages in each pallet position had the greatest damage, followed by the bottom-tier packages, while the middle-tier packages had the least damage	
Apple	To investigate the effect of vibration parameters (frequency and acceleration) and fruit properties (mass, curvature radius, and acoustical stiffness) on the bruising susceptibility of apple fruit	Not applicable	Increase in vibration acceleration increased bruise damage	Rostampour and Motlagh (2018)
			Increased vibration frequency decreased fruit bruising	
			Fruits with high acoustic stiffness and low curvature radius were more vulnerable to bruise damage	
Apple	To simulate the transport damage on paperboard cartons and evaluate the susceptibility of packed apples inside the packages to vibration bruise damage	Two types of ventilated paperboard cartons, both with internal packaging consisting of four layers of trays	The package type affected the bruise damage incurred by the fruit	Fadiji et al. (2016b)
			Fruit at the top trays had more damage compared to those at the bottom plate in both package types	
			No visible damage on the internal packaging (trays) in both package types	
Persimmon	To study the effects of different simulated vibration frequencies (5, 10, and 15 Hz) on the postharvest quality of persimmons	Persimmons were placed stem-side in polystyrene vials in a single layer, and then fruit vials were placed in plastic crates	Persimmon fruit showed a higher susceptibility to transport vibration	Mir and Shahbazi (2022)
			There was a correlation between mechanical damage on the fruit due to vibration force with faster quality degradation during shelf life storage	

Peach	To assess the protective performance of three cushioning materials and their packaging formats against mechanical damage to ripe peaches under simulated transportation	Polyurethane (PU) + corrugated fibreboard boxes (CFB)	Fruit packed with PU + CFB were less susceptible to vibration damage compared to those packed with EPE + CFB and CFB	Lin et al. (2020)
		Expandable polyethylene (EPE) + CFB		
		CFB		
Mushroom	To evaluate the effects of simulated transport vibration on the postharvest quality of mushrooms	Foam cushion material placed in a five-layer paper carton	Simulated transport vibration affected the deterioration of the quality of mushrooms during storage	Tao et al. (2021)
	To determine the protective performance of foam cushion material used in packaging during simulated transport		Foam cushion helped maintain the quality of the mushrooms	
Avocado	To study the effect of vibration on mechanical damage in avocados	Polyethylene foam(PF) mesh, stored in plastic baskets equally coated with the same material	Energy released by vibration during transportation is one of the primary causes of internal and superficial avocado damage	Ríos-Mesa et al. (2020)
			Fruit protected using PF had higher quality index than unprotected fruit	
Orange	To investigate the susceptibility to mechanical damage of orange fruit during simulated and road transportation	Packaging A—reusable plastic container (RPC)	Packaging C was suitable for transporting orange fruit compared to other packaging types	Zheng et al. (2022)
		Packaging B—cardboard box	Fruit damage rate at the rear position of the transport system (lorry) was significantly higher than those at the front position	
		Packaging C (specifically designed package)—comprised of a folding plastic basket, hard polypropylene (PP) board, hard plastic partition board, expandable polyethylene (EPE), and polypropylene (PP) board	Stack height affected the damage rate of the fruit	

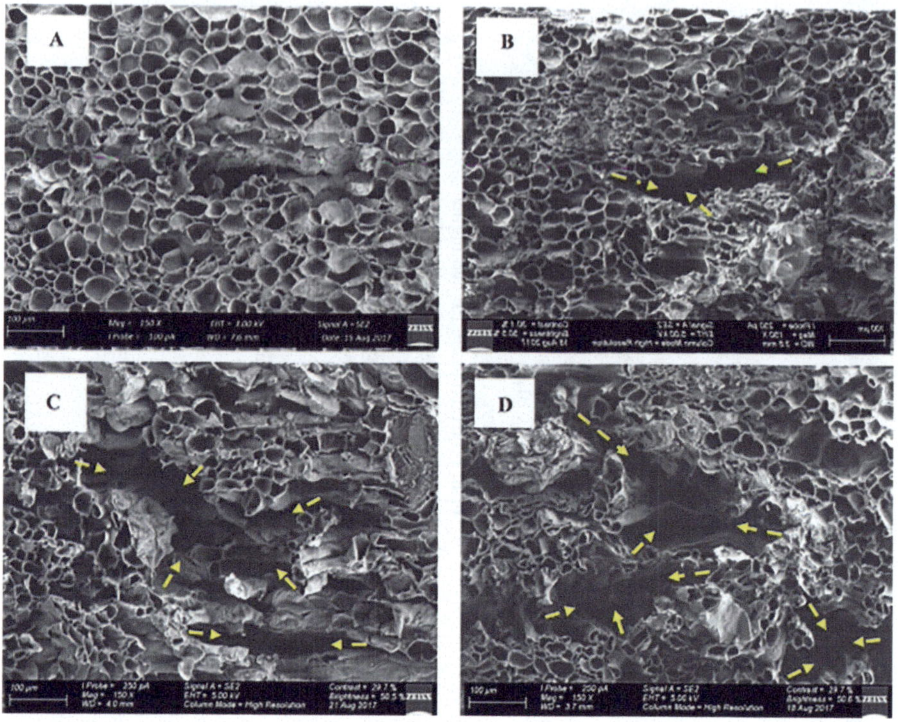

Fig. 15.4 Scanning electron images comparing the impact of bruising between in pomegranate fruit (**a**) unbruised fruit (**b**–**d**) bruised fruit from different drop heights, low (20 cm), medium (40 cm), and high (60 cm), respectively. Arrows show the bruised area. (Source: Hussein et al. (2019b). Reproduced with permission from © 2022 Springer)

tissues of the fresh produce (Hussein et al., 2019b). This process can occur inside the cell following cell membrane damage or outside the ruptured cell following the release of cell contents into the intercellular spaces and extends to the nearby intact cells during storage. Since oxygen is vital in the enzymatic oxidation process, the severity of enzymatic oxidation in fresh produce may be reduced by appropriate storage. Injury to the fresh produce tissues also results in oxidative stress, which facilitates the increased production of ethylene and reaction oxygen species (ROS), such as superoxide anion (O_2^-), hydroxyl radical (OH^-), and hydrogen peroxide (H_2O_2). High ROS results in lipid peroxidation activity in cells, producing radicals that damage lipids, proteins, and deoxyribonucleic acid (DNA). Hussein et al. (2019b) established that bruised pomegranates showed increased PPO activity, peel electrolyte leakage, and ROS production compared to non-bruised pomegranates. Nonetheless, treatment of bruised "Yali" pears with 1-methylcycloprope (1-MCP) lowered H_2O_2, O_2^- production rate, catalase, superoxide dismutase, and ascorbate peroxidase enzymes, when compared with the bruised 1-MCP untreated pears (Li et al., 2016).

15.3.2.3 Browning

Bruising induces enzymatic browning in fresh produce resulting in undesirable colour and flavour quality attributes (Ding & Ling, 2014). Chain reactions of enzymatic browning can occur outside the cells when cell contents are released into the intercellular spaces or within the cell upon rapture of the cell membrane (Mitsuhashi-Gonzalez et al., 2010; Maia et al., 2015). The *o*-quinines formed during the oxidation of phenolic compounds can be further oxidized and polymerized to form brown pigments, impairing the fresh produce's visual quality and resulting in loss of nutrients and flavour. Factors including the concentration and type of polyphenols, storage temperature, type of fresh produce, cultivar (membrane and cell wall integrity), and activity of the PPO have been implicated in the variation of the browning intensity (Hussein et al., 2019b). The variation of bruising-induced browning with cultivar was reported on green olives (Goldental-Cohen et al., 2019). The conversion of phenolic compounds to *o*-quinines may cause the softening of the damaged cells. Many types of phenolic compounds have been associated with the browning of fresh produce and differ with the type of fresh produce. Phenolic compounds, including dopamine and catechol, have been associated with browning in bananas and litchi. In contrast, chlorogenic acid has been implicated in the browning of pear, apple, tomato, and peach fruits (Li & Thomas, 2014). Segovia-Bravo et al. (2009) observed a reduction in phenolic compounds, including oleuropein, hydroxytyrosol, and verbascoside, in bruised olives, while phenolic compounds such as tyrosol glucoside, tyrosol, and hydroxytyrosol glucoside were not significantly affected. The browning process is heterogeneous because the content of polyphenols and PPO activity varies between the exocarp, mesocarp, and endocarp tissues. For instance, in litchi, the phenolic compounds, and PPO activity is greater in external tissues, which experience higher levels of browning, while in fruits including pear, tomatoes, and longan, a higher concentration of polyphenols, greater PPO activity is found in the internal tissues, which are associated with greater intensity of browning (Li & Thomas, 2014). Unlike the PPO activity, the activity of some cell wall enzymes, such as pectin methylesterase and polygalacturonase, have been reported to be unaffected by bruising over a short period. A summary of the effect of bruising on fresh produce is shown in Fig. 15.5.

15.4 How Does Bruise Damage Affect the Quality of Fresh Produce?

Generally, bruising is the most common form of mechanical damage known for downgrading the quality of any fresh produce, causing high economic losses (Manetto et al., 2017). One of the most common ways of bruising downgrade quality is by leaving visible undesirable marks on the surface of the fruit in the form of discolourations and injuries. In bananas, bruising is characterized by brownish and

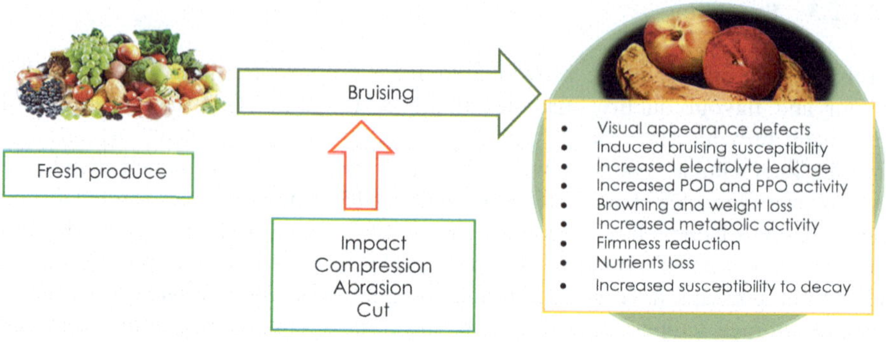

Fig. 15.5 A summary of the effects of bruising on fresh produce

dark-coloured marks in the epidermis rendering the fruit unsalable (Bugaud et al., 2014). Aside from the occurrence of enzymatic browning discussed in the previous section, other ways bruise damage affects the quality of fresh produce are highlighted below.

15.4.1 Colour

Bruise damage is a key contributing factor to fresh fruit discolouration during storage (Bodner & Scampicchio, 2020). Bruise damage significantly influenced skin colour parameters in pears (Pathare & Al-Dairi, 2021a). For example, the authors report a significant increase in the total colour difference of fruit with respect to drop height, storage temperature, and duration, being higher (21.2) in fruit impacted at a drop height of 60 cm and stored at 22 °C and lowest (17.1) in fruit impacted at 20 cm and stored at 10 °C. Similar results are reported in tomatoes, where the degree of bruise damage greatly influences the total colour changes taking place (Pathare & Al-Dairi, 2021b). In climacteric fruit such as tomatoes and pears, such an effect on overall colour change could be attributed to the acceleration of the ripening process by bruise damage (Al-Dairi & Pathare, 2021). Additionally, Mechanical damages were responsible for the higher enzyme activity and changes in colour and flavour observed in the wounded areas of banana fruit (Maia et al., 2015).

15.4.2 Moisture Loss

As observed in bananas and pomegranates, bruising increases moisture loss in fresh fruit (Hussein et al., 2020c; Maia et al., 2014; Pathare & Al-Dairi, 2022). The occurrence can be attributed to the injury and disruptions of the skin surfaces, exposing softer inter tissues and water pathways for accelerated evaporation of moisture to the surrounding atmosphere, as well as increased tissue permeability (Adi et al., 2019; Lufu et al., 2020). Bruising also removes the natural surface waxes from the fruit's skin, thus minimizing its protection against excessive moisture loss while increasing tissue permeability to water. In addition, the external and internal injury of tissues and disruptions of cells may cause physiological stress to the fruit, thus increasing the respiration rate, which contributes to moisture and mass loss of the fruit (Lufu et al., 2019). Bruising significantly influences the physiological processes of respiration and moisture loss across the damaged skin (Kumar et al., 2016).

15.4.3 Decay and Rot

Bruised fruit and vegetables are more susceptible to decay and rot than intact and undamaged produce during cold and shelf storage. Intact fruit peels offer barrier protection against microbial contamination. Bruising breaks this barrier and increases the susceptibility to microbial contamination across the wounded tissues leading to the decay and rotting of the rest of the produce (Pholpho et al., 2011). Therefore, surface bruising accelerates senescence and microbial spoilage (Li et al., 2012). The authors reported a higher decay incidence of 42.8% in bruised pears than in unbruised fruit at the end of 60 days of storage. Quite similarly, Hussein et al. (2020c) observed a 30% incidence and internal decay in high-impact bruised pomegranate fruit. The postharvest grey mould observed in pomegranate fruit originates from wounds and latent infections (Palou et al., 2013).

15.4.4 Respiration Rate, Ethylene Production, and Ripening

Though bruise damage may not be immediately visible in some fruit and vegetables, the symptoms of internal bruising could develop into severe external defects and can be accompanied by serious quality deterioration (Hussein et al., 2020a; Lee et al., 2005). The disruptions and injury of fruit skin caused by bruise damage affect physiological processes such as respiration (Kumar et al., 2016). Wounding of tissues triggers physical and metabolic reactions in the injured regions. This triggers ethylene production, resulting in accelerated ripening, senescence, and an overall deterioration in the shelf life of fruit (Ferreira et al., 2009; Wills & Kim, 1995). Bruise damage is shown to promote and expedite ripening and increase the rate of conversion of starch into total soluble sugars in fruit pulp (Maia et al., 2011), which consequently reduces the shelf life of fresh produce. Bruising increased ethylene production in bananas (Pathare & Al-Dairi, 2022).

15.5 Evaluating Damages to Packaging and Packed Fresh Produce

Mechanical damage, such as bruising, is of major concern in influencing postharvest losses and waste along the value chain of fresh produce, which could lead to poor consumer acceptance and marketability (Lin et al., 2020). Packaging is the most fundamental line of protection from bruising by providing cushioning against various forces such as impact, vibration, and compression experienced during packaging, handling, distribution, and transportation. The susceptibility of fresh produce to quality loss is primarily determined by mechanical contact and damage (Li et al., 2017). Often, the severity, incidence, and susceptibility of packaging and packed fresh produce to mechanical damage can be evaluated by examining the three most influentially contributing forces (impact, vibration, and compression) of action experienced along the value chain, coupled with methods of quantifying bruising (Fadiji et al., 2016b; Fernando et al., 2019). Some studies have investigated the protection of packaging for fresh produce against mechanical damage when subjected to the three most influential forces, as discussed in Sects. 15.3.1.1–15.3.1.3.

To assess the bruise damage on fresh produce, the most common and primary method is direct manual evaluations of bruising. This method uses empirical formulae based on a simple geometric assumption to estimate bruising, measured as bruise area, bruise volume, bruise depth, bruise diameter, bruise proportion as well as bruise index (Opara & Pathare, 2014; Li & Thomas, 2014; Lin et al., 2020). Furthermore, the recent advances in imaging techniques have fostered direct evaluation and detection of bruising damage in fresh produce. The imaging techniques include spectral techniques (biospeckle imaging, hyperspectral/multispectral imaging, fluorescence imaging, and structured-illumination reflectance imaging), nuclear magnetic techniques (magnetic resonance imaging and X-ray imaging), and thermal imaging (Du et al., 2020; He et al., 2021; Mahanti et al., 2022; Rashvand et al., 2022). While techniques such as the biospeckle, hyperspectral imaging, and thermal imaging can detect surface damages, they cannot identify internal ones. However, the X-ray imaging technique has great potential to detect internal damage to the fruit (Mahanti et al., 2022). The drawbacks of traditional approaches, such as their fallibility, tediousness, destructiveness, and time consumption, can be eliminated by these imaging techniques, which can detect bruised tissue and bruise volume in real-time (Du et al., 2020).

Additionally, with the advancement in computer simulation and modelling, finite element modelling (FEM) has been used to simulate the occurrence of mechanical damage in horticultural products such as fruit and vegetables (Zulkifli et al., 2020; Rashvand et al., 2022). FEM simulations can reduce experimentation time and provide important information that cannot be measured experimentally. Ashtiani et al. (2019) used FEM to simulate the quasi-static behaviour of grapefruit under compression loading and showed that the flesh tissue exhibited mechanical damage before the skin tissue. In addition, the internal texture reached yield points at a longitudinal displacement of 9.55 mm and a transverse displacement of 13.7 mm. The damage of kiwifruit under drop was successfully predicted using FEM by Du et al. (2019) and showed an increase in bruise susceptibility with ripening time. However, varying drop heights resulted in similar degrees of bruise susceptibility. Furthermore, the maximum errors obtained when simulation results were compared with experimental results were 17.1% and 18.3% for bruise volume and bruise susceptibility, respectively. An et al. (2022) analysed the textural mechanics of strawberry tissue and developed a finite element (FE) model to investigate the collision-damage sensitivity of the fruit. The authors reported a strong linear correlation between internal and surface damage in the strawberry fruit. A recent study by Liu et al. (2023) successfully predicted the internal mechanical damage of pineapple under compression using FEM. The study showed that the pulp tissue displayed mechanical damage before the peel and core tissues under the application of force. In addition, there was no damage to the pulp, peel, and core when the compression level was ≤5% (5.5 mm; 86 N); however, above this point, the fruit was damaged.

A good package should be able to protect the fruit and limit the amount of damage it sustains. As a vital and critical component of the distribution system, packaging requires acceptable damage at the lowest possible cost. A subjective pass/fail determination can be done in assessing package damage after performing laboratory experiments such as impact, compression, and vibration tests. For instance, the study by Fadiji et al. (2016b) used this method to evaluate the damage on packages (cartons) subjected to vibration force. A carton passed if it had no major gaps or tears, retained all products, and could still be manually handled. Conversely, a package was deemed to have failed if it had major holes or gaps, the contents spilled out or were exposed, or if the package could no longer be manually handled. In the study, no visible damage was observed externally on the carton; however, the internal packaging (trays) incurred damage in the form of cracks (Fig. 15.6).

Similar to detecting bruises in fruit, FEM can successfully predict different mechanical behaviours of packages such as corrugated paperboard cartons (Fadiji et al., 2018b, c). These include compression, buckling, collapse, stability, impact, drop, and the effects of complex designs on the performance of the carton (Ambaw et al., 2021). Fadiji et al. (2019) utilized FEM to investigate the role of geometrical designs of ventilated cartons on mechanical strength when under compression loading. The authors reported that the ventilation area and paperboard grades significantly affected the mechanical integrity of the cartons. Another study by Gong et al. (2020) used FEM to analyse the deformation and compressive capacity of the carton under the indentation condition of different shapes. Typically, FEM can be employed to investigate the packaging performances in the cushioning and damping of impact, compression, and vibration forces (Ambaw et al., 2021).

Fig. 15.6 Damage to internal packaging (trays) after a vibration test

15.6 Conclusions

Fresh produce, including fruit and vegetables, are vital food categories that assist in combating malnutrition diseases. However, they are prone to mechanical damage (bruising) during postharvest operations such as handling, transportation, storage, processing, packaging, and marketing. The presence of bruising reduces the quality of fresh produce, thereby reducing consumer acceptance and leading to postharvest and consequent economic losses. Packaging is one of the most effective methods for preserving the quality of fresh produce. It guarantees the safe and convenient delivery of fresh produce from the grower to the consumers at the lowest possible cost. In this chapter, we discussed the impact of packaging on the bruising of fresh produce, with an emphasis on fruit and vegetables. Packaging types, such as wood, bags, crates, hampers, baskets, corrugated paperboard cartons, plastics, bulk bins, and palletized containers, are available for handling, storing, transporting, and marketing fresh produce. However, corrugated paperboard cartons, combined with internal packaging (trays, liners, plastic bags, clamshells, etc.), are commonly used for fresh produce. As a result of the postharvest operations when handling fresh produce, the packaging and packed produce are subjected to different mechanical loadings, including impact, compression, and vibration/abrasion loadings. These loadings, singly and/or combined, could lead to mechanical injury of the produce exhibited as bruise damage. The occurrence of bruise damage in fresh produce results in cell damage, enzymatic oxidation, biochemical changes, and browning. The resultant physiological and biochemical processes could also cause colour changes, moisture loss, decay, increased respiration rate and ethylene production, and accelerated ripening. A quantitative and objective description of the degree of damage to packaging and packed produce is necessary for producers and consumers to assess their quality and for research into better handling operations. To this end, various methods have been used to evaluate damages to packaging and packed fresh produce. These include direct manual evaluations, imaging techniques, mathematical and numerical modelling, and simulation.

Acknowledgement This work is based on the research supported in part by the National Research Foundation of South Africa (Grant Numbers: 64813). The opinions, findings and conclusions or recommendations expressed are those of the author(s) alone, and the NRF accepts no liability whatsoever in this regard.

References

Adi, D. D., Oduro, I. N., & Tortoe, C. (2019). Physicochemical changes in plantain during normal storage ripening. *Scientific African, 6*, e00164.

Al-Dairi, M., & Pathare, P. B. (2021). Kinetic modeling of quality changes of tomato during storage. *Agricultural Engineering International: CIGR Journal, 23*(1), 183–193.

Al-Dairi, M., Pathare, P. B., Al-Yahyai, R., & Opara, U. L. (2022). Mechanical damage of fresh produce in postharvest transportation: Current status and future prospects. *Trends in Food Science and Technology, 124*, 195–207.

Aliasgarian, S., Ghassemzadeh, H. R., Moghaddam, M., & Ghaffari, H. (2013). Mechanical damage of strawberry during harvest and postharvest operations. *World Applied Sciences Journal, 22*(7), 969–974.

Alissa, E. M., & Ferns, G. A. (2017). Dietary fruits and vegetables and cardiovascular diseases risk. *Critical Reviews in Food Science and Nutrition, 57*(9), 1950–1962.

Ambaw, A., Fadiji, T., & Opara, U. L. (2021). Thermo-mechanical analysis in the fresh fruit cold chain: A review on recent advances. *Foods, 10*(6), 1357.

An, X., Liu, H., Fadiji, T., Li, Z., & Dimitrovski, D. (2022). Prediction of the temperature sensitivity of strawberry drop damage using dynamic finite element method. *Postharvest Biology and Technology, 190*, 111939.

Antala, D. K., Satasiya, R. M., & Chauhan, P. M. (2021). Design, development and performance evaluation of transportation container for sapota fruit. *Journal of Food Science and Technology, 58*(10), 4024–4033.

Ashtiani, S. H. M., Sadrnia, H., Mohammadinezhad, H., Aghkhani, M. H., Khojastehpour, M., & Abbaspour-Fard, M. H. (2019). FEM-based simulation of the mechanical behavior of grapefruit under compressive loading. *Scientia Horticulturae, 245*, 39–46.

Azam, M. M., Saad, A., & Amer, B. M. (2022). Assessment of the quality losses of cantaloupe fruit during transportation. *Processes, 10*(6), 1187.

Babarinsa, F. A., & Ige, M. T. (2012). Energy absorption capacity of packaged Roma tomatoes under compressive loading. *International Journal of Scientific and Engineering Research, 3*(10), 17–23.

Barchi, G. L., Berardinelli, A., Guarnieri, A., Ragni, L., & Fila, C. T. (2002). PH—Postharvest technology: Damage to loquats by vibration-simulating intra-state transport. *Biosystems Engineering, 82*(3), 305–312.

Belay, Z. A., Caleb, O. J., & Opara, U. L. (2019). Influence of initial gas modification on physicochemical quality attributes and molecular changes in fresh and fresh-cut fruit during modified atmosphere packaging. *Food Packaging and Shelf Life, 21*, 100359.

Bodner, M., & Scampicchio, M. (2020). Does bruising influence the volatile profile of pears? *Nutrition & Food Science, 51*, 643–652.

Bugaud, C., Ocrisse, G., Salmon, F., & Rinaldo, D. (2014). Bruise susceptibility of banana peel in relation to genotype and post-climacteric storage conditions. *Postharvest Biology and Technology, 87*, 113–119.

Celik, H. K., Ustun, H., Erkan, M., Rennie, A. E., & Akinci, I. (2021). Effects of bruising of 'Pink Lady' apple under impact loading in drop test on firmness, colour and gas exchange of fruit during long term storage. *Postharvest Biology and Technology, 179*, 111561.

Chang, S. K., Alasalvar, C., & Shahidi, F. (2016). Review of dried fruits: Phytochemicals, antioxidant efficacies, and health benefits. *Journal of Functional Foods, 21*, 113–132.

Chonhenchob, V., & Singh, S. P. (2003). A comparison of corrugated boxes and reusable plastic containers for mango distribution. *Packaging Technology and Science, 16*(6), 231–237.

Chonhenchob, V., Sittipod, S., Swasdee, D., Rachtanapun, P., Singh, S. P., & Singh, J. A. (2009). Effect of truck vibration during transport on damage to fresh produce shipments in Thailand. *Journal of Applied Packaging Research, 3*(1), 27–38.

Dagdelen, C., & Aday, M. S. (2021). The effect of simulated vibration frequency on the physico-mechanical and physicochemical properties of peach during transportation. *LWT, 137*, 110497.

Defraeye, T., Cronjé, P., Berry, T., Opara, U. L., East, A., Hertog, M., Verboven, P., & Nicolai, B. (2015). Towards integrated performance evaluation of future packaging for fresh produce in the cold chain. *Trends in Food Science & Technology, 44*(2), 201–225.

Ding, P., & Ling, Y. S. (2014). Browning assessment methods and polyphenol oxidase in UV-C irradiated Berangan banana fruit. *International Food Research Journal, 21*(4), 1667–1674.

Du, D., Wang, B., Wang, J., Yao, F., & Hong, X. (2019). Prediction of bruise susceptibility of harvested kiwifruit (Actinidia chinensis) using finite element method. *Postharvest Biology and Technology, 152*, 36–44.

Du, Z., Zeng, X., Li, X., Ding, X., Cao, J., & Jiang, W. (2020). Recent advances in imaging techniques for bruise detection in fruits and vegetables. *Trends in Food Science & Technology, 99*, 133–141.

Dunno, K., Stoeckley, I., & Hofmeister, M. (2021). Susceptibility of impact damage to whole apples packaged inside molded fiber and expanded polystyrene trays. *Foods, 10*(9), 1980.

Fadiji, T., Coetzee, C., Pathare, P., & Opara, U. L. (2016a). Susceptibility to impact damage of apples inside ventilated corrugated paperboard packages: Effects of package design. *Postharvest Biology and Technology, 111*, 286–296.

Fadiji, T., Coetzee, C., Chen, L., Chukwu, O., & Opara, U. L. (2016b). Susceptibility of apples to bruising inside ventilated corrugated paperboard packages during simulated transport damage. *Postharvest Biology and Technology, 118*, 111–119.

Fadiji, T., Berry, T. M., Coetzee, C. J., & Opara, U. L. (2018a). Mechanical design and performance testing of corrugated paperboard packaging for the postharvest handling of horticultural produce. *Biosystems Engineering, 171*, 220–244.

Fadiji, T., Coetzee, C. J., Berry, T. M., Ambaw, A., & Opara, U. L. (2018b). The efficacy of finite element analysis (FEA) as a design tool for food packaging: A review. *Biosystems Engineering, 174*, 20–40.

Fadiji, T., Ambaw, A., Coetzee, C. J., Berry, T. M., & Opara, U. L. (2018c). Application of finite element analysis to predict the mechanical strength of ventilated corrugated paperboard packaging for handling fresh produce. *Biosystems Engineering, 174*, 260–281.

Fadiji, T., Coetzee, C. J., Berry, T. M., & Opara, U. L. (2019). Investigating the role of geometrical configurations of ventilated fresh produce packaging to improve the mechanical strength—Experimental and numerical approaches. *Food Packaging and Shelf Life, 20*, 100312.

Fernando, I., Fei, J., Stanley, R., & Enshaei, H. (2018). Measurement and evaluation of the effect of vibration on fruits in transit. *Packaging Technology and Science, 31*(11), 723–738.

Fernando, I., Fei, J., & Stanley, R. (2019). Measurement and analysis of vibration and mechanical damage to bananas during long-distance interstate transport by multi-trailer road trains. *Postharvest Biology and Technology, 158*, 110977.

Fernando, I., Fei, J., Stanley, R., & Rouillard, V. (2020). Evaluating packaging performance for bananas under simulated vibration. *Food Packaging and Shelf Life, 23*, 100428.

Ferreira, M. D., Sargent, S. A., Brecht, J. K., & Chandler, C. K. (2009). Strawberry bruising sensitivity depends on the type of force applied, cooling method, and pulp temperature. *HortScience, 44*(7), 1953–1956.

Gołacki, K., Rowiński, P., & Stropek, Z. (2009). The determination of apples bruise resistance by the multiple impact method. *Technical Sciences, 12*, 29–39.

Goldental-Cohen, S., Biton, I., Many, Y., Ben-Sason, S., Zemach, H., Avidan, B., & Ben-Ari, G. (2019). Green olive browning differ between cultivars. *Frontiers in Plant Science, 10*, 1260.

Gong, G., Liu, Y., Fan, B., & Sun, D. (2020). Deformation and compressive strength of corrugated cartons under different indentation shapes: Experimental and simulation study. *Packaging Technology and Science, 33*(6), 215–226.

He, Y., Xiao, Q., Bai, X., Zhou, L., Liu, F., & Zhang, C. (2021). Recent progress of nondestructive techniques for fruits damage inspection: A review. *Critical Reviews in Food Science and Nutrition, 62*(20), 5476–5494.

Hervert-Hernández, D., García, O. P., Rosado, J. L., & Goñi, I. (2011). The contribution of fruits and vegetables to dietary intake of polyphenols and antioxidant capacity in a Mexican rural diet: Importance of fruit and vegetable variety. *Food Research International, 44*(5), 1182–1189.

Hou, J., Hu, W., Zhang, L., Ren, Z., Sun, Q., & Wang, W. (2021). Mechanical properties of mulberry fruit under compression and nuclear magnetic resonance tests. *Journal of Food Process Engineering, 44*(11), e13856.

Hussein, Z., Fawole, O. A., & Opara, U. L. (2019a). Bruise damage susceptibility of pomegranates (Punica granatum, L.) and impact on fruit physiological response during short term storage. *Scientia Horticulturae, 246*, 664–674.

Hussein, Z., Fawole, O. A., & Opara, U. L. (2019b). Determination of physical, biochemical and microstructural changes in impact-bruise damaged pomegranate fruit. *Journal of Food Measurement and Characterization, 13*(3), 2177–2189.

Hussein, Z., Fawole, O. A., & Opara, U. L. (2020a). Harvest and postharvest factors affecting bruise damage of fresh fruits. *Horticultural Plant Journal, 6*(1), 1–13.

Hussein, Z., Fawole, O. A., & Opara, U. O. (2020b). Bruise damage of pomegranate during long-term cold storage: Susceptibility to bruising and changes in textural properties of fruit. *International Journal of Fruit Science, 20*(2), S211–S230.

Hussein, Z., Fawole, O. A., & Opara, U. O. (2020c). Effects of bruising and storage duration on physiological response and quality attributes of pomegranate fruit. *Scientia Horticulturae, 267*, 109306.

Idah, P. A., Gana, Y. M., Ogbonnaya, C., & Morenikeji, O. O. (2012). Simulated transport damage study on fresh tomato (Lycopersicon esculentum) fruits. *Agricultural Engineering International: CIGR Journal, 14*(2), 119–126.

Jarimopas, B., Singh, S. P., Sayasoonthorn, S., & Singh, J. (2007). Comparison of package cushioning materials to protect post-harvest impact damage to apples. *Packaging Technology and Science, 20*(5), 315–324.

Kader, A. A. (2002). Quality parameters of fresh-cut fruit and vegetable products. In *Fresh-cut fruits and vegetables. Science, technology and market* (pp. 11–20). CRC Press.

Kitthawee, U., Pathaveerat, S., Srirungruang, T., & Slaughter, D. (2011). Mechanical bruising of young coconut. *Biosystems Engineering, 109*(3), 211–219.

Koutsimanis, G., Getter, K., Behe, B., Harte, J., & Almenar, E. (2012). Influences of packaging attributes on consumer purchase decisions for fresh produce. *Appetite, 59*(2), 270–280.

Kumar, V., Purbey, S. K., & Anal, A. K. D. (2016). Losses in litchi at various stages of supply chain and changes in fruit quality parameters. *Crop Protection, 79*, 97–104.

Lang, Y., Xie, Z., & Chen, L. (2022). Finite element simulation of mechanical properties of Rosa roxburghii under compression loading. *Journal of Food Process Engineering, 45*(1), e13939.

Lee, H. J., Kim, T. C., Kim, S. J., & Park, S. J. (2005). Bruising injury of persimmon (Diospyros kaki cv. Fuyu) fruits. *Scientia Horticulturae, 103*(2), 179–185.

Li, Z., & Thomas, C. (2014). Quantitative evaluation of mechanical damage to fresh fruits. *Trends in Food Science & Technology, 35*(2), 138–150.

Li, J., Yan, J., Cao, J., Zhao, Y., & Jiang, W. (2012). Preventing the wound-induced deterioration of Yali pears by chitosan coating treatments. *Food Science and Technology International, 18*(2), 123–128.

Li, J., Yan, J., Ritenour, M. A., Wang, J., Cao, J., & Jiang, W. (2016). Effects of 1-methylcyclopropene on the physiological response of Yali pears to bruise damage. *Scientia Horticulturae, 200*, 137–142.

Li, Z., Miao, F., & Andrews, J. (2017). Mechanical models of compression and impact on fresh fruits. *Comprehensive Reviews in Food Science and Food Safety, 16*(6), 1296–1312.

Li, D., Li, Z., & Tchuenbou-Magaia, F. (2021). An extended finite element model for fracture mechanical response of tomato fruit. *Postharvest Biology and Technology, 174*, 111468.

Lin, M., Chen, J., Chen, F., Zhu, C., Wu, D., Wang, J., & Chen, K. (2020). Effects of cushioning materials and temperature on quality damage of ripe peaches according to the vibration test. *Food Packaging and Shelf Life, 25*, 100518.

Liu, W., Liu, T., Zeng, T., Ma, R., Cheng, Y., Zheng, Y., Qiu, J., & Qi, L. (2023). Prediction of internal mechanical damage in pineapple compression using finite element method based on Hooke's and Hertz's laws. *Scientia Horticulturae, 308*, 111592.

Lu, Y., & Lu, R. (2017). Non-destructive defect detection of apples by spectroscopic and imaging technologies: A review. *Transactions of the ASABE, 60*(5), 1765–1790.

Lu, F., Ishikawa, Y., Kitazawa, H., & Satake, T. (2010). Impact damage to apple fruits in commercial corrugated fiberboard box packaging evaluated by the pressure-sensitive film technique. *Journal of Food, Agriculture & Environment, 8*(2), 218–222.

Lufu, R., Ambaw, A., & Opara, U. L. (2019). The contribution of transpiration and respiration processes in the mass loss of pomegranate fruit (cv. Wonderful). *Postharvest Biology and Technology, 157*, 110982.

Lufu, R., Ambaw, A., & Opara, U. L. (2020). Water loss of fresh fruit: Influencing pre-harvest, harvest and postharvest factors. *Scientia Horticulturae, 272*, 109519.

Mahanti, N. K., Pandiselvam, R., Kothakota, A., Chakraborty, S. K., Kumar, M., & Cozzolino, D. (2022). Emerging non-destructive imaging techniques for fruit damage detection: Image processing and analysis. *Trends in Food Science & Technology, 120*, 418–438.

Maia, V. M., Salomão, L. C. C., Siqueira, D. L., Puschman, R., Mota Filho, V. J. G., & Cecon, P. R. (2011). Physical and metabolic alterations in "Prata Anã" banana induced by mechanical damage at room temperature. *Scientia Agricola, 68*, 31–36.

Maia, V. M., Salomão, L. C. C., Siqueira, D. L., Aspiazú, I., & Maia, L. C. B. (2014). Physical and metabolic changes induced by mechanical damage in 'Dwarf-Prata' banana fruits kept under cold storage. *Australian Journal of Crop Science, 8*(7), 1029–1037.

Maia, L. C. B., Maia, V. M., Salomão, L. C. C., Salles, B. P. A., & Pegoraro, R. F. (2015). Metabolic alterations in banana induced by mechanical damage. *Revista Unimontes Científica, 17*(2), 27–34.

Manetto, G., Cerruto, E., Pascuzzi, S., & Santoro, F. (2017). Improvements in citrus packing lines to reduce the mechanical damage to fruit. *Chemical Engineering Transactions, 58*, 391–396.

Mir, H., & Shahbazi, F. (2022). Simulated transit vibration effects on the postharvest quality of persimmon during storage. *Postharvest Biology and Technology, 189*, 111918.

Mitsuhashi-Gonzalez, K., Pitts, M. J., Fellman, J. K., Curry, E. A., & Clary, C. D. (2010). Bruising profile of fresh apples associated with tissue type and structure. *Applied Engineering in Agriculture, 26*(3), 509–517.

Mukama, M., Ambaw, A., & Opara, U. L. (2020). Advances in design and performance evaluation of fresh fruit ventilated distribution packaging: A review. *Food Packaging and Shelf Life, 24*, 100472.

Okhuoya, J. A. (1995). Controlling post harvest losses in tomatoes and pepper. *Journal of Tropical Post Harvest, 2*, 136–142.

Onu, O. B., Ani, O. A., Asoiro, F. U., Uzoejinwa, B. B., Ugwu, S. N., Abu, N., & Uguru, M. I. (2021). Evaluation of different packaging types for adoption in safe handling and transportation of fresh tomato fruits in Nigeria. *Agricultural Engineering International: CIGR Journal, 23*(4), 309–318.

Onwude, D. I., Iranshahi, K., Martynenko, A., & Defraeye, T. (2021). Electrohydrodynamic drying: Can we scale-up the technology to make dried fruits and vegetables more nutritious and appealing? *Comprehensive Reviews in Food Science and Food Safety, 20*(5), 5283–5313.

Opara, L. U. (2007). Bruise susceptibilities of 'Gala' apples as affected by orchard management practices and harvest date. *Postharvest Biology and Technology, 43*(1), 47–54.

Opara, U. L., & Fadiji, T. (2018). Compression damage susceptibility of apple fruit packed inside ventilated corrugated paperboard package. *Scientia Horticulturae, 227*, 154–161.

Opara, U. L., & Pathare, P. B. (2014). Bruise damage measurement and analysis of fresh horticultural produce—A review. *Postharvest Biology and Technology, 91*, 9–24.

Palou, L., Taberner, V., Guardado, A., Del Río, M. Á., & Montesinos-Herrero, C. (2013). Incidence and etiology of postharvest fungal diseases of pomegranate (Punica granatum cv. Mollar de Elche) in Spain. *Phytopathologia Mediterranea, 52*, 478–489.

Pascall, M. A. (2020). The role and importance of packaging and labeling in assuring food safety, quality & compliance with regulations I: Packaging basics. In *Food safety and quality systems in developing countries* (pp. 261–283). Academic Press.

Pascua, A., Pascua, M. L., & Peralta, E. (2020). Compression damage threshold of young coconut (Cocos nucifera L.) at two maturity stages. *International Journal of Engineering Technologies and Management Research, 5*(1), 8–19.

Pathare, P. B., & Al-Dairi, M. (2021a). Bruise susceptibility and impact on quality parameters of pears during storage. *Frontiers in Sustainable Food Systems, 5*, 1–13.

Pathare, P. B., & Al-Dairi, M. (2021b). Bruise damage and quality changes in impact-bruised, stored tomatoes. *Horticulturae, 7*(5), 113.

Pathare, P. B., & Al-Dairi, M. (2021c). Effect of simulated vibration and storage on quality of tomato. *Horticulturae, 7*(11), 417.

Pathare, P. B., & Al-Dairi, M. (2022). Effect of mechanical damage on the quality characteristics of banana fruits during short-term storage. *Discover Food, 2*(1), 1–13.

Pathare, P. B., & Opara, U. L. (2014). Structural design of corrugated boxes for horticultural produce: A review. *Biosystems Engineering, 125*, 128–140.

Pathare, P. B., & Opara, U. L. (2018). Structural design of horticultural packaging. In *Reference module in food science*. Elsevier. https://doi.org/10.1016/B978-0-08-100596-5.22486-4. ISBN 9780081005965.

Pholpho, T., Pathaveerat, S., & Sirisomboon, P. (2011). Classification of longan fruit bruising using visible spectroscopy. *Journal of Food Engineering, 104*(1), 169–172.

Polychroniadou, C., Michailidis, M., Adamakis, I. D. S., Karagiannis, E., Ganopoulos, I., Tanou, G., Bazakos, C., & Molassiotis, A. (2022). Mechanical stress elicits kiwifruit ripening changes in gene expression and metabolic status. *Postharvest Biology and Technology, 194*, 112102.

Rashvand, M., Altieri, G., Genovese, F., Li, Z., & Di Renzo, G. C. (2022). Numerical simulation as a tool for predicting mechanical damage in fresh fruit. *Postharvest Biology and Technology, 187*, 111875.

Ríos-Mesa, A. F., Zuluaga Gallego, R., Osorio, M., Ciro-Velásquez, H. J., & Márquez Cardozo, C. J. (2020). Effect of vehicle vibration on the mechanical and sensory properties of avocado (Persea Americana Mill. Cv. Hass) during road transportation. *International Journal of Fruit Science, 20*(3), S1904–S1919.

Riva, S. C., Opara, U. O., & Fawole, O. A. (2020). Recent developments on postharvest application of edible coatings on stone fruit: A review. *Scientia Horticulturae, 262*, 109074.

Rostampour, V., & Motlagh, A. M. (2018). Evaluation of the bruising susceptibility of apple in transport conditions. *Bulgarian Journal of Agricultural Science, 24*(5), 902–908.

Ščetar, M., & Kurek, M. (2010). Trends in fruit and vegetable packaging—A review. *Hrvatski časopis za prehrambenu tehnologiju, biotehnologiju i nutricionizam, 5*(3–4), 69–86.

Segovia-Bravo, K. A., Jarén-Galán, M., García-García, P., & Garrido-Fernández, A. (2009). Browning reactions in olives: Mechanism and polyphenols involved. *Food Chemistry, 114*(4), 1380–1385.

Singh, S. P., Burgess, G., & Xu, M. (1992). Bruising of apples in four different packages using simulated truck vibration. *Packaging Technology and Science, 5*(3), 145–150.

Spricigo, P. C., Freitas, T. P., Purgatto, E., Ferreira, M. D., Correa, D. S., Bai, J., & Brecht, J. K. (2021). Visually imperceptible mechanical damage of harvested tomatoes changes ethylene production, color, enzyme activity, and volatile compounds profile. *Postharvest Biology and Technology, 176*, 111503.

Stropek, Z., & Gołacki, K. (2015). A new method for measuring impact related bruises in fruits. *Postharvest Biology and Technology, 110*, 131–139.

Stropek, Z., & Gołacki, K. (2022). Response of apple flesh to compression under the quasi-static and impact loading conditions. *Materials, 15*(21), 7743.

Subedi, G. D., Gautam, D. M., Baral, D. R., & Paudyal, K. P. (2017). Evaluation of packaging materials for transportation of apple fruits in CFB boxes. *International Journal of Horticulture, 7*(8), 54–63.

Sugino, N., Watanabe, T., Nakamura, N., & Kitazawa, H. (2021). Electrical and mechanical analysis to evaluate the cultivar difference in strawberries with respect to their bruising sensitivities and mass loss acceleration. *Postharvest Biology and Technology, 175*, 111489.

Sun, Y., Pessane, I., Pan, L., & Wang, X. (2021). Hyperspectral characteristics of bruised tomatoes as affected by drop height and fruit size. *LWT, 141*, 110863.

Tao, F., Chen, W., & Jia, Z. (2021). Effect of simulated transport vibration on the quality of shiitake mushroom (Lentinus edodes) during storage. *Food Science & Nutrition, 9*(2), 1152–1159.

Vursavuş, K. K., & Özgüven, F. (2004). Determining the effects of vibration parameters and packaging method on mechanical damage in Golden Delicious apples. *Turkish Journal of Agriculture and Forestry, 28*(5), 311–320.

Wills, R. B. H., & Kim, G. H. (1995). Effect of ethylene on postharvest life of strawberries. *Postharvest Biology and Technology, 6*(3–4), 249–255.

Wohner, B., Pauer, E., Heinrich, V., & Tacker, M. (2019). Packaging-related food losses and waste: An overview of drivers and issues. *Sustainability, 11*(1), 264.

Xia, M., Zhao, X., Wei, X., Guan, W., Wei, X., Xu, C., & Mao, L. (2020). Impact of packaging materials on bruise damage in kiwifruit during free drop test. *Acta Physiologiae Plantarum, 42*(7), 1–11.

Yeşiloğlu Cevher, E. (2022). Determination of bruise preventing capacity of the cushioning material in persimmon fruit under pendulum impact test. *Journal of Food Process Engineering, 45*(12), e14162.

Yousefi, S., Farsi, H., & Kheiralipour, K. (2016). Drop test of pear fruit: Experimental measurement and finite element modelling. *Biosystems Engineering, 147*, 17–25.

Zhang, B., Zhou, J., Meng, Y., Zhang, N., Gu, B., Yan, Z., & Idris, S. I. (2018). Comparative study of mechanical damage caused by a two-finger tomato gripper with different robotic grasping patterns for harvesting robots. *Biosystems Engineering, 171*, 245–257.

Zheng, D., Chen, J., Lin, M., Wang, D., Lin, Q., Cao, J., Yang, X., Duan, Y., Ye, X., Sun, C., Di Wu, J. W., & Chen, K. (2022). Packaging design to protect Hongmeiren orange fruit from mechanical damage during simulated and road transportation. *Horticulturae, 8*(3), 258.

Zulkifli, N., Hashim, N., Harith, H. H., & Shukery, M. F. M. (2020). Finite element modelling for fruit stress analysis—A review. *Trends in Food Science & Technology, 97*, 29–37.

Chapter 16
Importance of Bruise Assessment and Control in Fresh Produce Industry

Pankaj B. Pathare and Umezuruike Linus Opara

Abstract Bruising is the most common form of mechanical damage to fresh horticulture produce, lowering the quality for consumers and revenue for the fruit and vegetable industries. Bruising is a key physical defect that contributes to postharvest loss of fresh fruits and vegetables. Bruise detection is critical in fruit grading. A bruise detection system capable of detecting and eliminating damaged items on production lines will significantly enhance the quality of fresh produce for sale, and hence their economy. Bruise assessing methods for damage are investigated in this chapter, including manual evaluation, optical detection, computer vision, hyperspectral imaging, and X-ray computed tomography. Methods for measuring fruit susceptibility to mechanical damage based on the bruise threshold and the quantity of damage per unit of impact energy has been discussed. It is important to standardize the measurement and analysis of bruises so that study results can be compared for preharvest and postharvest treatments to reduce bruising, as well as to facilitate collaboration among laboratories.

Keywords Bruise assessment · Optical detection · Hyperspectral imaging · Postharvest handling · Impact test · Vibration damage

---

P. B. Pathare (✉)
Department of Soils, Water & Agricultural Engineering, College of Agricultural and Marine Sciences, Sultan Qaboos University, Al-Khod, Oman
e-mail: pankaj@squ.edu.om

U. L. Opara
SARChI Postharvest Technology Research Laboratory, Africa Institute for Postharvest Technology, Faculty of AgriSciences, Stellenbosch University, Stellenbosch, South Africa

UNESCO International Centre for Biotechnology, Nsukka, Enugu State, Nigeria

P. B. Pathare, U. L. Opara (eds.), *Mechanical Damage in Fresh Horticultural Produce*, https://doi.org/10.1007/978-981-99-7096-4_16

16.1 Introduction

Mechanical bruising to fruits and vegetables can be significant during postharvest handling. The appearance of bruises during harvest activities and postharvest processing reduces the quality of fruits and vegetables, resulting in significant economic losses. Fresh produce goes through a number of processes from the orchard to the market, including picking, sorting, packing, transportation, refrigeration, grading, and packaging. The produce is inevitably bruised during the postharvest value chain. After only a short period of storage, damaged tissue can easily encourage the growth of rot fungi, hastening the spread of rot to the surrounding undamaged fresh products. In fruits and vegetables, the bruise size increased when falling from specific height, unloading, and filling.

Bruise damage to fruits is common and recognized as a leading cause of fruit degradation and quality reduction. It is caused mostly by excess impacts as well as compressive pressures on concentrated areas of the fruit surface. The bruising reduce fresh product commercial acceptance rates and postharvest losses throughout the horticultural supply chain due to devaluation or outright refusal, increasing waste and the associated negative economic, social, and environmental implications.

In fresh horticultural produce, bruise damage is the most known and familiar mechanical damage. Manual evaluation and optical detection are the most two common methods to evaluate bruise damage, it measures bruise area (BA), bruise volume (BV), diameter and depth of the bruise (Opara & Pathare, 2014; Mitsuhashi-Gonzalez et al., 2010). Manual methods of evaluation are destructive, resulting in fruit and vegetables that are no longer available for storage or trading. Many technologies are being used in this field today, such as light reflection, light transmission, fluorescence, ultrasound, dielectric properties, X-rays, gamma rays, and magnetic resonance imaging (Opara & Pathare, 2014).

Fruit quality is a crucial factor for customer acceptance and market value. Furthermore, in recent years, there has been an increase in global demand for high-quality fruits and vegetables. Customers are reluctant to purchase things with unacceptable wholesomeness. As a result, detecting damage in fruits during preharvest and postharvest is a critical task. Measuring Fruit injury is an essential for the purpose of systematic improvement correction procedures. As a result, there is an urgent need to investigate novel methods for detecting bruising on fresh food in a timely and precise manner. Therefore, this chapter emphasizes the significance of bruise measurement for fresh fruits and vegetables. The chapter also discusses how to assess mechanical damage susceptibility to reduce the incidence of the damage of fresh fruits. The role of nondestructive techniques for bruise measurements also discussed.

16.2 Why Assess Bruise Damage?

Fruit quality is mainly judged by consumers based on visual appearance, size, shape, and slight visible damage resulting in lower quality, price, and financial revenue. Consumers prefer fresh fruits and vegetables of high quality. However, how can farmers, handlers, transporters, distributers, and retailers measure the quality attributes without destroying them. Also, how can researchers analyze the damage of the fruit in order to understand mechanisms and forecast harm caused by mechanical handling? Furthermore, mechanical damage can cause physiological disorders which can result in off-flavor, nutritional value loss, decay, and spoiling (Opara & Pathare, 2014). Damage from mechanical forces can result in bruises, cuts, punctures, splits, and abrasions (Opara & Pathare, 2014); bruises damage epidermal and hypodermal tissues without ruptures, involving living and dead cells (Opara & Pathare, 2014; Mitsuhashi-Gonzalez et al., 2010). A minor injury can result in bacterial and fungal infection, thereby shortening the fresh produce's shelf-life (Van Zeebroeck et al., 2007).

The costs of fruit bruising are enormous, and the annual return to fruit firms and consumers in decreasing bruising might be in the billions of dollars worldwide. Understanding and knowing the effect of bruising is very important and can help in the reduction of fruit bruises. Damage susceptibility studies are an important experimental tool in this field of study. The general goal of any susceptibility test is to determine how fruits respond to various types and intensities of loads. The specific goal is to establish when fruits begin to exhibit mechanical damage symptoms. This is known as the bruise threshold (Ruiz-Altisent & Moreda, 2011).

Bruise resistance is an important feature of fruit that ensures product quality. Bruise resistance has both qualitative and quantitative aspects (Brusewitz & Bartsch, 1989). The ability to test bruise resistance would be extremely useful in assessing harvest readiness, the effect of storage methods, selecting fruit/vegetable varieties for planting, and exploring methods to reduce fruit bruising during harvest and handling. There is no clear, worldwide accepted standard technique for testing bruise resistance at the moment. The degree of bruising is directly connected to the energy absorbed by the fruit, which is of particular importance in evaluating materials handling equipment, package design, and measuring the effects of storage and temperature on bruising. As a result, damage is primarily connected to the stress created within the produce during postharvest handling and processing.

16.3 Measurements of Mechanical Damage

Using different methods to measure bruising or damage in fresh fruit is critical for the efficient improvement of preventative measures. A precise evaluation of mechanical damage in fresh fruit gives immediate evidence for understanding the process of damage occurrence during postharvest handling practices. Once damage is shown, it is vital to quantify the degree of damage thoroughly and objectively to sort and grade the damage in fruit and estimate postharvest losses in damaged fruit, thus reducing economic losses. Nowadays, the recognition and detection of damage in fresh produce relies on direct techniques like manual inspection as well as optical detection (Opara & Pathare, 2014; Li & Thomas, 2014).

There are several quantitative analyses used to describe fruit damage including, bruise area (BA), bruise volume (BV), bruise susceptibility (BS), specific bruise susceptibility (SBS), and bruise score (Opara & Pathare, 2014). The main and direct measurements used to assess the bruise size are depth and diameter of the bruise. The bruise susceptibility is identified by the ratio of bruise volume (BV) to impact energy (Ei) (Opara & Pathare, 2014; Pathare & Al-Dairi, 2021a; Pathare et al., 2023). Other approaches aim to determine fruit as damaged or not and show "probability of damage" based on chosen handling conditions or to correlate fresh fruit mechanical attributes to damage or probability of damage.

The ability of a fruit to resist exterior loads is known by its susceptibility/sensitivity to mechanical damage. Postharvest handling of fresh produce and its susceptibility to mechanical injury is an important subject of study (Opara, 2007; Bugaud et al., 2014). In the postharvest supply chain, farmers, processors, and operators must understand the factors that influence fruit susceptibility. Studies on bruising thresholds are going to provide fresh understandings into the mechanism of bruise damage. Susceptibility of fruit to mechanical damage can be measured to offer important facts for postharvest handling operations in to take preventive and protective actions to reduce the incidence of damage in fruit. To get controlled bruising damage on different fresh fruit and vegetables, impact, drop, or compression tests are commonly used shown in Fig. 16.1.

16.3.1 Impact Test for Bruise Measurement

Bruise damage in fresh fruit can be simulated in the laboratory by performing several methodologies, most of which are aimed to imitate various forms of dynamic loadings that may be encountered during real-time harvesting and handling processes. Several approaches are used in the laboratory to imitate fruit bruising and test the influence on various agricultural products. The drop impact test (Pathare et al., 2021; Hussein et al., 2019) and the pendulum impactor method (Pathare et al., 2023) are the most commonly used methods. To investigate the damage susceptibility of fruit to damage in many fruits, impact experiment which includes dropping a uniform steel-ball with a known mass from different heights into the fruit or by dropping the fruit on a firm surface (Pathare et al., 2021). The impact energy (Ei) was estimated utilizing $Ei = mgh$, where m is the mass of the produce (kg), g is the gravitational constant (9.81 m/s^2), and h is the drop height (m). The fruit was stored at room temperature after impact testing. A drop impact test has been applied to examine bruise damage measurements of numerous fruit like pear (Pathare & Al-Dairi, 2021a), tomato (Pathare & Al-Dairi, 2021b), banana (Pathare &

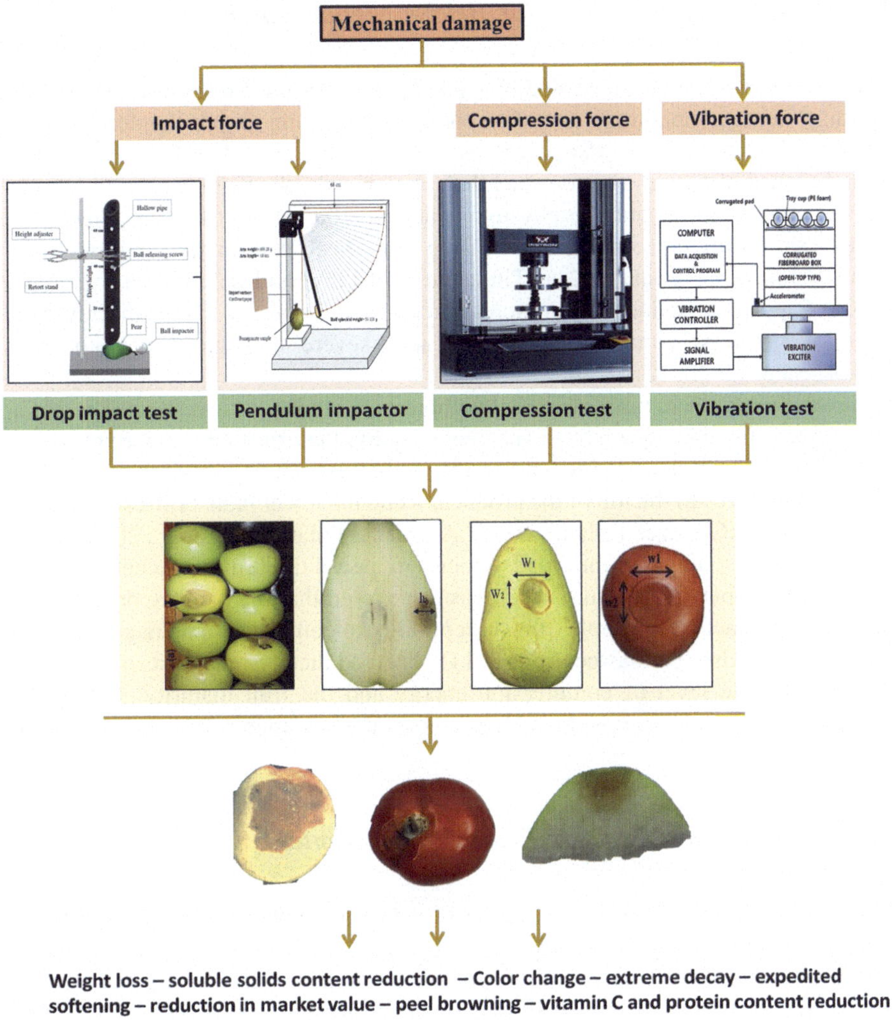

Fig. 16.1 Experimental setup for impact, drop and compression tests

Al-Dairi, 2022), guava (Htike et al., 2023). The other popular method is dropping a pendulum arm with a specific weight (impactor) from a chosen angle into the targeted fruit, that has been used to analyze bruising measurements of various fruits such as apple (Stropek & Golacki, 2018), pear (Azadbakht et al., 2019a), persimmon (Yeşiloğlu Cevher, 2022), nectarine (Polat et al., 2012), and pomegranate (Pathare et al., 2023).

16.3.2 Compression Test for Bruise Measurement

When fresh produce are overfilled or stacked, compression force occurs during the loading, storage, handling, and transportation, resulting in mechanical damage (Opara & Pathare, 2014). Also, compression force can be applied to fruits by other fruits because of extreme carton stack height or bin depth (Kitthawee et al., 2011). The structure of tissue and the physiological properties of the fruits are affected by compression. Fruit detachment and bruising caused cell deformation and loss under compression (Guo et al., 2021). Titratable acid (TA) and soluble solid contents (SSC) of apples are reduced by compression forces (Li et al., 2018). At the advanced stage of ripeness, compression forces cause severe damage to tomato fruits (Babarinsa & Ige, 2012).

Several researchers have also used compression studies to investigate the possibility of bruising damage on fresh vegetables. Individual fruits are squeezed between two flat steel plates at a specific loading rate in this test. The upper steel plate applies compression force to the top of the produce, while the lower plate bruises the bottom of the produce (Opara & Pathare, 2014). Aktas and Polat (Aktas et al., 2008) used a compression test at 1 mm/min load velocity to assess the bruise susceptibility and mechanical properties of tomatoes. It was discovered that the packing design had a substantial impact on the apple fruit susceptibility to compression damage (Opara & Fadiji, 2018). Also, Kitthawee et al. (2011) observed that coconut susceptibility was impacted by the level of compression energy and the fruit maturity, stating that young coconuts were more vulnerable to impact damage than compression damage.

16.3.3 Vibration Test for Bruise Measurement

Mechanical damage caused by vibration is one of the most common quality issues with fresh fruit during transit (Al-Dairi et al., 2022). The fruits susceptibility to vibration damage is a result of various parameters and is often investigated utilizing in-transit, simulation, and transit-simulation experiments (Al-Dairi et al., 2022). The purpose of the in-transit experiment is to explore the relationship between various factors that affect the vibration level during transportation and to directly correlate these factors with the resulting change and damage in fresh produce quality. The in-transit experiment examined the relationship of several factors that influence the level of vibration during transport process and correlate them with the resulting alteration and injury in the quality of fresh produce. In-transit investigations are being conducted to determine the influence of vibration on the quality metrics of fresh fruits and vegetables (Soleimani & Ahmadi, 2014). Simulated transport are considered as laboratory examinations where a known vibration intensity values based on a restricted frequency or standard vibration profile is applied to drive the vibration system or shaker (Fernando et al., 2018). Inside the truck, the vibration intensity is always determined in terms of acceleration (G), power spectral density

(PSD), and root mean square acceleration (Grms) (Lu et al., 2010). A PSD plot can be utilized in most circumstances to deliver transport environment with real-time monitoring, which is a crucial tool for evaluating simulated vibration in laboratories (Singh et al., 2006).

Visual defects in fresh produce caused by mechanical damage throughout the transit process from the farm gate to markets may significantly reduce the visual quality of product's affecting producer and retailer profit (Opara & Pathare, 2014; Chonhenchob et al., 2009). Mechanical damage caused by vibrational stress result in severe morphological and physiological alterations in fresh produce (Mahanti et al., 2021). Vibration generated by vibration resulted in many internal and external changes and damage to fresh horticultural products. Walkowiak-Tomczak et al. (2021) found that the mechanical vibration caused by simulated transport influenced the firmness stated of "Idared" and "Gala" apples after 14 days of storage by 13% and 9%, respectively. The weight loss rate of vibrated tomato fruit (2.88 Hz for 2 h) was greater than that tomatoes with no vibration after 10 days of storage (Pathare & Al-Dairi, 2021c). A more bruise % incidence was observed in vibrated tomato fruit. It was shown that raising the frequency of simulated transport vibration of mushrooms to 16 Hz improved cap development, pH, weight loss %, and electrolyte leakage while lowering firmness, water activity, and whitening index (WI) (Wu & Wang, 2014). The reduction of fresh produce quality during transportation mostly occurs because of physical and biological damages/changes caused by vibration (Wu & Wang, 2014; Al-Dairi et al., 2022). Vibration level measurements and analysis during transit are important for improving fruit quality (Ríos-Mesa et al., 2020).

16.4 Nondestructive Techniques for Bruise Damage Assessment

Bruise identification is generally accomplished subjectively via manual assessment. Utilizing modern sorting and grading applications is performed by personnel expert on the qualitative attributes of the fresh produce. Visual inspection to identify the incidence of bruises is regularly followed by a decision to degrade, accept, or reject the fresh produce. Additionally, graders quickly identify and eliminate products with bursts, cracks and punctures during quality control, but fresh bruises may not be detected due to delayed bruising, defective produce might not be detected. Simple inspection is adequate to identify if the injured tissue is visible and observed by the naked eye. However, one significant downside of subjective visual identification of bruise damage is that after a period of time from nonstop inspection, the effectiveness of produce quality examiners rapidly declines, potentially leading to inaccurate and inconsistent grading results. Manual evaluation decreases the efficiency and objectivity of bruise detection (Opara & Pathare, 2014). It can be especially challenging to detect mechanical damage located under the produce skin. Bruise detection accuracy is influenced by a number of factors such as the bruise type, bruise age, bruise severity, pre- and postharvest fruit conditions, and fresh produce type and variety (Lu, 2003). Damage detection in mass-produced fruit necessitates quick and non-invasive procedures. Nondestructive approaches are gaining popularity since they may swiftly gather damage-related information while not compromising the integrity of crops. Nondestructive methods are extremely efficient, with capable consequences for fruit quality assessment. Nondestructive techniques are utilized to evaluate fresh fruit and vegetable surfaces without impacting their freshness (Pathare & Rahman, 2022).

There has been a growing interest in exploring fast, nondestructive methods of detecting fresh produce damage over the last few years. Among them, spectroscopy and optical imaging techniques, computer vision, nuclear magnetic techniques, hyperspectral imaging, X-ray imaging, and thermal imaging used to detect mechanical damage (Htike et al., 2023; Mahanti et al., 2021; Diels et al., 2017; Zhang et al., 2020; Lorente et al., 2015; Jiménez-Jiménez et al., 2012; Al-Dairi et al., 2023).

16.4.1 Computer Vision Technology

For fresh fruit and vegetables quality evaluation, computer vision technology (image, interpretation) mimics human eyesight. Image analysis is important and valuable nondestructive approach for evaluating fruit quality. It is faster, more effective, less expensive, and more consistent than human evaluation, which is prone to fatigue and loss of attention (Pace et al., 2011) . Computer vision is economical, quick, hygienic, and consistent (Mahendran et al., 2012; Narendra & Hareesha, 2010). Fruit attributes can be assessed using image features such as color, size, shape, and texture. Color characteristics can be utilized to assess mechanical, physical, and chemical qualities of fruits such as ripening, index of maturity, and fruit defects (Pace et al., 2011). Htike et al. (2021) found that fractal dimension (FD) was a sufficient indicator of impact bruising damage and browning of pulp in "Glom Sali" guava than bruise measurements (bruise area or bruise volume). Besides, normalized FD difference ($\Delta FD/FD_0$) was more repeatable and reliable than the bruise area (BA) from image analysis in identifying vibration bruising of "Glom Sali" guava fruit resulted from simulated transportation (Htike et al., 2021). Al-Dairi et al. (2023) recently investigated the exterior quality of impact damaged bananas after 25 days of storage at three temperature conditions using a computer vision system (CVS).

16.4.2 Hyperspectral Imaging

A hyperspectral imaging (HSI) technique combines image and spectroscopic technology. The HSI system collects both spectral and spatial data. HSI can detect both outward faults and interior chemicals in samples. Hypersecretion imaging technology may identify and determine the degree of bruises through selecting relevant information, such as hypersecretion near-infrared information or imagine passing. Keresztes et al. (2017) summarized four-phase experimental protocols for detecting fruit bruises using hyperspectral imaging techniques, which includes sample supply, mechanical damage procedure (optimal design of experiment), hyperspectral model building (stepwise/multiclass bruise damage prediction), and nondestructive bruise damage monitoring (monitor browning development for accuracy of prediction).

Several studies (Pandiselvam et al., 2020; Ansari & Nirala, 2013; Arefi et al., 2017) employed hyperspectral imaging to assess impact bruising on fresh fruits. Zhang et al. (2020) investigated the use of hyperspectral transmittance imaging to detect early interior bruising in blueberries. Sun et al. (2021) sed partial least squares discriminant analysis models to categorize bruised tomatoes, with a 90.93% total classification accuracy. The spectra of injured tomatoes were lower than those of healthy tomatoes, with the features related to the spectra around 810 nm playing a significant role.

16.4.3 Biospeckle Laser Technique

Food material are examined using biospeckle, a nondestructive optical approach (Pandiselvam et al., 2020). When laser light impacts a substance, it reveals a variety of biological and non-biological dynamic processes or activities. This is referred to as a biospeckle (Ansari & Nirala, 2013). Arefi et al. (2017) employed a nondestructive biospeckle approach to measure mealiness in apple fruit. The biospeckle approach can be used to examine fruit injury induced by handling, storage, transit, and package design, as well as to compare distinct affects in different varieties of fruits. The biospeckle approach, together with the autocorrelation coefficient, was used to detect bruising in apples and pears (Vega & Torres, 2013). They observed that the BA of fresh pears and apple fruits was higher than that of bruised fruits, and that it decreased further after 1 day.

16.4.4 Thermal Imaging

The concept of thermal imaging (TI) involves measuring the released infrared radiation, not the reflected light, as all materials emit infrared radiation (Gowen et al., 2010). Different materials have different physicochemical properties as well as variances in thermal diffusivity (Varith et al., 2007). Thermal diffusivity differences serve as contrast parameters in the creation of thermograms. What distinguishes TI from spectral imaging techniques and nuclear magnetic techniques is that it does not require external lighting to eliminate the effects of uneven illumination and scattering, but it does require a constant ambient temperature, which requires a long period of heating or cooling treatment to control (Lu & Lu, 2017; Baranowski et al., 2009). The rate of cooling and heating between sound and damaged tissue can be used to detect bruises in their early stages (Vadivambal & Jayas, 2011). Zeng et al. (2020) used thermal images to detect and classify of bruises of pears and obtained prediction accuracy 99.25%.

16.4.5 X-Ray CT Imaging

X-ray computed tomography (CT) is a novel nondestructive machine vision technology with a significant potential for detecting internal faults that other machine vision technologies are incapable of detecting. Because they can pass through most objects, X-rays offer a competitive edge. It could be used to detect interior defects in vegetables and fruits in the food processing sectors (Azadbakht et al., 2019b). Diels et al. (2017) used X-ray CT scans to develop a method for automatically detecting and quantifying bruise volumes in the equatorial region of apples. Azadbakht et al. (2019b) used an X-ray imaging approach to determine the degree of bruising in pears during transportation in a recent study. The pressure deformation device and pendulum were used to model the static and moving loads that occurred during transportation. They confirmed that the bruises that appeared were not visible at lower loads. X-ray CT is better suited for detecting damage within the fresh produce.

16.4.6 Near-Infrared Spectroscopy

Near-infrared spectroscopy (NIRS) is another nondestructive approach that can be used in combination with molecular functional groups that contain polar groups like –OH, C–O, and N–H bonds (Blanco & Villarroya, 2002). Based on internal quality parameters, NIRS has become one of the most extensively utilized technologies for authenticating and detecting adulteration in fresh fruits and vegetables (Fan et al., 2009; Arendse et al., 2017). Furthermore, a number of studies have shown that VIS/NIR spectroscopy can be used to detect fruit damage, such as the differentiation of bruise levels in table olives (Jiménez-Jiménez et al., 2012), bruised apple detection (Luo et al., 2012), microbial infection evaluation in citrus (Lorente et al., 2015), bruise susceptibility predictions for table olives (Jiménez-Jiménez et al., 2013), and apples (Jian et al., 2017).

Nondestructive methods advancements have enabled the demonstration of a variety of practical applications for detecting the mechanical damage of fresh fruits and vegetables. Unlike traditional methodologies, nondestructive procedures can handle vast amounts of data independent of the diversity and complexity of biomaterial features encountered during the development of such applications.

16.5 Conclusions

Several studies have investigated mechanical damages using various experimental methodologies to determine their influence and link to quality changes in fresh fruit during postharvest handling. Mechanical damage has been shown to change and reduce the shelf life and edible quality of fresh fruits and vegetables. Mechanical damage in fresh commodities is a major concern in the harvest and postharvest supply chain, resulting in significant economic loss due to reduced shelf-life and loss of quality appearance. Bruise susceptibility studies are quite useful in reducing damage during handling processes. Bruise damage can be reduced by managing handling and postharvest conditions, which reduces the susceptibility of horticultural products to damage. Bruise damage could be reduced by decreasing bruise susceptibility of the horticultural produce by modifying growth, handling and postharvest factors. Bruise susceptibility of fresh horticultural produce could be minimized by modifying growth and postharvest handling practices. Bruise prevention is required to manage the quality of fresh horticultural produce, and efficient prevention is only achievable when the components that cause bruise formation are identified. Nondestructive techniques such as computer vision, hyperspectral imaging, thermal imaging, X-ray computed tomography, and near-infrared spectroscopy are featured. Measurement standards are also necessary to ensure that bruise measurements can be traced and transferred from one laboratory to another, facilitating for inter-laboratory comparisons.

Acknowledgments This work is based on the research supported in part by Sultan Qaboos University under the project codes: CL\SQU\QU\AGR\23\02 and IG/AGR/SWAE/19/03. This work is based on the research supported in part by the National Research Foundation of South Africa (Grant Numbers: 64813). The opinions, findings and conclusions or recommendations expressed are those of the author(s) alone, and the NRF accepts no liability whatsoever in this regard.

References

Aktas, T., Polat, R., & Atay, U. (2008). Mechanical properties and bruise susceptibility of tomatoes related to impact direction and drop-height. *The Philippine Agricultural Scientist, 91*(2), 180–186.

Al-Dairi, M., et al. (2022). Mechanical damage of fresh produce in postharvest transportation: Current status and future prospects. *Trends in Food Science & Technology, 124*, 195–207.

Al-Dairi, M., Pathare, P. B., & Al-Yahyai, R. (2023). Applying a computer vision system to monitor external quality attributes of damaged banana fruit during storage. In *Proceedings of the 3rd International Conference on Sustainable Agriculture for Rural Development (ICSARD 2022)*. Springer Nature.

Ansari, M. Z., & Nirala, A. K. (2013). Biospeckle activity measurement of Indian fruits using the methods of cross-correlation and inertia moments. *Optik, 124*(15), 2180–2186.

Arefi, A., et al. (2017). Towards real-time speckle image processing for mealiness assessment in apple fruit. *International Journal of Food Properties, 20*(Suppl 3), S3135–S3148.

Arendse, E., et al. (2017). Development of calibration models for the evaluation of pomegranate aril quality by Fourier-transform near infrared spectroscopy combined with chemometrics. *Biosystems Engineering, 159*, 22–32.

Azadbakht, M., Mahmoodi, M. J., & Vahedi Torshizi, M. (2019a). Effects of different loading forces and storage periods on the percentage of bruising and its relation with the qualitative properties of pear fruit. *International Journal of Horticultural Science and Technology, 6*(2), 177–188.

Azadbakht, M., Vahedi Torshizi, M., & Mahmoodi, M. J. (2019b). The use of CT scan imaging technique to determine pear bruise level due to external loads. *Food Science & Nutrition, 7*(1), 273–280.

Babarinsa, F., & Ige, M. (2012). Young's modulus for packaged Roma tomatoes under compressive loading. *International Journal of Scientific Engineering and Research, 3*(10), 314–320.

Baranowski, P., et al. (2009). Detection of early apple bruises using pulsed-phase thermography. *Postharvest Biology and Technology, 53*(3), 91–100.

Blanco, M., & Villarroya, I. (2002). NIR spectroscopy: A rapid-response analytical tool. *Trends in Analytical Chemistry, 21*(4), 240–250.

Brusewitz, G., & Bartsch, J. (1989). Impact parameters related to post harvest bruising of apples. *Transactions of the ASAE, 32*(3), 953.

Bugaud, C., et al. (2014). Bruise susceptibility of banana peel in relation to genotype and post-climacteric storage conditions. *Postharvest Biology and Technology, 87*, 113–119.

Chonhenchob, V., et al. (2009). Effect of truck vibration during transport on damage to fresh produce shipments in Thailand. *Industrial Technology, 3*, 27–38.

Diels, E., et al. (2017). Assessment of bruise volumes in apples using X-ray computed tomography. *Postharvest Biology and Technology, 128*, 24–32.

Fan, G., et al. (2009). Determination of soluble solids and firmness of apples by Vis/NIR transmittance. *Journal of Food Engineering, 93*(4), 416–420.

Fernando, I., et al. (2018). Measurement and evaluation of the effect of vibration on fruits in transit. *Packaging Technology & Science, 31*(11), 723–738.

Gowen, A., et al. (2010). Applications of thermal imaging in food quality and safety assessment. *Trends in Food Science & Technology, 21*(4), 190–200.

Guo, W., et al. (2021). Effect of mechanical bruises on optical properties of mature peaches in the near-infrared wavelength range. *Biosystems Engineering, 211*, 114–124.

Htike, T., et al. (2021). Investigation and evaluation of impact bruising in guava using image processing and response surface methodology. *Horticulturae, 7*(10), 411.

Htike, T., et al. (2023). Fractal image analysis and bruise damage evaluation of impact damage in guava. *Information Processing in Agriculture.*

Hussein, Z., Fawole, O. A., & Opara, U. L. (2019). Bruise damage susceptibility of pomegranates (Punica granatum, L.) and impact on fruit physiological response during short term storage. *Scientia Horticulturae, 246*, 664–674.

Jian, Y., Jiyu, G., & Qibing, Z. (2017). Predicting bruise susceptibility in apples using Vis/SWNIR technique combined with ensemble learning. *International Journal of Agricultural and Biological Engineering, 10*(5), 144–153.

Jiménez-Jiménez, F., et al. (2012). Non-destructive determination of impact bruising on table olives using Vis–NIR spectroscopy. *Biosystems Engineering, 113*(4), 371–378.

Jiménez-Jiménez, F., et al. (2013). Table olive cultivar susceptibility to impact bruising. *Postharvest Biology and Technology, 86*, 100–106.

Keresztes, J. C., et al. (2017). Glare based apple sorting and iterative algorithm for bruise region detection using shortwave infrared hyperspectral imaging. *Postharvest Biology and Technology, 130*, 103–115.

Kitthawee, U., et al. (2011). Mechanical bruising of young coconut. *Biosystems Engineering, 109*(3), 211–219.

Li, Z., & Thomas, C. (2014). Quantitative evaluation of mechanical damage to fresh fruits. *Trends in Food Science & Technology, 35*(2), 138–150.

Li, P., et al. (2018). Effects of static pressure injury on respiration intensity, quality and aroma components of apple. *Storage and Process, 18*(2), 1–8.

Lorente, D., et al. (2015). Visible–NIR reflectance spectroscopy and manifold learning methods applied to the detection of fungal infections on citrus fruit. *Journal of Food Engineering, 163*, 17–24.

Lu, R. (2003). Detection of bruises on apples using near–infrared hyperspectral imaging. *Transactions of the ASAE, 46*(2), 523.

Lu, Y., & Lu, R. (2017). Non-destructive defect detection of apples by spectroscopic and imaging technologies: A review. *Transactions of the ASABE, 60*(5), 1765–1790.

Lu, F., et al. (2010). Effect of vehicle speed on shock and vibration levels in truck transport. *Packaging Technology and Science, 23*(2), 101–109.

Luo, X., et al. (2012). Wavelength selection in Vis/NIR spectra for detection of bruises on apples by ROC analysis. *Journal of Food Engineering, 109*(3), 457–466.

Mahanti, N. K., et al. (2021). Emerging non-destructive imaging techniques for fruit damage detection: Image processing and analysis. *Trends in Food Science & Technology, 120*, 418–438.

Mahendran, R., Jayashree, G., & Alagusundaram, K. (2012). Application of computer vision technique on sorting and grading of fruits and vegetables. *Journal of Food Processing & Technology, 10*, 2157–7110.

Mitsuhashi-Gonzalez, K., et al. (2010). Bruising profile of fresh apples associated with tissue type and structure. *Applied Engineering in Agriculture, 26*(3), 509–517.

Narendra, V., & Hareesha, K. (2010). Prospects of computer vision automated grading and sorting systems in agricultural and food products for quality evaluation. *International Journal of Computer Applications, 1*(4), 1–9.

Opara, L. U. (2007). Bruise susceptibilities of 'Gala' apples as affected by orchard management practices and harvest date. *Postharvest Biology and Technology, 43*(1), 47–54.

Opara, U. L., & Fadiji, T. (2018). Compression damage susceptibility of apple fruit packed inside ventilated corrugated paperboard package. *Scientia Horticulturae, 227*, 154–161.

Opara, U. L., & Pathare, P. B. (2014). Bruise damage measurement and analysis of fresh horticultural produce—A review. *Postharvest Biology and Technology, 91*, 9–24.

Pace, B., et al. (2011). Relationship between visual appearance and browning as evaluated by image analysis and chemical traits in fresh-cut nectarines. *Postharvest Biology and Technology, 61*(2–3), 178–183.

Pandiselvam, R., et al. (2020). Biospeckle laser technique—A novel non-destructive approach for food quality and safety detection. *Trends in Food Science & Technology, 97*, 1–13.

Pathare, P. B., & Al-Dairi, M. (2021a). Bruise susceptibility and impact on quality parameters of pears during storage. *Frontiers in Sustainable Food Systems, 5*, 658132.

Pathare, P. B., & Al-Dairi, M. (2021b). Bruise damage and quality changes in impact-bruised, stored tomatoes. *Horticulturae, 7*(5), 113.

Pathare, P. B., & Al-Dairi, M. (2021c). Effect of simulated vibration and storage on quality of tomato. *Horticulturae, 7*(11), 417.

Pathare, P. B., & Al-Dairi, M. (2022). Effect of mechanical damage on the quality characteristics of banana fruits during short-term storage. *Discover Food, 2*(1), 4.

Pathare, P. B., & Rahman, M. S. (2022). Nondestructive techniques for fresh produce quality analysis: An overview. In *Nondestructive quality assessment techniques for fresh fruits and vegetables* (pp. 1–10). Springer.

Pathare, P. B., Al-Dairi, M., & Al-Mahdouri, A. (2021). Bruise damage susceptibility and fruit quality assessment of pear. *The Open Agriculture Journal, 15*(1), 82–90.

Pathare, P. B., et al. (2023). Physiological response of stored pomegranate fruit affected by simulated impact. *Foods, 12*(6), 1122.

Polat, R., Aktas, T., & Ikinci, A. (2012). Selected mechanical properties and bruise susceptibility of nectarine fruit. *International Journal of Food Properties, 15*(6), 1369–1380.

Ríos-Mesa, A. F., et al. (2020). Effect of vehicle vibration on the mechanical and sensory properties of avocado (Persea Americana Mill. Cv. Hass) during road transportation. *International Journal of Fruit Science, 20*(Suppl 3), S1904–S1919.

Ruiz-Altisent, M., & Moreda, G. P. (2011). Fruits, mechanical properties and bruise susceptibility. In J. Gliński, J. Horabik, & J. Lipiec (Eds.), *Encyclopedia of agrophysics* (pp. 318–321). Springer.

Singh, J., Singh, S. P., & Joneson, E. (2006). Measurement and analysis of US truck vibration for leaf spring and air ride suspensions, and development of tests to simulate these conditions. *Packaging Technology and Science, 19*(6), 309–323.

Soleimani, B., & Ahmadi, E. (2014). Measurement and analysis of truck vibration levels as a function of packages locations in truck bed and suspension. *Computers and Electronics in Agriculture, 109*, 141–147.

Stropek, Z., & Golacki, K. (2018). Viscoelastic response of apple flesh in a wide range of mechanical loading rates. *International Agrophysics, 32*(3), 335–340.

Sun, Y., et al. (2021). Hyperspectral characteristics of bruised tomatoes as affected by drop height and fruit size. *LWT - Food Science and Technology, 141*, 110863.

Vadivambal, R., & Jayas, D. S. (2011). Applications of thermal imaging in agriculture and food industry—A review. *Food and Bioprocess Technology, 4*, 186–199.

Van Zeebroeck, M., et al. (2007). Impact damage of apples during transport and handling. *Postharvest Biology and Technology, 45*(2), 157–167.

Varith, J., et al. (2007). Design of multimode-circular microwave cavity for agri-food processing. In *2007 Asia-Pacific Microwave Conference*. IEEE.

Vega, F., & Torres M. (2013). Automatic detection of bruises in fruit using biospeckle techniques. In *Symposium of Signals, Images and Artificial Vision-2013: STSIVA-2013*. IEEE.

Walkowiak-Tomczak, D., et al. (2021). The effect of mechanical vibration during transport under model conditions on the shelf-life, quality and physico-chemical parameters of four apple cultivars. *Agronomy, 11*(1), 81.

Wu, G., & Wang, C. (2014). Investigating the effects of simulated transport vibration on tomato tissue damage based on Vis/NIR spectroscopy. *Postharvest Biology and Technology, 98*, 41–47.

Yeşiloğlu Cevher, E. (2022). Determination of bruise preventing capacity of the cushioning material in persimmon fruit under pendulum impact test. *Journal of Food Process Engineering, 45*(12), e14162.

Zeng, X., et al. (2020). Detection and classification of bruises of pears based on thermal images. *Postharvest Biology and Technology, 161*, 111090.

Zhang, M., et al. (2020). Fully convolutional networks for blueberry bruising and calyx segmentation using hyperspectral transmittance imaging. *Biosystems Engineering, 192*, 159–175.

The manufacturer's authorised representative in the EU is Springer Nature Customer Service Centre GmbH, Europaplatz 3, 69115 Heidelberg, Germany. If you have any concerns regarding our products, please contact ProductSafety@springernature.com

Printed and bound by CPI Group (UK) Ltd, Croydon, CR0 4YY
23/09/2026
02221546-0014